OPTICAL WAVEGUIDE SCIENCES

Academic Council of the Symposium

Co-Chairmen

Huang Hung-chia
Allan W. Snyder

INTERNATIONAL MEMBERS

J. Arnaud
Limoges University, France

H. Blok
Delft University of Technology, The Netherlands

Charles K. Kao
ITT Corp., USA

W. A. Gambling
University of Southampton, England

A.K. Ghatak
Indian Institute of Technology, India

I. P. Kaminow
Bell Laboratories, USA

J. E. Midwinter
British Telecom, England

T. Okoshi
University of Tokyo, Japan

A. A. Oliner
Polytechnic of New York

Harrison E. Rowe
Bell Laboratories, USA

C.G. Someda
University of Pavada, Italy

Allan W. Snyder
Australian National University

E. Spitz
Thomson CSF, France

H. G. Unger
Braunschweig University of Technology
West Germany

NATIONAL MEMBERS OF PRC

Fang Jun-xin
Jiao-Tong University, Shanghai

Huang Hung-chia
Shanghai University of Science and Technology

Lin Wei-gan
Chengdu Institute of Radio Engineering

Ma Xun
Kweilin Institute of Optical Communications

Yan En-ze
Wuhan Institute of Post and Telecommunications

Ye Pei-da
Peking Institute of Post and Telecommunications

Optical Waveguide Sciences

Proceedings of the International Symposium, held at Kweilin, People's Republic of China (PRC), June 20-23, 1983

Sponsored by The Chinese Institute of Electronics (CIE)

Edited by

HUANG HUNG-CHIA
Member, Academia Sinica
Vice President, Shanghai University of Science and Technology

ALLAN W. SNYDER
Head, Department of Applied Mathematics
School of Advanced Studies
Australian National University

1983 **MARTINUS NIJHOFF PUBLISHERS**
a member of the KLUWER ACADEMIC PUBLISHERS GROUP
THE HAGUE / BOSTON / LANCASTER

Distributors

for the United States and Canada: Kluwer Boston, Inc., 190 Old Derby Street, Hingham, MA 02043, USA
for all other countries: Kluwer Academic Publishers Group, Distribution Center, P.O.Box 322, 3300 AH Dordrecht, The Netherlands

Library of Congress Cataloging in Publication Data

Library of Congress Cataloging in Publication Data
Main entry under title:

Optical waveguide sciences.

Includes index.
1. Fiber optics--Congresses. 2. Optical wave guides--Congresses. I. Huang, Hung-chia. II. Snyder, Allan W., 1940- . III. Chinese Institute of Electronics.
TA1800.O69 1983 621.36'92 83-8143
ISBN 90-247-2848-7

ISBN 90-247-2848-7 (this volume)

FOREWORD

Over the past decade or more, the art relating to modern optical waveguides has evolved as a highly focused interdisciplinary field, so attractive, stimulating and full of far-reaching promise that no parallel could be found earlier in other branches of applied sciences except, perhaps, in solid-state electronics and computer technology.

This Proceedings, with a selection of 53 papers and briefs by 96 authors of 16 countries, is of real international dimensions. It is the outgrowth of the International Symposium held June 20-23, 1983, the first international meeting on this subject taking place in China. Since almost every country in the world, Eastern or Western, large or small, has been involved in the study and promotion of this technological revolution, it is only natural that China should be happy to serve as host country of the international meeting to promote an interchange of experiences and ideas conducive to greater achievements in the future.

At present, several outstanding international meetings are being held annually or biannually in the United States, in Europe, or in other regions, on topics relevant to optical communications. The scope covered by those meetings is usually very comprehensive, including optical fibers and integrated optics, laser sources and detectors, systems problems, etc. With the rapid growth of the science and technology, we consider it both practical and feasible to restrict the scope of scientific discussions at an international meeting to optical waveguide sciences only, as the subject itself is already broad enough to frame a modern branch of applied science in its own right.

A unique feature of the Symposium is its intent to integrate the educational, industrial and research aspects relating to optical waveguides. The inclusion of the educational aspect, in particular, lends an orientation not found at other technical meetings on this and similar subjects. The educational feature is made manifest by the inclusion in the Proceedings of a paper which proposes a comtemporary course on optical waveguides. The same feature is also reflected in papers such as source-solution approach to optical waveguides, microwave-approach to optical waveguides, etc, which attempt in-depth treatment of optical waveguides from different viewpoints and by way of different mathematical procedures.

One of the reasons why special emphasis has been placed on the educational aspect of optical waveguide sciences at this Symposium is that our understanding of optical waveguides has, in a sense, reached a sufficient degree of maturity to permit all concerned to benefit from discussions on effective methods of education and the application of already acquired knowledge in this field. By education, we do not mean narrowly the teaching of college students alone, for the importance of transferring the wide variety of transmission theories and technical information of optical waveguides to the enormous field of industry should not be underestimated, i.e. to the designers, manufacturers, technicians, etc, who are the actual makers and/or users of optical waveguides, but who may not share a great interest in mathematics and abstract concepts.

Especially in China, who lost an entire decade through 1966-1976, and in some other countries under different situations, the educational aspect of this science may be of particular importance in reducing the technology gap and in making it practicable to catch up with the most advanced countries in the West as speedily as possible.

We do not pretend that we have solved at this Symposium the problems relating to education in optical waveguide sciences. We have only made a start. A satisfactory solution of these educational problems presents a hard challenge to the world-wide scientific circles concerned. But we do hope that an integration of the afore-mentioned three aspects relating to optical waveguide sciences, which has been devised as the Symposium theme, is adequately reflected in the present Proceedings.

Another feature of the Proceedings is reflected in papers which deal with the highly focused research areas relating to the subject concerned. In some sense, a certain degree of maturity has been attained by the optical waveguide sciences, as we so estimated in connection with the educational feature of the Symposium papers. In a broader sense, however, this science is far from mature in view of the fact that a multitude of theoretical and technological problems still await answers, such as the eventual solution of the making of long-length polarization-stable single mode optical fibers, the design and manufacture of various waveguide devices for coherent transmission systems, the new fibers working at ultra-long wavelengths, the nonlinear phenomena in optical waveguides. etc. In this sense, optical waveguide as a branch of modern science is still in its infancy, but with promises of rapid growth. We have the satisfaction to note that a variety of papers by scientists of different nationalities have been presented at the Symposium and included in this Proceedings, dealing with forefront research subjects with fruitful results.

There are also a number of contributed papers dealing with a variety of specific problems relating to optical waveguide science and technology. While papers on single-mode fibers occupy a prominent position in this volume, several interesting papers on multimode fibers are selected which indicate an unfading interest in this more conventional fiber version. Included in the present volume are also papers on film waveguides and planar waveguide devices, relevant to the developing field of integrated optics. In reviewing the papers, the editors consider it prudent to allow immature or sometimes controversial views, especially when the art concerned is not yet in a state of full development.

We are very proud to acknowledge, and honored to have, the sponsorship of the Chinese Institute of Electronics (CIE), who is fully aware of the necessity to encourage and accelerate research in optical waveguide sciences, in general, and optical fiber in particular. We are also happy to acknowledge the valuable cooperation of many institutions and organizations, including especially the Chinese Microwave Society of the CIE, the Shanghai University of Science and Technology, the Institute of Transmission Lines in Shanghai, and the Institute of Optical Communications in Kweilin. Special acknowledgement is expressed to the School of Advanced Studies of the National University of Australia, whose kind assistance has been indispensible in organizing the Symposium.

Among the large number of individuals to whom we feel grateful, special thanks go to Miss Shi Zhen-juan, who typed and proof-read many of the papers with admirable patience and unfailing cheerfulness.

During the preparation of the Proceedings, it has been a great pleasure to work with the Martinus Nijhoff Publishers, whose enthusiasm in promoting the cause of modern science and technology has made possible the publication of this book in time for the conference.

It is to be regretted that several outstanding scholars who have been so generous as to have given, on various occassions, many constructive suggestions for improvement of the Symposium were not able to come to China and participate in the Symposium. To all of them we feel deeply indebted.

Owing to the far distance that separates the two co-editors and the members of the Academic Council of the Symposium, closer collaboration in editing the papers has at times been difficult. Any deficiencies or problems which may occur in this volume will, therefore, be the responsibility of the first undersigned, and not of the overseas editor and the academic members.

HUANG HUNG-CHIA

ALLAN W. SNYDER

March, 1983

CONTENTS

PART THREE: OPTICAL WAVEGUIDE SYSTEMS

PART FOUR: OPTICAL WAVEGUIDE THEORY AND PRACTICE

PART FIVE: PLANAR OPTICAL WAVEGUIDES

PART SIX: SYMPOSIUM BRIEFS

PART ONE

EDUCATION-ORIENTED PAPERS

PULSE PROPAGATION IN OPTICAL FIBERS

J. ARNAUD* and C. FROEHLY**

INTRODUCTION

In this paper, we have no attempt to present some results which, basically, have not been known before. What we wish to emphasize is the analogy that exists between space and time concepts, pulse spreading in dispersive media being analogous to beam diffraction.

TIME-HARMONIC PLANE WAVES

Let us first consider a time-harmonic plane wave propagating in the x-z plane. The field has the form

$$\psi(x,z,t) = \cos(k_x x + k_z z - \omega t)$$

$$= \text{Real Part of } \{\exp[i(k_x x + k_z z)] \exp(-i\omega t)\} \qquad (1)$$

to within arbitrary amplitude and phase factors. The numbers k_x, k_z can be considered the components of a wave vector $\vec{k}$. The angular frequency ω is taken to be a constant here. The crests of the field are clearly given at $t = 0$ by the straight lines

$$k_x x + k_z z = 2m\pi \quad ; \quad m = 0, \pm 1, \pm 2 \ldots \qquad (2)$$

shown on top of Fig.1 (a), these lines being perpendicular to the $\vec{k}$ vector.

---

* University of Montpellier, Place Eugène Bataillon
MONTPELLIER - 34060 - cédex, France.

** Laboratory of Optics, University of Limoges, 123, rue A. Thomas
LIMOGES - 87060 - cédex, France.

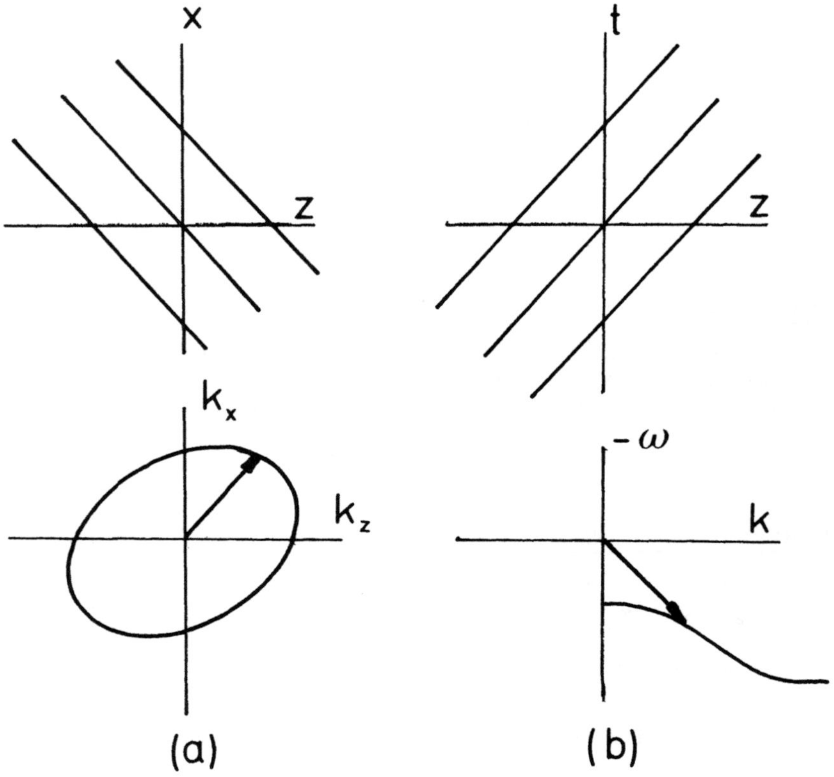

Fig.1

The distance between adjacent crests in the x direction at a given time is $\lambda_x = 2\pi/k_x$, and, similarly $\lambda_z = 2\pi/\lambda_z$. We define $k = |\vec{k}| = (k_x^2 + k_z^2)^{1/2}$ and the wavelength in the medium (not to be confused with the free-space wavelength λ_o) is $\lambda = 2\pi/k$. If λ depends on the direction of $\vec{k}$, the medium is called anisotropic. In isotropic media the curve described by the tip of the $\vec{k}$ vector is a circle. The nonzero curvature of that circle is responsible for beam diffraction as we shall see.

Let us now consider phenomena depending on time (t) and one spatial coordinate (z). The field has the form

$$\psi(z,t) = \cos(kz - \omega t) = \text{Real Part of } \exp[i(kz - \omega t)] \quad (3)$$

We shall omit the subscript z on k when there is no risk of ambiguity. The crests are given by the straight lines

$$kz - \omega t = 2m\pi \quad ; \quad m = 0, \pm 1, \pm 2 \ldots \tag{4}$$

shown on top of Fig.1(b). If T denotes the wave period, that is the interval of time separating two successive maxima of the field, then $\omega = 2\pi/T$. Similarly, $k = 2\pi/\lambda$. Clearly from Eq.(3), the wave crests in space-time are perpendicular to the $\vec{k}$ vector with components k, $-\omega$. The $\vec{k}$ curve at the bottom of Fig.1b is a straight line going through the origin when the medium is non dispersive, k being proportional to ω, or $k = (\omega/c)n$, with n a constant. Alternatively one can say that the phase velocity $v_\phi = c/n$ is independent of frequency. But a general medium is dispersive. For example, if we restore the dependence of ψ on the x-coordinate and maintain k_x a constant (this is the case when a plane wave is launched on a transmission grating of period p, then $k_x = \pm 2\pi/p$, and for a metallic waveguide or an oversized optical fiber of width d : $k_x = \pi/d$), k_z is no longer proportional to ω. In that sense, free-space can be viewed as dispersive. Indeed, in free space

$$k_z = \left[(\omega/c)^2 - k_x^2\right]^{1/2} , \; k_x = \text{constant}. \tag{5}$$

implies a nonlinear dependence of k_z on ω.

An optical fiber (of axis z) is dispersive also because silica glass is a Lorentz medium that can be modeled by microscopic oscillators in the ultraviolet (UV) range and in the infrared (IR) range. Low-loss propagation occurs in between (from, say, $\lambda_o \simeq 0.8$ µm to 1.7 µm). In such a medium the refractive index squared is, neglecting losses,

$$n^2 = 1 - \frac{A_1}{\omega^2 - \omega_1^2} - \frac{A_2}{\omega^2 - \omega_2^2} \tag{6}$$

where A_1, A_2 are oscillator strengths and ω_1, ω_2 are resonant frequencies in the UV and IR, respectively.

We have the correspondence

x,z	⟷	z,t
k_x, k_z	⟷	k(or k_z), $-\omega$
λ_x, λ_z	⟷	λ, T

(ω fixed) (k$_x$ fixed)

RAY TRAJECTORIES

Next, let us consider the average direction, or speed, of wave packets. These quantities involve now the slopes of the $\vec{k}$ curves shown at the bottom of Fig.1, in the neighborhood of the carrier wavevector $\vec{k} = \vec{k}_o$.

In the x-z space domain, we need consider transversely limited beams whose widths comprise however many wavelengths and we are concerned with the direction in which the beam has maximum energy. This is the ray direction. In the space-time domain, we are considering wave packets of duration much larger than the period T. The average speed of a wave packet is called the group velocity, and its trajectory in space-time is called a worldline (or space-time ray).

As we shall see in the next section beams and wave packets in general spread out and distort about their average motion, and one may wonder if the wave packet center is a well-defined concept. This is so because, as we make the propagation length L longer and longer, the transit time $t = L/v_g$ increases in proportion to length, while the pulse spreading increases only in proportion of $\sqrt{L}$ if we are careful to increase the input pulse duration in proportion to $\sqrt{L}$ in order to minimize its frequency spectrum. Thus, the accuracy with which one can measure the group velocity can be increased without limit, in spite of the wavepacket deformation.

To summarize this section, we have the correspondance :

ray trajectory in space	⟷	world line
space-space	⟷	space-time

Finally, we shall be concerned with the distortion of a beam about its average path and the spreading of a wave packet about its world line, with the correspondence

diffraction	⟷	pulse spreading
space-space	⟷	space-time

These latter effects are dependent on the curvature of the k-curves shown at the bottom of Fig.1. We could go on step further and consider geometrical aberrations in space and time, but this will not be discussed here.

Because we shall be concerned only with a small neighborhood of the ω, k curve, centered at ω_o, k_o, we restrict ourselves to the following representation of the $k(\omega)$ curve

$$k = k_o + a(\omega-\omega_o) + \frac{1}{2} b(\omega-\omega_o)^2 \qquad (8)$$

were $a = dk/d\omega$, $\omega = \omega_o$ and $b = d^2k/d\omega^2$, $\omega = \omega_o$. For a given material and a given carrier frequency (or free-space wavelength λ_o), k_o, a and b are fixed quantities. It is useful to introduce in place of k_o, a, b dimensionless quantities n, D and M as was done in Ref.1

$$k = (\omega/c)n$$
$$d = D-1 = (\omega/k)\ dk/d\omega - 1$$
$$M = (\omega^2/k)d^2k/d\omega^2 = (\lambda_o^2/n)(d^2n/d\lambda_o^2) \qquad (9)$$

where all quantities are evaluated at $\omega = \omega_o$.
For silica at $\lambda_o = 0.9$ µm, for example, we have :

$$n = 1.45 \quad ; \quad d = 8.8 \times 10^{-3} \quad ; \quad M = 12 \times 10^{-3}$$

As λ_o increases, the d factor increases slightly, while M goes to zero around 1.3 µm, and then becomes negative. For other materials of practical use, the wavelength at which M is zero may be larger, e.g., $\lambda_o = 1.7$ µm for pure germania. The physical significance of these results will be appreciated later.

OPTICAL LENGTH OF A SPACE-TIME RAY

We wish to evaluate the field generated at z and time t by a short pulse (not a Dirac function) with carrier frequency ω_o, emitted at $z = 0$ and $t = 0$. Within the approximation of Eq.(8), this is the Van Vleck propagator (1928) discussed in Ref.1 for square-law anisotropic media. We restrict ourselves here to homogeneous media, and we will give only the rapidly varying phase term $\exp[iS(z,t)]$. It is understood that, at the end of the calculations, one must take the real part of the quantity obtained, since the field, being a measurable quantity, must be real at some z, t point (in many optical experiments, however, it is sufficient to know the complex field, or more precisely, the analytic signal).

Because the medium is homogeneous and time-invariant, the world lines are straight with slope

$$\frac{t}{z} = \frac{dk}{d\omega} = a + b\ (\omega-\omega_o) \qquad (10)$$

where a and b are the constants introduced in Eq. (8). This slope is the reciprocal of the group velocity v_g. It takes the value a precisely at $\omega=\omega_o$. Unless b = 0, the world lines fan out from the origin z = t = 0. The initial pulse has a frequency spectrum centered about $\omega=\omega_o$ with a width reciprocal to the (as yet unspecified) pulse duration. The frequency that reaches some specified point z, t in space-time is clearly, from Eq. (10)

$$\omega=\omega_o + (t/z - a)/b \qquad (11)$$

Because the pulse spectrum is fairly narrow, most of the pulse energy is to be found whenever t/z is close to a (reciprocal of the group velocity).

The spatial analog of the situation depicted above are rays emitted from a small spot (not quite a point source). The ray directions are normal to the surface of wave normals $x/z = dk_x/dk_z$. This is only for isotropic media that we have $x/z = k_x/k_z$ (rays directed along the $\vec{k}$ vectors).

Next, we calculate the phase shift [see Eq. (3)]

$$S(z,t) = k(\omega)\, z - \omega t \qquad (12)$$

Substituting $\omega-\omega_o$ from Eq. (11) into Eq. (8) and Eq. (12), we obtain, after rearranging,

$$- S = \omega_o t - k_o z + \frac{1}{2} (t - az)^2/bz \qquad (13)$$

In space-space, at some fixed frequency ω, $S/2\pi$ (if this is an integer) corresponds physically to the number of crests of the field from x = z = o to the x, z point, at some given time. Alternatively, we can view the S(x,z) = constant curve as the wavefront emitted by the radiating spot. This curve differs vastly from the surface of wave normals ($\vec{k}$ curve), except if the latter is circular.

In space-time $S/2\pi$ (if this is an integer) is the number of crests from t = z = 0 to the t, z point. (For the analogy to be complete, one should restore the x coordinate and keep k_x a constant, in the same way as we kept ω a constant previously).

In connection with the hamiltonian - lagrangian formalism developed in Ref.1, note incidentally that Eq. (10) is one of the two

Hamilton's equations, the Hamilton characteristic function being taken as $\omega = \omega(k)$. S in Eq.(13) is the product L(z, dz/dt) x t, L being the Lagrangian function, easily deducible from the Hamiltonian function.

Note also that the result in Eq. (13) can be obtained alternatively by expanding ω up to second order in $k - k_o$ (instead of the opposite), provided z is replaced in the denominator of the last term by t/a. This is a permissible approximation because the power is significant only in the neighborhood of z = t/a.

REPRESENTATIONS OF PULSE SPREADING

It is often pointed out that pulse spreading in dispersive media is analogous to beam diffraction. In particular, a gaussian pulse of carrier frequency ω_o and initial duration σ_o (defined at the 1/e point of the field intensity) spreads out after a propagation length z to σ given by

$$\sigma^2 = \sigma_o^2 + b^2 z^2/\sigma_o^2 \qquad (14)$$

where $b = d^2k/d\omega^2$, $\omega = \omega_o$, as before. This result can be obtained by multiplying the pulse spectrum

$$\exp\left[-\frac{1}{2}\sigma^2(\omega-\omega_o)^2\right] \qquad (15)$$

by the phase function $\exp\left[ik(\omega)z\right]$ and performing a Fourier transform back to the time-dependent field. It is quicker to use the complex-ray representation of gaussian beams proposed by one of us in 1968 for isotropic[3,4] media and shortly there after for anisotropic media[1].

The half-width of a gaussian beam propagating in a medium with wave number k_o on the other hand spreads out according to the law[1]

$$\xi^2 = \xi_o^2 + z^2/k_o^2\,\xi_o^2 \qquad (16)$$

whose similarity with Eq.(14) is obvious. The space-time analogy is of course not limited to gaussian beam pulses. More generally, one can set up a vertical axis $t - z/v_g$, where v_g is the group velocity, and describe pulse spreading as if it were a diffraction problem in medium with wave number b^{-1}. This procedure, however, is incomplete and provides only pulse envelopes. The reason why the above description is uncompleteis that pulse spreading is analogous to diffraction in anisotropic media, while Eq.(16) as written, is applicable only to isotropic media. It takes a coordinate transformation to go from anisotropic to isotropic media. The gaussian beam described by Eq.(16) for example, has zero average transverse spatial frequency at $z = 0$, while pulses have always non zero carrier frequencies. It is not sufficient to subtract z/v_g from t to take that fact into account.

It is however possible to find coordinates ξ, ζ such that the expression for the phase shift S in Eq.(13) be equal to $\varpi(\xi^2 + \zeta^2)^{1/2}$, as it would be the case for isotropic media, in the neighborhood of the group delay, up to second order. We will give only the result : one must set on the vertical axis, instead of t the quantity

$$\xi = t - \left(\frac{\cos^2\Theta}{v_g} + \frac{\sin^2\Theta}{v_\phi}\right) z \tag{17}$$

On the horizontal axis, instead of z, one must set

$$\zeta = \left(\frac{1}{v_\phi} - \frac{1}{v_g}\right) \sin\Theta \cos\Theta\, z \tag{18}$$

and the effective angular frequency $\overline{\omega}$ is

$$\overline{\omega} = \omega_o/\sin\Theta \tag{19}$$

In these expressions, Θ defines the direction of a pulse at the mean group velocity $v_g = a^{-1}$; $v_\phi = \omega_o/k_o$ is the phase velocity. We have

$$\tan\Theta = (M/d)^{1/2} \tag{20}$$

if we introduce the dimensionless parameters M and d. This representation is restricted to positive M/d ratios (in silica, $\lambda_o < 1.3$ μm).
Note, however, that the field modulus is unchanged if M is changed to $-M$ (as an example see Eq.(14) where only b^2 enters). This observation extends the range of validity of the proposed representation. For the values of n, d and M given earlier (Silica at $\lambda_o = 0.9$ μm), we find from Eqs.(17) to (20)

$$\Theta = 50°$$

$$\xi = t - 1.45\, z/c$$

$$\zeta = 0.0063\ z/c$$

$$\bar{\omega} = 435 \text{ Terahertz}$$

where c is the velocity of light in free space. Using these coordinates, pulse spreading can indeed be accurately described if it were beam diffraction in free space.

GENERALIZATIONS

The complex ray representation of gaussian pulses alluded to before enables us to deal with media that are not only dispersive but also have quadratic spatial temporal variations of the refractive index. For example, the pulse spreading of tubular modes measured in Ref. 5 by interferometric techniques can be dealt with accurately using such a representation.

Nonlinear (Kerr) effects can be dealt with also, and solitary waves discovered, with the above formalism, provided one assumes that the refractive index is a logarithmic (rather than linear) function of the optical intensity. Whether this assumption is academic or has some physical significance remains to be seen.

REFERENCE

1 - J. ARNAUD "Beam and Fiber Optics" - Academic Press (N.Y), 1976.

2 - C. FROEHLY "Temporally Coherent Fiber Optics" - NATO course in Cargèse, Corsica, july 1982.

3 - J. ARNAUD "Representation of Gaussian Beams by Complex Rays", unpublished Bell Labs memorandum, oct. 10, 1968.

4 - J. ARNAUD "Degenerate Optical Cavities" - Applied Optics, 8, pp. 189 and 1909 (1969). See the foot notes in these papers describing the complex ray representation used.

5 - A. BARTHELEMY "Dispersion of Tubular Modes Propagating in Multimode Optical Fibers". Opto and Quantum Electronics, (to appear).

SOURCE-TYPE REPRESENTATION FOR FIELDS IN OPTICAL WAVEGUIDE SECTIONS AND THEIR CONSEQUENCES FOR THE SPECTRAL FIELD CONSTITUENTS

H. BLOK and A.T. DE HOOP

Laboratory of Electromagnetic Research
Department of Electrical Engineering
Delft University of Technology
P.O. Box 5031, 2600 GA Delft
The Netherlands

ABSTRACT

A consistent theory is developed by which the transmission properties of a uniform optical waveguide section of arbitrary cross section are characterized by the modal constituents as they are excited by equivalent surface-current distributions located in the end planes of the section. The representation leads in a natural way to orthogonality and completeness relations as well as to the appropriate normalization conditions.

1. INTRODUCTION

The waveguiding part of an optical communication system generally consists of a cascading of a number of uniform waveguide sections. The excitation and transmission properties of the system as a whole are fully known once the transmission properties of each uniform section and the transfer properties of the junction between two successive sections are known. In the present paper we investigate the excitation and transmission properties of a single uniform waveguide section with special emphasis on the axial spectral decomposition of the field.

2. THE WAVEGUIDE FIELDS IN THE SPECTRAL DOMAIN

Figure 1 shows an open waveguide section of isotropic material that is uniform in the z-direction of a Cartesian coordinate system. The cross section consists of a bounded waveguiding part $\mathcal{D}_1$ in which the permittivity and or the permeability differ from the values of the homogeneous surroundings $\mathcal{D}_2$. The transmission properties of the waveguide section are commonly specified in terms of their modal constituents. These follow upon subjecting the electromagnetic field equations to a Fourier transform with respect to time with zero initial conditions and a Fourier transform with respect to the axial coordinate over $\mathcal{Z}$. For example

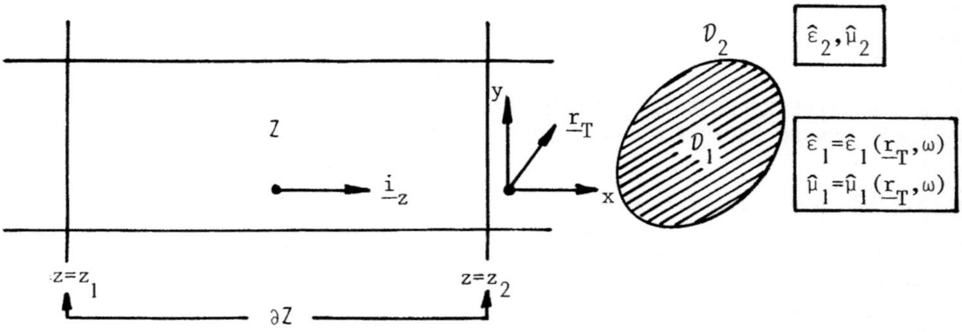

FIGURE 1. Uniform open waveguide section.

$$(2.1)\qquad \underline{\hat{E}}(\underline{r}_T,z,\omega) = \int_0^\infty \exp(-j\omega t)\underline{E}(\underline{r}_T,z,t)dt,$$

with $\omega+j\Delta\in\mathbb{R}$, where $\Delta>0$ ensures causality, and

$$(2.2)\qquad \underline{\tilde{E}}(\underline{r}_T,k_z,\omega) = \int_{z_1}^{z_2} \exp(jk_z z)\underline{\hat{E}}(\underline{r}_T,z,\omega)dz,$$

with $k_z\in\mathbb{R}$. The electromagnetic field equations then transform into

$$(2.3)\qquad \underline{\tilde{\nabla}} \times \underline{\tilde{H}} - j\omega\hat{\varepsilon}\underline{\tilde{E}} = \underline{\tilde{J}},$$

$$(2.4)\qquad \underline{\tilde{\nabla}} \times \underline{\tilde{E}} + j\omega\hat{\mu}\underline{\tilde{H}} = -\underline{\tilde{K}},$$

where $\underline{\tilde{\nabla}} = \underline{\nabla}_T - jk_z\underline{i}_z$, $\hat{\varepsilon} = \hat{\varepsilon}(\underline{r}_T,\omega)$, $\hat{\mu} = \hat{\mu}(\underline{r}_T,\omega)$,

$$(2.5)\qquad \underline{\tilde{J}} = (\underline{i}_z\times\underline{\hat{H}})_{z_2}\exp(jk_z z_2) - (\underline{i}_z\times\underline{\hat{H}})_{z_1}\exp(jk_z z_1),$$

$$(2.6)\qquad \underline{\tilde{K}} = -(\underline{i}_z\times\underline{\hat{E}})_{z_2}\exp(jk_z z_2) + (\underline{i}_z\times\underline{\hat{E}})_{z_1}\exp(jk_z z_1),$$

and 'T' stands for transverse. Inversion of the Fourier transform with respect to z yields

$$(2.7)\qquad (2\pi)^{-1}\int_{k_z\in\mathbb{R}} \exp(-jk_z z)\underline{\tilde{E}}(\underline{r}_T,k_z,\omega)dk_z = \{1,\tfrac{1}{2},0\}\underline{\hat{E}}(r_T,z,\omega) \text{ when } z\in\{Z,\partial Z,Z'\},$$

which shows that in the interior of the section the field is recovered, while outside the section the left-hand side leads to destructive interference (ONSEEN extinction theorem). From (2.3)-(2.7) it follows that the field

$\{\underline{\hat{E}},\underline{\hat{H}}\}$ in the waveguide section can be considered as generated by the transverse, end-plane, current distributions $\underline{\hat{J}} = \underline{J}_{T,2}\delta(z-z_2) + \underline{\hat{J}}_{T,1}\delta(z-z_1)$, $\underline{\hat{K}} = \underline{\hat{K}}_{T,2}\delta(z-z_2) + \underline{\hat{K}}_{T,1}\delta(z-z_1)$, where $\underline{\hat{J}}_{T,2} = (\underline{i}_z\times\underline{\hat{H}})_{z_2}$, $\underline{\hat{J}}_{T,1} = -(\underline{i}_z\times\underline{\hat{H}})_{z_1}$, $\underline{\hat{K}}_{T,2} = -(\underline{i}_2\times\underline{\hat{E}})_{z_2}$, $\underline{\hat{K}}_{T,1} = (\underline{i}_z\times\underline{\hat{E}})_{z_1}$. The modal constituents of the field now show up when the excitation by a transverse current-source distribution, located in a single transverse plane is further analyzed. This is done in the next section.

3. THE FIELD EXCITED BY A PLANAR TRANSVERSE CURRENT DISTRIBUTION

The field $\{\underline{\hat{E}}^e,\underline{\hat{H}}^e\}$ generated by the transverse electric current-source distribution $\underline{\hat{J}} = \underline{\hat{J}}_T\delta(z)$ follows from

$$(3.1)\qquad \underline{\tilde{\nabla}} \times \underline{\tilde{H}}^e - j\omega\hat{\varepsilon}\underline{\tilde{E}}^e = \underline{\hat{J}}_T,$$

$$(3.2)\qquad \underline{\tilde{\nabla}} \times \underline{\tilde{E}}^e + j\omega\hat{\mu}\underline{\tilde{H}}^e = \underline{0}.$$

Now, k_z enters as a parameter in the coefficients of the spectral-domain field equations in an analytic manner. Consequently, $\{\underline{\tilde{E}}^e,\underline{\tilde{H}}^e\}$ will be analytic functions of k_z in some domain of the complex k_z-plane around the real axis. Further they show the following singularities: a finite collection of simple poles $\{\kappa_m^e, m=1, \dots, M\}$ and a branch point $k_z = k_2$, where $k_2 = \omega(\hat{\varepsilon}_2\hat{\mu}_2)^{\frac{1}{2}}$ in the fourth quadrant of the k_z-plane, while, due to the reflection symmetry, in the second quadrant the singularities are a collection of simple poles $\{-\kappa_m^e\}$ and the branch point $k_z = -k_2$. In order to show the relevance of these singularities to the representation of the field $\{\underline{\hat{E}}^e,\underline{\hat{H}}^e\}$, we close the path of integration in

$$(3.3)\qquad \{\underline{\hat{E}}^e,\underline{\hat{H}}^e\} = (2\pi)^{-1}\int_{k_z\in\mathbb{R}} \exp(-jk_z z)\{\underline{\tilde{E}}^e,\underline{\tilde{H}}^e\}dk_z,$$

in the upper (lower) half plane when $z<0(z>0)$. The supplementing circular arcs, at infinity yield a vanishing contribution for all $\underline{r}_T$, provided that the branch cuts $\mathcal{B}^-(\mathcal{B}^+)$ are introduced joining $-k_2(k_2)$ and infinity such that $\mathrm{Im}(k_2^2 - k_z^2)^{\frac{1}{2}} = 0$; in the cut k_z-plane, we then have $\mathrm{Im}(k_2^2 - k_z^2)^{\frac{1}{2}} < 0$. When $z > 0$, the resulting expression is written as

$$(3.4)\qquad \{\underline{\hat{E}}^e,\underline{\hat{H}}^e\} = \sum_{m=1}^{M} \{\underline{E}_m^e,\underline{H}_m^e\}\exp(-j\kappa_m^e z) + \int_{\kappa\in\mathcal{B}^+} \{\underline{E}_\kappa^e,\underline{H}_\kappa^e\}\exp(-j\kappa z)d\kappa \quad (z > 0),$$

where $\{\underline{E}^e_m, \underline{H}^e_m\}$ denotes $j \times \mathrm{Res}_{k_z=\kappa^e_m}\{\tilde{\underline{E}}^e, \tilde{\underline{H}}^e\}$ and $\{\underline{E}^e_\kappa, \underline{H}^e_\kappa\}$ denotes the jump in $\{\tilde{\underline{E}}^e, \tilde{\underline{H}}^e\}$ across $\mathcal{B}^+$ in accordance with the choosen direction of integration along $\mathcal{B}^+$(see Figure 2). The summation on the right-hand side of (3.4)

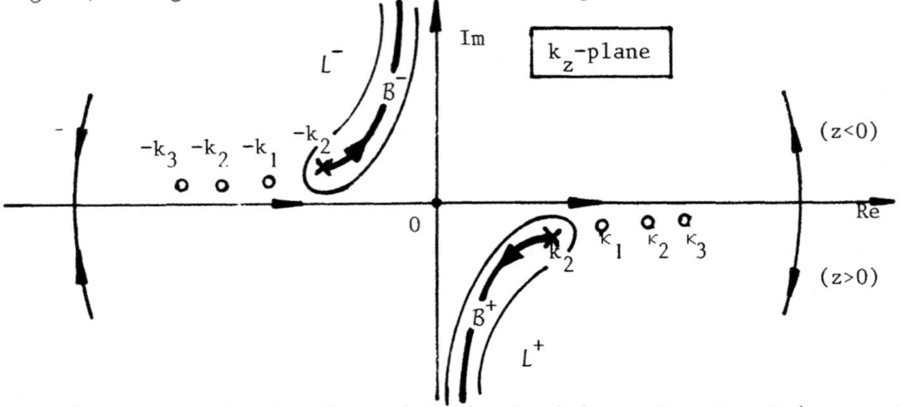

FIGURE 2. The complex k_z-plane with singularities and paths of integration.

represents the contribution of the surface-wave modes to the field; the integration represents the contribution of the radiation modes to the field. When $z < 0$ the relevant expressions follow from (3.4) in view of the symmetry properties

$$\begin{aligned} \hat{\underline{E}}^e_T &\to +\hat{\underline{E}}^e_T, & \hat{E}^e_z &\to -\hat{E}^e_z, \\ \hat{\underline{H}}^e_T &\to -\hat{\underline{H}}^e_T, & \hat{\underline{H}}^e_z &\to +\hat{\underline{H}}^e_z, \end{aligned} \qquad \text{when } z \to -z. \tag{3.5}$$

From (3.4) and the corresponding expression for $z < 0$, we recover the general strucure of $\{\tilde{\underline{E}}^e, \tilde{\underline{H}}^e\}$ by subjecting the field representation to the Fourier transform with respect to z over the entire range $z \in \mathbb{R}$. The result is

$$\{\tilde{\underline{E}}^e_T, \tilde{H}^e_z\} = \sum_{m=1}^{M} \{\underline{E}^e_{m,T}, H^e_{m,z}\} \frac{-2\kappa^e_m}{j(k_z^2 - \kappa^{e2}_m)} + \int_{\kappa\in\mathcal{B}^+} \{\underline{E}^e_{\kappa,T}, H^e_{\kappa,z}\} \frac{-2\kappa}{j(k_z^2 - \kappa^2)}\, d\kappa, \tag{3.6}$$

$$\{\tilde{\underline{H}}^e_T, \tilde{E}^e_z\} = \sum_{m=1}^{M} \{\underline{H}^e_{m,T}, E^e_{m,z}\} \frac{-2k_z}{j(k_z^2 - \kappa^{e2}_m)} + \int_{\kappa\in\mathcal{B}^+} \{\underline{H}^e_{\kappa,T}, E^e_{\kappa,z}\} \frac{-2k_z}{j(k_z^2 - \kappa^2)}\, d\kappa. \tag{3.7}$$

Equations (3.6) and (3.7) clearly show that the branch-cut contribution can be envisaged as a continuous version of the summation of discrete modes. The choice of our branch cut is in accordance with this point of view, as has been stipulated by FELSEN and MARCUVITZ [1].

4. ORTHOGONALITY, NORMALIZATION AND COMPLETENESS OF THE MODAL CONSTITUENTS

The field distributions in the transverse plane $\{\underline{E}^e_m,\underline{H}^e_m\}$ and $\{\underline{E}^e_\kappa,\underline{H}^e_\kappa\}$ occurring in (3.4) are, of course, dependent on $\hat{\underline{J}}_T$ in (3.1). To express this (linear) dependence, we introduce properly normalized field distributions of the type $\{\underline{e}^e_m,\underline{h}^e_m\}\exp(-j\kappa^e_m z)$ and $\{\underline{e}^e_\kappa,\underline{h}^e_\kappa\}\exp(-j\kappa z)$ with $\kappa\in\mathcal{B}^+$, and corresponding expressions for $z < 0$ (cf. (3.5)). Each of them satisfy the source-free electromagnetic field equations

$$(4.1)\qquad \underline{\tilde{\nabla}}^e_m \times \underline{h}^e_m - j\omega\hat{\varepsilon}\underline{e}^e_m = \underline{0},$$

$$(4.2)\qquad \underline{\tilde{\nabla}}^e_m \times \underline{e}^e_m + j\omega\hat{\mu}\underline{h}^e_m = \underline{0},$$

with $\underline{\tilde{\nabla}}^e_m = \underline{\nabla}_T - j\kappa^e_m\underline{i}_z$, and simular equations for $\{\underline{e}^e_\kappa,\underline{h}^e_\kappa\}$. With the aid of (4.1) and (4.2), applied to two different surface-wave modes propagating in opposite directions, we then arrive at

$$(4.3)\qquad \int_{\mathcal{D}}(\underline{e}^e_m \times \underline{h}^e_m)\cdot\underline{i}_z d\underline{r}_T = \tfrac{1}{2}\, N^e_n\delta_{m,n},$$

where $\mathcal{D}$ denotes the entire cross section of the waveguide and N^e_n is a properly choosen normalization constant. When $m \neq n$, (4.3) expresses the orthogonality of two different surface-wave modes. In a similar way we arrive at

$$(4.4)\qquad \int_{\mathcal{D}}(\underline{e}^e_{\kappa'} \times \underline{h}^e_{\kappa''})\cdot\underline{i}_z d\underline{r}_T = \tfrac{1}{2}\, N^e_{\kappa'}\delta(\kappa' - \kappa''),$$

with $\kappa'\in\mathcal{B}^+$, $\kappa''\in\mathcal{B}^+$. If $N^e_n = 1$ and $N^e_\kappa = 1$ for all modes, we call the modes Lorentz normalized.

We now write

$$(4.5)\qquad \{\underline{E}^e_m,\underline{H}^e_m\} = \alpha^e_m\{\underline{e}^e_m,\underline{h}^e_m\},\quad \{\underline{E}^e_\kappa,\underline{H}^e_\kappa\} = \alpha^e_\kappa\{\underline{e}^e_\kappa,\underline{h}^e_\kappa\},$$

where α^e_m and α^e_κ are directly related to $\hat{\underline{J}}_T$. To arrive at the corresponding expressions, we observe that from the frequency-domain counterpart of (3.1) it follows that

$$(4.6)\qquad \lim_{\zeta\downarrow 0}\,[\underline{i}_z \times \hat{\underline{H}}^e]_{z=-\zeta}^{z=\zeta} = \hat{\underline{J}}_T .$$

Using (3.4) and (4.5) in (4.6) we arrive at

$$(4.7)\qquad 2\sum_{m=1}^{M} \alpha_m^e(\underline{i}_z \times \underline{h}_m^e) + 2\int_{\kappa\in B^{+\kappa}} \alpha_\kappa^e(\underline{i}_z \times \underline{h}_\kappa^e)d\kappa = \hat{\underline{J}}_T .$$

Applying the respective operations $\int_D \underline{e}_n^e d\underline{r}_T\cdot$ and $\int_D \underline{e}_{\kappa'}^e d\underline{r}_T\cdot$ to (4.7) and using (4.3) and (4.4) we arrive at

$$(4.8)\qquad \alpha_n^e = N_n^e \int_D \underline{e}_n^e\cdot\hat{\underline{J}}_T d\underline{r}_T,\ \alpha_\kappa^e = N_\kappa^e\int_D \underline{e}_\kappa^e\cdot\hat{\underline{J}}_T d\underline{r}_T .$$

Substitution of (4.8) in (4.7) shows that the sets $\{\underline{e}_m^e,\underline{h}_m^e\}$ and $\{\underline{e}_\kappa^e,\underline{h}_\kappa^e\}$ with $\kappa\in B^+$, form a complete, orthogonal set of functions in which any excitation by transverse electric-current distributions located in a single transverse plane can be expressed. The simplest normalization occurs when the Lorentz normalization is used; this normalization will be understood in the subsequent analysis. Writing

$$(4.9)\qquad \hat{\underline{E}}^e = \int_D \underline{\underline{G}}^{ee}\cdot\hat{\underline{J}}_T d\underline{r}_T' ,\ \hat{\underline{H}}^e = \int_D \underline{\underline{G}}^{me}\cdot\hat{\underline{J}}_T d\underline{r}_T',$$

we obtain by inspection

$$(4.10)\qquad \underline{\underline{G}}^{ee} = \sum_{m=1}^{M} \underline{e}_m^e(\underline{r}_T)\otimes\underline{e}_m^e(\underline{r}_T')\exp(-j\kappa_m^e z) + \int_{\kappa\in B^{+\kappa}} \underline{e}_\kappa^e(\underline{r}_T)\otimes\underline{e}_\kappa^e(r_T')\exp(-j\kappa z)d\kappa,$$

$$(4.11)\qquad \underline{\underline{G}}^{me} = \sum_{m=1}^{M} \underline{h}_m^e(\underline{r}_T)\otimes\underline{e}_m^e(\underline{r}_T')\exp(-j\kappa_m^e z) + \int_{\kappa\in B^{+\kappa}} \underline{h}_\kappa^e(\underline{r}_T)\otimes\underline{e}_\kappa^e(r_T')\exp(-j\kappa z)d\kappa.$$

A similar analysis holds for the fields generated by the transverse magnetic-current distribution $\hat{\underline{K}} = \hat{\underline{K}}_T\delta(z)$. Writing this field as

$$(4.12)\qquad \hat{\underline{E}}^m = \int_D \underline{\underline{G}}^{em}\cdot\hat{\underline{K}}_T d\underline{r}_T',\ \hat{\underline{H}}^m = \int_D \underline{\underline{G}}^{mm}\cdot\hat{\underline{K}}_T d\underline{r}'_T,$$

we obtain

$$(4.13)\qquad \underline{\underline{G}}^{em} = \sum_{m=1}^{M} \underline{e}_m^m(\underline{r}_T)\otimes\underline{h}_m^m(\underline{r}_T')\exp(-j\kappa_m^m z) + \int_{\kappa\in B^{+\kappa}} \underline{e}_\kappa^m(\underline{r}_T)\otimes\underline{h}_\kappa^m(\underline{r}_T')\exp(-j\kappa z)d\kappa,$$

$$(4.14)\qquad \underline{\underline{G}}^{mm} = \sum_{m=1}^{M} \underline{h}_m^m(\underline{r}_T)\otimes\underline{h}_m^m(\underline{r}_T')\ \exp(-j\kappa_m^m z) + \int_{\kappa\in\mathcal{B}} \underline{h}_{\mp\kappa}^m(\underline{r}_T)\otimes\underline{h}_\kappa^m(\underline{r}_T')\ \exp(-j\kappa z)d\kappa.$$

Equations (4.10), (4.11), (4.13) and (4.14) can be used to construct the Lorentz normalized orthonormal modal constituents from the tensorial point-source fields $\underline{\underline{G}}^{ee}$, $\underline{\underline{G}}^{me}$, $\underline{\underline{G}}^{em}$ and $\underline{\underline{G}}^{mm}$.

5. CONCLUSION

In the previous sections we have arrived at the complete modal expansions of the tensorial Green's functions that fully describe the transmission properties of a uniform open waveguide section. As could be expected, the modal expansion contains the discrete surface-wave modes as well as the continuous radiation modes. It is shown that the Lorentz normalized orthogonal transverse field distributions of the modal constituents show up in the expressions for the tensor Green's functions in a natural way. In the literature the discussion on the modal expansion of the tensor Green's functions is restricted to separable geometries [e.g. 1, 2, 3, 4], the present analysis shows that this restriction is unnessesary.

REFERENCES

1. Felsen, L.B. and N. Marcuwitz, *Radiation and scattering of waves*, Prentice-Hall, Inc., Englewood Cliffs, N.J., 1973, Chapter 3.
2. Snyder, A.W., Continuous mode spectrum of a circular dielectric rod, IEEE Trans. Microwave Theory Techn., Vol. MIT-19, 1971, pp. 720-727.
3. Shevchenko, V.V., *Continuous transitions in open waveguides*, The Golem Press, Boulder, 1971.
4. Vassallo, C., Radiating normal modes of lossy planar waveguides, J.O.S.A., Vol. 69, 1979, pp. 311-316.

MICROWAVE APPROACH TO OPTICAL WAVEGUIDES

Huang Hung-chia

Shanghai University of Science and Technology

ABSTRACT

This paper proposes an educational approach to optical waveguides from the viewpoint of a microwave theorist. The substance of the text is based on the published literature, including especially some papers by the present author. Emphasis is laid on the coupled-mode theories early developed in microwave science, which can be profitably borrowed, or extended in a natural way, in structuring an educational scheme for understanding optical waveguides in general, and optical fibers in particular.

INTRODUCTION

Modern optical waveguides have become a common area for optics and microwaves research. In this new branch of science, microwaves and optics have a particularly intimate union for optimum solutions of problems relating to a wide variety of optical waveguides.

The union of microwaves and optics dates back to the very beginning of Maxwell's theory. A highlight of this union is the reformulation of the entire classical optics by the use of Maxwell equations. At the early time of Hertz, microwave technology borrowed heavily from light optics, to which man had accumulated a wealth of experimental and theoretical contributions since antiquity, and such achievements in optics were completely independent of the Maxwell's theory of electromagnetic wave, and were well established in their own right.

Both microwaves and optics have played important roles in research and development of fiber optics, but on the whole, it seems that microwaves appears more prominent. If it is correct to say, as stated above, that microwaves borrowed heavily from optics at the early stage of their union, then, it is now justifiable to say that microwaves has returned, or partially returned, its debt to optics.

The purpose of this paper is to present some basic viewpoints relevant to the educational aspect of optical waveguide sciences. It will be seen that the microwave approach is preferrable particularly in the study of single-mode optical fiber, which is one of the highly focused research areas of the scientific discipline concerned.

1. MICROWAVE GUIDES AND OPTICAL FIBERS

In this paper, we are speaking of microwave guides in a narrow sense, referring to closed metallic structures of different cross-sectional shapes. In the long run, a "sheltered" transmission medium will probably be more attractive than radio in many respects for both continental and transoceanic communications. About two or three decades ago, waveguide transmission using H_{01} mode at millimeter wavelengths was considered the most promising technology for long-haul high-capacity communications. Intensive world-wide study in this subject area for many years has structured a consistent and self-contained theoretical frame-work which is relevant and remains useful for the study of modern optical fibers.

From a theoretical point of view, or from the standpoint of Maxwell equations, optical waveguides, especially the single-mode versions, do not differ much from microwave guides, even though there exists an enormous disparity in their operating wavelengths. A careful comparison of the theoretical frame-work adopted in the early microwave literature[(1)] and that adopted in the theory of dielectric optical waveguides[(2)] discloses that both begin with the study of normal modes, then with couplings of ideal modes, and then with couplings of local modes and so on. A distinctive feature of optical fibers is the possible existence of a continuum of radiation modes and a continuum of evanescent fields, in addition to the set of discrete guided modes.

Owing to the above difference, a crucial step necessary for structuring a unified theoretical frame-work applicable both to microwave guides and optical fibers is to discretize the continuum of mode spectrum associated with optical fibers.

Physical intuition suggests that the continuum of modes may be considered a sequence of discrete modes, in the same manner that an integral is replaced by a summation for calculation purposes. In a previous paper[3], we considered the discretization of the continuum of modes under certain restrictions, including especially the weak coupling restriction. In a later unpublished work by Qian[4], it was shown that discretization of the continuum of modes can be done on a mathematical basis, without recourse to the weak-coupling restriction.

Thus, we discretize the electric field $\overline{E}$ of the radiation-mode continuum as

$$\overline{E} = \sum_{\ell} \int_{-n_2k}^{n_2k} \xi(\beta)\overline{E}_{\ell}(\beta)d\beta = \sum_{\ell} \sum_{k=1}^{M} A_{\ell k} \overline{E}_{\ell k} \tag{1}$$

where ℓ is the circumferential index of radiation modes, and $k=1,2\cdots M$ indicates the sequence of the discretized segments, M being sufficiently large to ensure the required accuracy. The meanings of n_1, n_2, k and β are in comformity with any standard text, and need no expalanation. The normalized electric field of radiation modes is

$$\overline{E}_{\ell k} = 1/(\Delta\beta_k) \int_{\beta_{k-1}}^{\beta_k} E_{\ell}(\beta)\ d\beta \tag{2}$$

with $\Delta\beta_k = \beta_k - \beta_{k-1}$, and the mode amplitude is

$$A_k = \xi_{\ell}(\beta_k)/(\Delta\beta_k) \tag{3}$$

Note that Eqs.(1) and (2) are valid also for the magnetic field if $\overline{E}$ in the equations is replaced by $\overline{H}$.

Introduction of the factor $1/(\Delta\beta_k)$ in the above equations is necessary for expressing the orthonormalization for the discretized modes in the form

$$\int_{\Omega} \overline{E}_{\ell i} \times \overline{H}^{*}_{\ell k} \cdot \overline{i}_z\ d\Omega = \delta_{ik}\beta_i/|\beta_i| \tag{4}$$

where the Kronecker delta takes the place of the Dirac delta function in the usual orthonormal expression for the rediation continuum.[2]

With the aid of the above mathematical artifice, all formulas established in microwave theory, notably the coupled-mode formulation and the modal transformations, become equally applicable to solving the fiber-optic problems.

2. EIGEN-MODES AND COUPLED MODES

The concept of mode is a topic of frequent discussion, both in microwaves and in fiber-optics. In our papers [3,5] we illustrate, with taper and bend as specific examples, the concepts of ideal modes, of local modes and of super-local modes. The three sets of modes apply equally well to microwave guides and optical fibers. As a matter of fact, nonconventional versions of optical fibers [6], such as twisted and spun fibers, provide excellent examples for illustration of the distinctive properties of the super-local modes. (Refer pp.57 - 68 of this Proceedings)

It seems that the importance of the existence of a variety of mode sets, especially the above three, has not yet received sufficient appreciation. Early in mid-'50s, it was pointed out without proof that eigen-modes and coupled modes are related by a linear transformation [7]. Towards the end of that decade [5], the author obtained by way of the matrix theory the result that if the matrices associated with the ideal modes A, local modes W and super-local modes $\overline{W}$ are, respectively, K, N and $\overline{N}$ (i.e. $A' = KA$, $W' = NW$, $\overline{W}' = \overline{NW}$), then K, N and $\overline{N}$ are similar in the sense of Lowey, such that

$$N = O_k^{-1} K O_k - O_k^{-1} O_k' \tag{5}$$

where O_k is the modal matrix of K. The expression of $\overline{N}$ takes exactly the same form, if k is replaced by N.

In his classic treatise on dielectric waveguides [2], D. Marcuse considered the local-mode expansion as well as the ideal-mode expansion separately, both starting from Maxwell equations. The lengthy mathematical derivations with all the vector notations involved look rather embarrassing. We prefer the use of matrix algebra which relates different sets of modes by similarity properties of the matrices concerned. Matrix algebra is a straightforward mathematics, but is tedious for the general case involving matrices of an order higher than 2. Fortunately, in the weak-coupling regime of multimode fibers, and in the general case of single-mode fibers involving only two polarized modes, the relevant matrix equation is not only succint in form, but also admirably simple in performing actual derivations and calculations.

The above line of approach appears mathematically elegant, and is probably an adequate approach as far as the educational aspect of optical waveguides is concerned.

3. MODE COUPLINGS AND ARTIFICIAL SOURCES

Coupling of modes are produced by any deviations of the actual waveguide structure from a simpler waveguide structure chosen as the reference, for which the complete set of modes is available, or can be found more easily. For our immediate purpose, we put the coupled-mode equations in the form

$$A_i' = K_i A_i + K_{ij} A_j, \qquad i = 1,2 \cdots n \tag{6}$$

where A_i is the amplitude of mode i, K_i is its propagation coefficient, and K_{ij} $(i \neq j)$ is the coupling coefficient between modes i and j. Eqs.(6) are homogeneous for source-free cases. It has been well recognized that such equations are extremely useful in studying fiber-optics just as they are in studying microwave guide transmission.

Alternatively, if we consider the medium perturbation $(\bar{\bar{\varepsilon}}' - \bar{\bar{\varepsilon}})$ as a fictitious current source

$$\bar{J} = i\omega(\bar{\bar{\varepsilon}}' - \bar{\bar{\varepsilon}}) \cdot \bar{E} \tag{7}$$

where $\bar{\bar{\varepsilon}}'$ and $\bar{\bar{\varepsilon}}$ are respectively the tensor dielectric constant of the unperturbed medium and of the perturbed medium, then the relevant differential equations become inhomogeneous, with the presence of some source functions:

$$A_i' = K_i A_i - \tfrac{1}{2}(\beta_i/|\beta_i|) \int_\Omega \bar{E}_i \cdot \bar{J}\, d\Omega \quad i = 1,2 \cdots n \tag{8}$$

where A_i are coefficients of the field expansions:

$$\bar{E} = \sum_k A_i \bar{e}_i \,, \qquad \bar{H} = \sum_k A_i \bar{h}_i \tag{9}$$

Each summation includes all guided modes and the discretized radiation modes. $\bar{e}_i$, $\bar{h}_i$ are vectors satisfying the orthonormal expression (4).

Substitution of Eq.(7) in Eqs.(8), which are the so-called excitation equations, leads to the coupled-mode equations(6). The expressions for coupling coefficients thus found agree with the results derived from conversion of Maxwell equations[(2)].

In our early work[(1)], we showed the equivalence of Eqs.(6) and (8), with microwave guides taken as illustrative examples. Inasmuch as most microwave guides are described by their wall impedances, we naturally focused our attention on wall irregularities,[(8)] for which we introduced the concept of wall-impedance perturbations as fictitious sources causing couplings of modes. In optical fibers, dielectric medium perturbations possess a pre-eminent position, yet the equivalence of the above two formulations remains valid.

4. PHYSICAL INTUITION AND EARLY APPROXIMATIONS

It appears that there is a continuing need for simplification of mathematical procedures as well as clarification of concepts in the study of complicated boundary-value problems involved in nonideal microwave and optical waveguides.

A casual look at the relevant literature, including some early publications by the present author, reveals that almost all the papers begin with Maxwell equations, and arrive at the required answers only after a lengthy mathematical manipulation. The basic concepts and the key points crucial to the analysis are often obscured under the mask of the multitude of complicated mathematical equations.

Admittedly, mathematics is indispensable in almost every branch of modern science. However, as our understanding of the problem grows more mature, it becomes possible to construct a theoretical frame-work on the known facts with the least effort at mathematical derivation.

We again take coupled-mode equations as an example. Physical intuition will help to establish the coupled-mode equations in a simple phenomenological way. In every infinitesmally short section of the waveguide each mode A_i undergoes a small change in amplitude ΔA_i, partially as a result of wave propagation in this mode along the short section, and partially due to energy exchanges with all the coupled modes, yielding $\Delta A_i = K_i A_j \Delta z + K_{ij} A_j \Delta z$ which, when divided by Δz and passing to the limit $\Delta z \to 0$, is exactly the set of coupled-mode equations frequently used in microwave and optical waveguides.

While expressions for the coupling coefficients are obtained as a direct consequence of converting Maxwell equations into coupled-mode equations, they can also be derived in a very simple manner by adopting early approximations. On the analogy of low-frequency coupled circuits, the coupling coefficients can be defined as $P_{ji}/(P_i P_j)^{\frac{1}{2}}$, P_i and P_j being the power carried by mode i and of mode j, respectively, and P_{ji}, the power converted from mode i to mode j. With reference to Eq. (7), perturbation of the medium permittivity $\Delta\bar{\bar{\varepsilon}} = (\bar{\bar{\varepsilon}}' - \bar{\bar{\varepsilon}})$ can be replaced by a fictitious source with current density $i\omega(\Delta\bar{\bar{\varepsilon}}) \cdot \bar{E}_i$, where $\bar{E}_i$ is the vector electric field of the incident mode i. Early approximation is adopted by assuming that couplings between the incident and spurious modes are sufficiently weak so that the power carried by the incident mode is nearly constant, and the power generated by this fictitious source is reasonably interpreted as a conversion to the spurious mode j. Let a_i and a_j be the amplitudes of $\bar{E}_i$ and $\bar{E}_j$, so normalized that $|a_i|^2 = P_i$, and $|a_j|^2 = P_j$, then substitution of the above results in the aforesaid formula for low frequency coupled oscillation leads to the expression for the coupling coefficient[9]

$$K_{ji} \simeq \pm \frac{i\omega}{2} \int \Delta\bar{\bar{\varepsilon}}\ \bar{E}_i \cdot \bar{E}_j \, ds \tag{10}$$

with + and - referring, respectively, to forward- and backward-mode coupling.

Other well-known examples may be referred to for which judicious early approximations yield useful results, such as the introduction of weakly guiding approximation by Snyder[10] and Gloge[11], and the Gaussian approximation of the fields in conventional single-mode optical fiber[12]

CLOSING REMARKS

In this paper, we have stressed the importance of the simplified theories for education of optical waveguide sciences. But it is by no means our intention to underestimate the coherence and beauty as well as the amazingly useful function that mathematics has displayed in all branches of science, including of course the modern fiber-optics.

Among the different lines of mathematical approach, the function-theoretical technique is perhaps most elegant in its capacity to define at one blow all kinds of modes existing in an open waveguide structure like the optical fiber. Derivations involved in this approach are convincing enough, mathematically, but it is almost impossible to associate each mathematical step with a physical meaning.

Mathematics often goes a long way without any apparent physical meaning attached to it, but eventually yields the expected or unexpected results, which can hardly be found, or even imagined, by physical comtemplation alone. It reflects the reality exactly as physics does, though in an implicit and sometimes rather intriguing way. From the educational standpoint, the author believes that, while the mathematical approach is never to be neglected, it may be appropriate to include this approach at a later stage in the curriculum.

The present paper attempts to present a consistent analytic approach to optical waveguides from the standpoint of microwave theories, which is not difficult in mathematics, but is easily accessible to physical interpretation. As the author understands, modal analysis well developed in microwaves is capable of dealing with a wider subject area of fiber-optics than the ray method. However, it is quite possible that the ray-optic technique, if only sufficiently extended, can possibly solve almost the same problems which are solvable by microwave methods, especially for multimode optical fibers. Furthermore, a composite method combining the advantages of both microwave and optic methods may prove useful for certain problems. Indeed, the forming of a satisfactory scheme for education of optical waveguide sciences remains a challenge to the large scientific cicle concerned.

ACKNOWLEDGEMENT

The author is particularly indebted to Prof. Qian Jing-ren of the University of Science and Technology of China (Hefei) for very many times of fruitful and valuable discussions.

REFERENCES

1. Huang Hung-chia, Microwave Principles, vol. I & II, Academia Sinica Press, 1963 & 1964.
2. D. Marcuse, Theory of Dielectric Optical Waveguides, Academic Press, New York & London, 1974.
3. Huang Hung-chia, Weak coupling theory of optical fiber and film waveguides, Radio Science, vol.16, No.4, p.495, July-Aug. 1981.
4. Qian Jing-ren, unpublished work.
5. Huang Hung-chia, Generalized theory of coupled local normal modes in multi-wave guides, Scientia Sinica, vol.9, p.142, Jan. 1960.
6. Huang Hung-chia and Qian Jing-ren, Theory of imperfect nonconventional single-mode optical fibers, pp. - , this publication.
7. Proceedings of the Symposium on Modern Advances in Microwave Techniques, New York, N. Y., Sponsored by Microwave Research Institute of Polytechnic Institute of Brooklyn, Nov. 8-10, 1954 (pp.481-492).
8. Huang Hung-chia, General theory of nonconventional waveguides for long-distance transmission, Scientia Sinica, vol.11, p.761, June 1962.
9. Huang Hung-chia, Coupled Modes and Nonideal Waveguides (Collected papers), Microwave Research Institute (MRI), Polytechnic Institute of New York, November, 1981.
10. A.W. Snyder, Asymptotic expressions for eigenfunctions and eigenvalues of a dielectric or optical waveguide, IEEE Trans. MTT-17, p.1130, Dec. 1969.
11. D. Gloge, Weakly guiding fibers, Appl. Opt., vol.10, p.2252, Oct. 1971.
12. A• W• Snyder, Understanding monomode optical fibers, Proc. IEEE, vol.69, No.1, p.6, Jan. 1981.

A COURSE IN OPTICAL WAVEGUIDES*

B.P. PAL+ and R.TH. KERSTEN

Fraunhofer-Institut für Physikalische Meßtechnik
Heidenhofstraße 8, D-7800 Freiburg, West Germany

1. INTRODUCTION

In the last decade the fields of fiber and integrated optics have seen vigorous research and developmental efforts throughout the world. The basic of this emerging descipline of science and technology is an optical waveguide. Several books /1-16/ and reviews /17-20/ are available now on these topics. However, to our opinion it is difficult to use any of these straightaway to develop an unified syllabus for class room teaching of optical waveguides. In the following we have tried to outline a course structure on optical waveguides with the aim of making it a basic course on the subject so that the students are well prepared to undertake detailed R&D work on one or more areas of its special applications. In part I, after a discussion of the course approach, we outline theory course topics while in part II we attempt to provide guidelines for some experiments which should supplement the theoretical knowledge.

2. PART I: THEORY

2.1. Course approach:

In introducing a new course it is instructive to devote some time on the motivation behind its introduction. Introductory lectures should therefore discuss topics like: potentiality of optical waveguides to yield optical devices in microstructures owing to their dimensions being comparable to optical wavelengths /20/, large information carrying capacity at optical frequencies /21/, advantages of optical fibers and integrated optics /20, 21/, development of high purity materials and composites for realizing low loss waveguides, semiconductor lasers, and photodetectors /8/.

---

*Work supported by Alexander-von-Humboldt-Stiftung, Bonn, FRG.

+on leave from Indian Institute of Technology Delhi, Physics Department, New Delhi-110016, India.

Descriptions of propagation effects in multimode wavegudies are documented in several books. However, the following discussions are often missing in these books: utility of the knowledge of modal fields in power coupling estimations, physical interpretation of integers identifying a mode, cladding modes and the necessity of stripping these modes in fiber characterization experiments, the consequence of the existence of a narrow boron-doped inner cladding and axial index dip in real CVD-fibers, utility of differential mode delay characterization of a multimode fiber for correcting preform fabrication process, multiple α-profiled fibers, and so on.

In view of the recent emergence of monomode fibers as i) most suitable media for longhaul high capacity optical transmission systems and ii) being most suited for a majority of fiber sensors, it is necessary to include a detailed description of waveguidance in monomode fibers. For example, the rigorous calculation of the $d^2\beta/d\omega^2$ term for predicting the minimum dispersion wavelength /22/, which is an important monomode fiber design parameter. Due to the deviation of practical monomode fibers from an ideal step index profile, the concept of equivalent step index profile (ESI) /23, 24/ and the variational method /25, 26/ for the fundamental mode field have assumed great importance in the prediction of monomode fiber performance. In the same context the two-mode fibers also have assumed some importance because of the possibility of achieving zero dispersion even in these fibers /27/ in addition to the benefit of greater splicing tolerances /28/.

Another topic which has attracted attention is the technology of fiber optic sensors. A number of fiber optic sensors /29/ have already been demonstrated e.g. hydrophones, magnetometers, gyroscopes, liquid level, bubble and pressure sensors. Broadly all these sensors are based on phase or intensity modulation of light propagating through a fiber by the external stimulus to be sensed. Polarization properties of monomode fibers, e.g. birefringence, beat length, polarization maintainance etc. require tutorial level discussions to form the foundations for undertaking R&D studies on fiber sensors relying on phase measurements.

2.2. Course topics:

2.2.1. Introduction: definition of optical waveguides, advantages of optical waveguide devices, and large information carrying capacity at optical frequencies.

2.2.2. Pre-requisite optics and electromagnetic wave theory: Maxwell's equations, vector and scalar wave equations, reflection and refraction of plane waves, Bragg angle, evanescent waves and Goos-Hähnchen shift, interference, diffraction and spatial filtering, polarization.

2.2.3. Slab and rectangular strip waveguides: ray optics of waveguidance, exact solutions to pure TE- and TM-modes in homogeneous waveguides, power flow in planar structures, rectangular waveguides, graded index waveguides, examples of some passive and active waveguiding devices e.g. directional couplers etc.

2.2.4. Circular waveguides: weakly guiding fibers and linearly polarized modes, exact scalar wave solution for step-index fiber, cut-off frequencies, power flow calculations and necessity of low loss cladding for monomode fibers, Snyder-Young modes, homogeneous w-fibers.

2.2.5. Graded index optical fibers: Laguerre-Gauss modes of an infinitely extended parabolic medium, WKB-theory of cladded power-law index, brief outlines of approximate approaches: perturbation and variational analyses, power series expansion, finite element and staircase approximation.

2.2.6. Dispersion: broadening of a gaussian pulse in a fiber, Gloge-Marcatili's theory of impulse response, differential mode delay, influence of excitation conditions on impulse response measurements, modal noise.

2.2.7. Monomode and two-mode fibers: intramodal dispersion in step-index monomode fibers, wavelength of minimum dispersion, graded index monomode fibers, gaussian approximation for the fundamental mode and ESI formalism, variational analyses of graded core monomode and two-mode fibers, polarization properties of monomode fibers and their importance in sensors.

2.2.8. Ray optics of waveguidance in multimode fibers: derivation of ray equation, ray classifications, power excitation and pulse dispersion under ray optics, examples on: fiber endoscopy, faceplates, image converters, multimode fiber optic sensors etc.

2.2.9. Fundamentals of coupling: Liouville's theorem and its application to source-waveguide coupling, fiber-to-fiber coupling.

2.2.10. Bending loss and mode coupling effects: bending loss in optical waveguides, coupled mode theory, impulse response of mode coupled fibers.

2.2.11. Materials for integrated and fiber optical waveguides: sources of attenuation in waveguides, fabrication of low loss I.O. and fiber waveguides.

3. PART II: PLANNING A TEACHING LABORATORY ON OPTICAL WAVEGUIDES

A dark room laboratory is required to carry out experiments on optical waveguides although for a few experiments it is not essential e.g. for those with phase sensitive detection by means of lock-in amplifiers. However, we recommend a well cleaned dark room. We give below an almost complete list of equipments for planning teaching and research level experiments on optical waveguides. It will be possible to perform a lot of experiments even if not all the equipments are available:

Few to several optical benches supported on vibration free legs (e.g. on inflated tubes) or granite tables; large number of stackable x-y-z translation stages (of 10 μm and a few with 1 μm resolution); 360° rotation table with resolution of about 1 minute of arc (for high precision about 30 seconds); microscope objectives with different NA (0.1-0.6) and magnification (4x to 40x); optical components: mirrors, beamsplitters, lenses, prisms, polarizers, babinet compensator, neutral density and interference filters, $\lambda/2$ and $\lambda/4$ plates etc.; infrared viewer; monochromator; spatial filter with different pinhole sizes; HeNe-lasers; LEDs and semiconductor lasers (some with fiber pigtail) with emission around 0.8 or 1.3 μm; tungsten lamps with highly stabilized power supply; PIN and avalanche detectors (Si and Ge); optical power meter; general purpose laboratory equipments (power supplies, oscilloscopes, pulse generators etc.); lock-in amplifier and mechanical chopper; fusion splicing kit, several fibers: step & graded multimode, monomode (at 0.63, 0.8, 1.3 μm), highly birefringent, single polarization etc.; chemicals like $AgNO_3$, polystyrene, modestrippers, optical epoxies etc.

In the following we briefly discuss some experiments which should supplement the theory syllabus outlined in Part I. For brevity we avoid details on waveguide characterization available in books /11, 14, 15/.

In order to demonstrate large information transmission capacity of an optical fiber the best solution is to buy an inexpensive commercial low bit rate time division fiber system (available from several companies like Hewlett Packard, ITT etc.) and demonstrate the system to the students.

A homogeneous planar waveguide can be fabricated e.g. by coating a well cleaned microscope glass slide with a higher refractive index material e.g. polystyrene dissolved in toluene by dipping the slide into the solution followed by a gradual withdrawal /30/. Waveguidance in this structure can be demonstrated by the prism-film evanescent wave coupling of the laser light

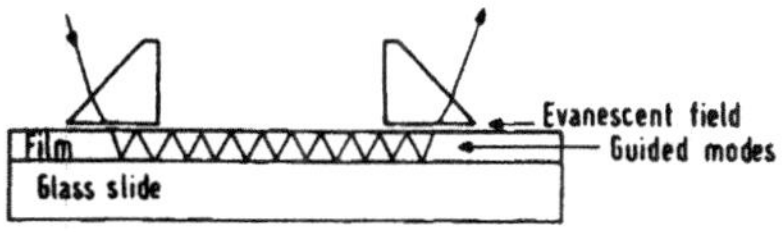

FIGURE 1. Principle of prism-film coupling.

into the waveguide (cf. Fig.1). Each guided mode appears as a band (so called m-lines) when projected on a screen by placing a decoupling prism in reverse orientation on the guide. Several physical phenomena/parameters corresponding to a planar optical waveguide can be described/determined: evanescent wave, waveguidance, propagation constant of a mode from a measurement of the exit angles of the m-lines, TE- and TM-modes being two distinct polarization states by rotating a sheet polarizer in front of the screen, approximate attenuation of the waveguide by dividing 27 dB which is the average dynamic range for human eye in darkness with the measured distance of scattered light observable on the waveguide, and so on. Fabrication of inhomogeneous waveguides can be made e.g. by thermal diffusion of $AgNO_3$ into a glass slide and above experiments can be repeated. The distinction between a homogeneous and an inhomogeneous waveguide can be demonstrated by noting the spacing between the m-lines which will be different in the two cases.

The different optical regions (core, cladding) of an optical fiber can be displayed by injecting white light into one end of a small piece (about 1 meter) of multimode fiber and projecting a magnified image of its near field pattern (NFP experiment) on a screen. This experiment will also give an idea of the circularity of the fiber, the quality of the cleaved fiber endface, presence of axial index dip which is a characteristic of CVD fibers, and so on.

The conventional experiment to determine the numerical aperture (NA) of a fiber can be used to demonstrate that in a step-index multimode fiber a guided mode is uniquely formed by rays propagating in the fiber at an angle (to the axis) characteristic of that mode. The far-field power at an angle θ_p would correspond to a mode having a modal parameter

$$u = u_p = a n_{co} k_o \sin \theta_p .$$

Since $V > u > 0$, accordingly a measure of the far-field angular distribution of power yields the fiber NA.

Existence of cladding modes can be assessed by measuring the NA of the fiber once without and then with a cladding mode stripper. In the first case the measured NA will be more than that in the latter.

The physical meaning of the radial (μ) and azimutal (ν) mode numbers which identify a particular mode can be demonstrated by selective excitation /7/ of one mode in a multimode fiber by means of a prism coupler. The near field intensity pattern of each mode will exhibit 2ν maxima along the circumference of the mode pattern and μ number of maxima in the radial direction.

To demonstrate the power acceptance of a fiber for leaky rays, it is instructive to make NFP measurements on a step-index fiber using an apertured launch microscope objective. Measuring NFP once overfilling and then by restricting the launch NA equal or less to the fiber NA will show in the former case humps on the xy recorder scan at the core-cladding edges which disappear in the latter case (cf. Fig.2). In a graded core fiber such

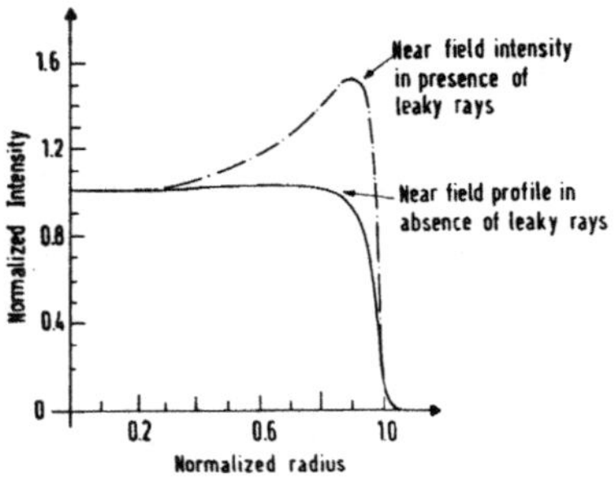

FIGURE 2. Near field profile (NFP) measurement on a near step-index profile fiber in the presence and absence of leaky rays (after ref. 34).

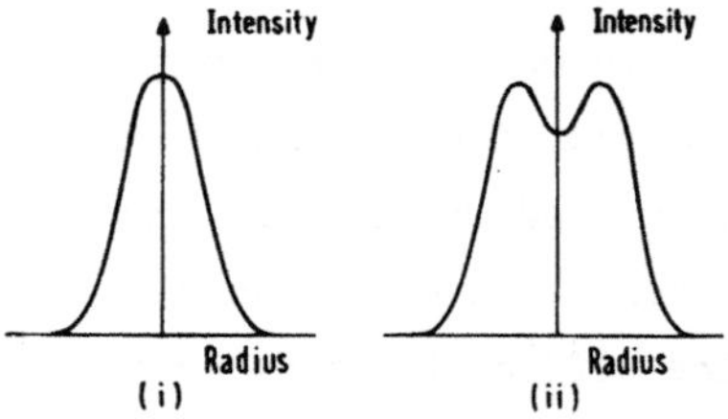

FIGURE 3. Intensity distribution of i) LP_{01}-mode, ii) LP_{01}- and LP_{11}-modes combined.

demonstration of leaky rays is difficult in a single experiment. However, individual tunneling leaky modes can be demonstrated /31/ in a graded core Selfoc fiber by coupling unfocused HeNe-laser light into these modes through the side of a short piece of fiber using a prism coupler and then photographing its NFP. NFP measurement often indicates that it is hard to fit it with a single value for the index exponent (α) of the power law profile. On the contrary the index profile /32/ requires fitting with two or more values of α's for different regions of the core, a fact which demonstrates the importance of studying multiple α-profiled fibers.

Spreading of power of a mode into the cladding with decrease in V-number of a fiber can be demonstrated by repeating the NFP experiment once by launching very low order modes on a highly multimoded fiber (large V-number) and second time on a very low V-number fiber and in each case measuring the power corresponding to the cladding region.

Influence of excitation conditions on conventional dispersion measurements /11/ can be shown by making changes on the excitation conditions e.g. imaging laserdiode spot at various points on the fiber by micromovements, making use of mode scramblers and so on. Differential mode delay measurements can be made in a multimode fiber by measuring time delay at different points across its diameter. In low bandwidthfibers the plot of delay time against mode number would show strong dependence on mode number whereas in optimum profiled fibers it will be almost independent of mode number.

The importance of ESI and variational trial field approximation for characterizing a monomode fiber can be demonstrated by determining the ESI profile from a measurement of the λ-dependence of the mode spot size /24/ and the modal field from the fiber's far-field /26/ followed by a correlation of these results to the actual index profile determined independently by the refracted near-field technique.

The field profile of a two-mode fibers can be demonstrated by scanning automatically the NFP of a monomode fibers as a function of wavelength. As the wavelength drops below the cut-off wavelength one will observe a dip instead of a pure peak (characteristic of LP_{01}-mode) at the centre of the field profile due to the propagation of both LP_{01}- and LP_{11}-modes (cf. Fig.3). The same experiment also determines the cut-off wavelength.

Source-fiber and fiber-fiber coupling efficiency can be carried out in the conventional way /33/. The effect of an intermediate coupling element e.g. lens, taper etc. can also be demonstrated by repeating these measurements in the presence of these coupling elements.

Bending induced loss in a fiber can be damonstrated by winding a piece of fiber on spools of different radii and measuring the loss as a function of the radii without changing the launching. Similarly microbending induced fiber losses can be measured by sandwiching some length of fiber between sand papers or some deforming device that introduces small bends in the fiber. Such experiments on microbending induced losses find applications in sensor studies /29/: by varying the pressure e.g. by means of an acoustic field on the deformer device and monitoring the corresponding intensity change of the transmitted light, one can get an idea of the measure of the acoustic field (cf. Fig.4).

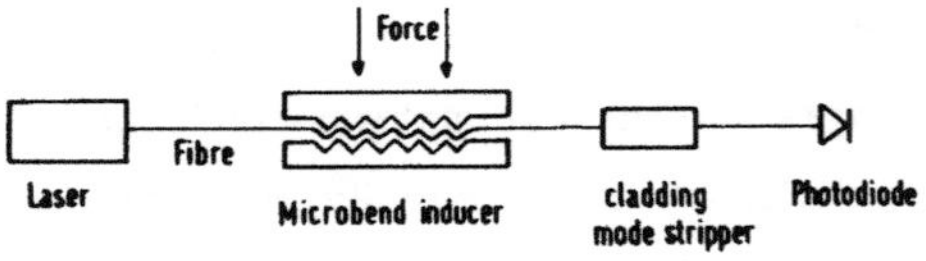

FIGURE 4. A fiber sensor based on microbending effect (after ref. 29).

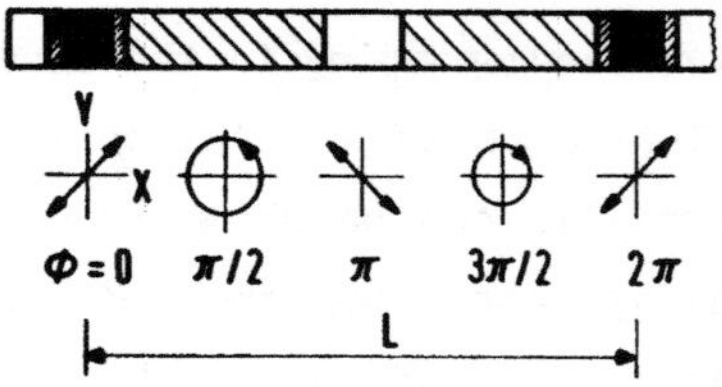

FIGURE 5. Evolution of polarization in a fiber due to birefringence (after ref. 29).

To demonstrate birefringence effects in monomode fibers /29/ a linearly polarized light is launched into an elliptical core single mode fiber, the light being polarized at 45° to the axes of the elliptical core. Since an elliptically cored fiber has an inherent birefringence $B = (\beta_x - \beta_y)/k_o$, the injected light will continuously change ist polarization as it propagates through the fiber. If the room is dark enough one will abserve (cf. Fig.5) a cyclic variation in intensity of the scattered light along the fiber with the darkest regions periodically spaced by a distance of L, the so called beat length ($= \lambda/B$). A measure of beat length has important consequences in determining performance of fiber optic sensors.

4. CONCLUSION

We have tried to provide a contemporary course structure in Optical Waveguides alongwith some guidelines for experiments which should supplement the theory. It may be used either as a two semester or as an one semester course, the latter will obviously require skipping too much details.

5. ACKNOWLEDGEMENT

The authors wish to thank all colleagues as well as their students for helpful discussions and encouragement to build up such courses at their institutions (IIT Delhi, India, Optical Waveguide Group; Technical University of Berlin, FRG, Institute of High Frequency Techniques). B.P. PAL is, in particular, indebted to Professors M.S. SODHA and A.K. GHATAK and Dr. K.

THYAGARAJAN of IIT Delhi for their critical comments.

REFERENCES

1. Kapany NS, Burke JJ. 1972. Optical Waveguides, Academic Press, New York.
2. Marcuse D. 1972. Light Transmission Optics, Van Nostrand Reinhold, New York; ibid. 1974. Theory of Dielectric Optic Waveguides, Academic Press, New York.
3. Tamir T. 1975. Integrated Optics, Springer-Verlag, Berlin.
4. Arnaud JA. 1976. Beam and Fiber Optics, Academic Press, New York.
5. Unger HG. 1977. Planar Optical Waveguides and Fibers, Clarendon Press, Oxford.
6. Sodha MS, Ghatak AK. 1977. Inhomogeneous Optical Waveguides, Plenum Press, New York.
7. Midwinter JE. 1979. Optical Fibers for Transmission, Wiley Interscience, Somerset.
8. Miller SE, Chynoweth AG. 1979. Optical Fiber Telecommunications, Academic Press, New York.
9. Sharma AB, Halme SJ, Butusov MM. 1981. Optical Fiber Systems and their Components, Springer-Verlag, Berlin.
10. Grau G. 1981. Optische Nachrichtentechnik, Springer-Verlag, Berlin.
11. Marcuse D. 1981. Principles of Optical Fiber Measurements, Academic Press, New York.
12. Snyder AW, Love JD. 1981. Optical Waveguide Theory, Chapman and Hall, London.
13. Adams MJ. 1982. An Introduction to Optical Waveguides, Wiley, Chichester.
14. Hunspurger RG. 1982. Integrated Optics: Theory and Technology, Springer-Verlag, Berlin.
15. Kersten RTh. 1983. Einführung in die Optische Nachrichtentechnik, Springer-Verlag, Berlin.
16. Okoshi T. 1982. Optical Fibers, Academic Press, New York.
17. Clarricoats PJB. 1976. Optical Fiber Waveguides: A Review, Progress in Optics, Vol.XIV, North Holland, Amsterdam.
18. Ghatak AK, Thyagarajan K. 1980. Graded Index Optical Waveguides: A Review, Progress in Optics, Vol.XVIII, North Holland, Amsterdam.
19. Pal BP. 1979. Optical Communication Fiber Waveguide Fabrication: A Review, Fiber and Integrated Optics, vol.2, 195-252.
20. Tien PK. 1977. Integrated Optics and New Wave Phenomena in Optical Waveguides, Revs. Modern Physics, vol.49, 361-420.
21. Thyagarajan K, Ghatak AK. 1981. Lasers: Theory and Applications, Plenum Press, New York.
22. Pal BP, Kumar A, Ghatak AK. 1981. J.Opt.Commun. vol.2, 97; Sammut RA. 1979. Electron.Lett. vol.15, 590.
23. Snyder AW, Sammut RA. 1979. JOSA vol.69, 1663; Pask C, Sammut RA. 1980. Electron.Lett. vol.16, 310.
24. Matsumura H, Suganuma T. 1980. Appl.Opt. vol.19, 3151; Miller CA. 1981. Electron.Lett. vol.17, 458.
25. Sharma A, Ghatak AK. 1981. Opt.Commun. vol.36, 22; Hussain SI, Sharma A, Ghatak AK. 1982. Appl.Opt. vol. 21, 2716.
26. Tewari R, Thyagarajan K, Pal BP, Ghatak AK. 1982. Opt.Commun. (in press); Ghatak AK, Srivastava R, Faria IF, Thyagarajan K, Tewari R. 1983. Electron.Lett. vol.19, 97.

27. Kitayama K, Kato Y, Seikai S, Uchida N, Ikeda M. 1979. IEEE J. Quant.Electron. vol.QE-15, 6.
28. Thyagarajan K, Gupta A, Pal BP, Ghatak AK. 1982. Opt.Commun. vol.42, 92.
29. Fiber Optic Sensor Handbook. 1982. Dynamic Systems, Inc., Virginia.
30. Batchman TE, Mohoney JW. 1977. IEEE Trans. on Education, vol.E-20, 87.
31. Zemon S, Fellows D. 1976. Appl.Opt. vol.15, 1936.
32. Pal BP. 1977. Electron.Res.Lab.Report No.STF44 A77208, NTH, Trondheim, Norway.
33. Technical Staff of CSELT. 1980. Optical Fiber Communication, McGraw-Hill Book Company, New York, Chapter 3.
34. Sladen FME, Payne DN, Adams MJ. 1976. Appl.Phys.Lett. vol.28, 257.

PART TWO

SINGLE-MODE OPTICAL FIBERS

THEORY OF IMPERFECT NONCONVENTIONAL SINGLE-MODE OPTICAL FIBERS

Huang Hung-chia
Shanghai University of Science and Technology

Qian Jing-ren
University of Science and Technology of China

ABSTRACT

This paper presents a theoretical frame-work for analyzing, in a general way, the imperfect nonconventional optical fibers met with in practice. Spun and twisted fibers with various imperfections are taken as specific examples to illustrate the general theory. Super-local modes of such fibers are found to be quasi-circularly polarized. Some advantageous features of spun fibers are viewed from the standpoint of mode conversion-reconversions. Couplings between the super-local modes caused by various imperfections are much weakened as compared with couplings between the ordinary (or ideal) modes, thus facilitating the statistical calculation of random imperfections in nonconventional fibers of long lengths. On the basis of this statistical study, an attempt is made to deal with quantitatively, perhaps for the first time, the all-important tolerance problem relating to long-length nonconventional fibers.

INTRODUCTION

While conventional single-mode optical fibers, featured by a simple round geometry and without any intentional deformation of the fiber structure, are recognized as impractical for long-length optical transmission requiring polarization-preserving, low birefringence and other related properties, nonconventional optical waveguides such as spun and twisted fibers appear increasingly attractive in recent years. The purpose of this paper is to structure a theoretical framework for describing the transmission characteristics of imperfect nonconventional fibers from the viewpoint of perturbed super-local modes, and for dealing with certain problems of practical importance hitherto not adequately considered, or not satisfactorily solved, in the published literature.

1. COUPLINGS OF MODES

Any deviations from the chosen reference waveguide cause couplings between the normal modes, or independent fields, of the reference structure. If, as usual, we consider the normal modes of an ideal conventional single-mode optical fiber, i.e. the two orthogonally polarized HE_{11} modes, then such nonconventional features like cross-sectional ellipticity, bending, twisting, spinning, as well as the externally applied electric and/or magnetic fields, etc, all originate mode couplings. For succintness, we put the coupled mode equations in matrix form:

$$\underline{A}' = \underline{K}\ \underline{A} \tag{1}$$

where $\underline{A}$ for a single-mode fiber is a column matrix with two elements, $\underline{A}'$ is the derivative of A with respect to the transmission direction z and $\underline{K}$ is a square matrix of the following form:

$$\underline{K} = j \begin{bmatrix} -\beta_1 + \Delta\beta_1 & K_1 + jK_2 \\ K_1 - jK_2 & -\beta_2 + \Delta\beta_2 \end{bmatrix} \tag{1a}$$

where we ignore attenuations of the modes caused by the coupling mechamisms, and consider β_1 equal to β_2 (i.e. $\beta_1=\beta_2 =\beta$).

By nonconventional but otherwise perfect optical fibers, it is meant that the nonconventional features of the fibers are invariant along the longitudinal direction z. Couplings of ideal modes in nonconventional fibers are generally of two kinds[1]. For the first kind, the nonconventional features appear symmetrical on the two sides of a fiber cross-section, such as cross-sectional ellipticity, bending, transverse applied fields, etc, which are associated with an orientation or a direction over the fiber cross-section. Couplings of this kind are characterized by imaginary coupling coefficients (i.e. jK_1, K_1 real) as well as by a change in phase constants (i.e. $\Delta\beta_1$ and $\Delta\beta_2$). For the second kind, the nonconventional features appear anti-symmetrical on the two sides of a fiber cross-section, such as twist, spin and longitudinal applied fields. They are not associated with an orientation or a direction over the fiber cross-section, and are characterized by real coupling coefficients (i.e. K_2) and by invariance of phase constants in the usual case.

For our immediate interest, we consider one specific example of the first kind, e.g. elliptical deformation of the fiber cross-section. Following reference (2,3), we have

$$K_1 = \frac{b}{2} \sin 2\,\theta \qquad \Delta\beta_1 = -\Delta\beta_2 = \frac{b}{2} \cos 2\,\theta \tag{2}$$

where b is a parameter descriptive of the shape of the fiber cross-section (for given values of index difference and the normalized frequency), and θ designates the orientation of the ellipse. For twisted or spun fibers, we have $\theta = \tau z$, where the coefficient τ is a measure of the twisting or spinning speed.

With the aid of Eq.(2), Eq.(1a) for the case of elliptical deformation assumes the form:

$$\underline{K} = j \begin{bmatrix} -\beta + \frac{b}{2}\cos(2\tau z) & \frac{b}{2}\sin(2\tau z) + jK_2 \\ \frac{b}{2}\sin(2\tau z) - jK_2 & -\beta - \frac{b}{2}\cos(2\tau z) \end{bmatrix} \tag{1c}$$

In the simple case where the nonconventional features are longitudinally invariant, b, K_2 and τ are constants. But generally they are functions of z.

2. SUPER-LOCAL MODES IN NONCONVENTIONAL FIBERS

In our previous papers[4,5] we indicate that the coupled-mode formulation is by no means unique, as the choice of the reference waveguide is in a sense arbitrary. With reference to a taper or a curved waveguide, we consider the ideal modes, the local modes and the super-local modes as three natural choices among all the possibilities.

In nonconventional optical fibers, the above three sets of modes remain useful, although they refer to different field structures for different nonconventional features. For definiteness, we consider a twisted or spun fiber with an elliptical core. The ideal modes, as afore-said, refer to the normal modes in a round conventional fiber taken as the reference waveguide. The local modes refer to the normal modes in a fiber with an elliptical cross-section identical to the local cross-section of the twisted or spun fiber, but invariant along the direction z. The super-local modes are distinct from the local modes in that they refer to the normal modes in a structure with the same rate of change of cross-section as the local rate of change of cross-section of the twisted or spun fiber.

Obviously, the super-local modes are normal modes of the actual structure for the simple case of constant parameters (b, K_2 and τ), while they become coupled in case one or more parameters vary with z.

We first consider the simple case of constant b, K_2 and τ. As was shown previously[4], different sets of modes are related by a transformation of the form:

$$\underline{A} = \underline{O}\ \underline{W} \tag{3}$$

where $\underline{W}$ denotes a matrix for the super-local modes, and $\underline{O}$ is a modal matrix which, for the present case, represents three successive transformations:

$$\underline{O} = \underline{O}_1\ \underline{O}_2\ \underline{O}_3 \tag{4}$$

where

$$\underline{O}_1 = \begin{bmatrix} \cos\theta & -\sin\theta \\ \sin\theta & \cos\theta \end{bmatrix} \tag{4a}$$

$$\underline{O}_2 = \begin{bmatrix} \cos\Psi & -j\sin\Psi \\ -j\sin\Psi & \cos\Psi \end{bmatrix} \tag{4b}$$

in which $\Psi = \frac{1}{2}\, tg^{-1} \frac{2(\tau - K_2)}{b}$, and

$$\underline{O}_3 = \begin{bmatrix} \exp(-j\tau z) & 0 \\ 0 & \exp(j\tau z) \end{bmatrix} \tag{4c}$$

With Eq. (4) introduced in Eq. (3), we arrive at the matrix equation for the super-local modes:

$$\underline{W}' = \underline{N}\ \underline{W} \tag{5}$$

where $\underline{N}$ is a diagonal matrix for constant parameters (b, K_2, τ):

$$\underline{N} = \begin{bmatrix} -j(\beta+g-\tau) & 0 \\ 0 & -j(\beta-g+\tau) \end{bmatrix} \tag{6}$$

and

$$g = \{(\tfrac{b}{2})^2 + (\tau-K_2)^2\}^{\frac{1}{2}} \tag{6a}$$

The super-local modes for a twisted or spun fiber with constant parameters, being elliptically polarized, or quasi-circularly polarized if $\tau-K_2 \gg b$, represent the eigenstates of the fiber, whose elliptical birefringence is easily found to be

$$B = \frac{\lambda(g-\tau)}{\P} = \frac{\lambda\{(b/2)^2 + (\tau-K_2)^2\}^{\frac{1}{2}} - \lambda\tau}{\P} \tag{7}$$

The above formula applies of course to the special case of a fiber which is otherwise conventional except for its elliptical core, by setting $\tau=0$. On the other hand, for the special case $b = 0$, $\tau \neq 0$, we have

$$B = \frac{\lambda K_2}{\P} \tag{7a}$$

in which τ does not appear, implying that twisting or spinning of a conventional round fiber does not change its birefringence, except indirectly via the stress parameter K_2.

For a spun fiber with $K_2 = 0$, the birefringence as expressed by Eq.(7) is reduced to

$$B = \frac{\P}{\lambda}(b^2/2\tau) \tag{7b}$$

if $\tau \gg b/2$, and is very small under this condition.

3. MODE CONVERSION-RECONVERSIONS

Nonconventional fibers met with in practice can hardly be described by a set of constant parameters (i.e. b, K_2, and τ do not vary). Small imperfections always exist, and are expressible as perturbations of these parameters, i.e. Δb, ΔK_2, $\Delta\tau$, which are functions of the longitudinal coordinate z.

Intuitive physical insight reveals that such perturbations are responsible to couplings between the super-local modes, and strightforward matrix algebra yields simple formulas for various coupling coefficients.

Consider the case of nonuniform twisting or spinning. If we replace $\Theta = \tau z$ by $\Theta = (\tau + \Delta\tau)z$, an additional term will appear on the right side of Eq.(6) in the form:

$$\underline{N} = \begin{bmatrix} -j\Delta\tau\sin2\Psi & \cos2\Psi\exp(j2\tau z) \\ -\Delta\tau\cos2\Psi\exp(-j2\tau z) & j\Delta\tau\sin2\Psi \end{bmatrix} \quad (8)$$

in which the diagonal terms represent perturbations of phase constants of the super-local modes, and are negligible if $\Delta\tau \ll \tau$. The off-diagonal terms, $N_{12} = -N^*_{21} = \Delta\tau\cos2\Psi\exp(j2\tau z)$, represent coupling coefficients for the super-local modes.

By similar approach, coupling coefficients caused by a perturbation of K_2 assume the form: $N_{12} = -N^*_{21} = -\Delta K_2\cos2\Psi \cdot \exp(j2\tau z)$. Like the above case, change in phase constants of the super-local modes is negligible if $\Delta K_2 \ll \tau$.

By adding Δb to b in Eq.(1a), derivation of $\underline{N}$ lead to the results that the super-local modes are coupled with a coefficient

$$N_{12} = -N^*_{21} = \frac{b}{2}\sin2\Psi\exp(j2\tau z) \quad (9)$$

Perturbations of phase constants are $\pm\frac{\Delta b}{2}\cos2\Psi$ for the pair of super-local modes concerned.

It is apparent from Eqs.(8) and (9) that mode couplings for the nonconventional fibers concerned are periodic. With reference to the literature[6], the total coupling capacity is given by

$$\bar{Q} = \frac{2(\Delta\tau - \Delta K_2)\cos2\Psi + \Delta b\sin2\Psi}{2g} \quad (10)$$

where the denominator, $2\tau-(2g+2\tau) = 2g$, is in accordance with Eq.(13) on p.188 of ref.6.

Eq.(10) leads to a very important result. For conventional

fibers with non-periodic couplings, low birefringence is associated with large coupling capacity, and hence large conversion-reconversions between the coupled modes. However, for nonconventional optical fibers, the coupling capacity (not coupling coefficients) is small because the denominator of Eq.(10) never vanishes, even though the relevant birefringence is extremely small. Each of the pair of super-local modes, being quasi-circularly polarized, is therefore capable of preserving its polarization in such nonconventional optical fibers of long lengths.

When couplings exist, super-local modes no longer represent the eigenstates of the nonconventional single-mode fibers in the strict sense. Under the usual initial condition $W_1(0) = 1$ and $W_2(0) = 0$, iterative solution of the relevant matrix equation gives(7)

$$W_1(L) = \exp[-j(\beta+g-\tau)L]\ \{1 - \int_0^L [(\Delta\tau - \Delta K_2)\cos 2\Psi + \frac{\Delta b}{2}\sin 2\Psi\ \exp(j2gz')\cdot \int_0^{z'} [(\Delta\tau - \Delta K_2)\cos 2\Psi + \frac{\Delta b}{2}\sin 2\Psi]\exp(-j2gz'')dz''dz'\}$$

$$W_2(L) = -\exp[-j(\beta-g+\tau)L]\cdot \int_0^L [(\Delta\tau - \Delta K_2)\cos 2\Psi + \frac{\Delta b}{2}\sin 2\Psi]\cdot \exp(-j2gz)dz \tag{11}$$

in which $\cos 2\Psi = \frac{b}{2g}$, $\sin 2\Psi = \frac{\tau - K_2}{2g}$, and

$$g = \{(\frac{b}{2})^2 + (\tau - K_2)^2\}^{\frac{1}{2}}$$

as given previously by Eq.(6a).

It now becomes clear from Eq.(10) that a spun fiber with vanishing K_2 experiences less conversion-reconversions between the super-local modes as compared with a twisted fiber for which $K_2 \neq 0$. Thus we view this favorable feature of spun fibers from a new point based on strict analysis.

4. STATISTICS AND TOLERANCES

Iterative solutions of the coupled-mode equations converge more rapidly if the modes concerned refer to super-local modes, instead of ideal or local modes, for imperfect nonconventional optical waveguides.

For fibers of long lengths, the imperfections are randomly distributed along their lengths. As far as the imperfections are small ($\Delta K_2 << \tau$, $\Delta\tau << \tau$, $\Delta b << b$), statistical calculations for different kinds of perturbations can be done separately. The statistical sum of these calculations gives the overall result.

For example, we take ΔK_2 into account. With the aid of Eq.(11), the ensemeble average of $W_2(L)$ is expressed by

$$<|W_2(L)^2|> = \cos^2 2\Psi < \Delta K_2^2 > [\int_{-\infty}^{\infty} \rho(u) \exp(j2gu) du] L \qquad (12)$$

where u=z-z' and the corelation function is, as usual, assumed to be exponential[8]

$$\rho(u) = \exp(-2\pi |u| / L_C)$$

$$\int_{-\infty}^{\infty} \rho(u) \exp(j2gu) du = \frac{L_C}{\pi} \frac{1}{1+(L_C \frac{g}{\pi})^2} \qquad (13)$$

in which L_C is the correlation distance.

With the aid of Eqs.(12)-(13), calculation of the tolerances for given values of extinction ratio can be carried out in a straightforward manner. Curves in Fig.1, which apply either to $<\Delta K_2^2>$ or to $<\Delta\tau^2>$, show normalized extinction radio vs the correlation distance. They are applicable also to $<\Delta b^2>$, if the readings on the ordinate are multiplied by $\frac{1}{4} tg^2 2\psi = (\frac{\tau - K_2}{b})^2$.

It is seen from each curve in Fig.1 that, for a given extinction ratio, the tolerance is most severe at a certain value of L_C. Thus, $L_C = \pi / g$, according to Eq.(13), which corresponds to the condition that the correlation distance coincides with the beat length of the pair of super-local modes.

Figs. 2 and 3 show, respectively for $<\Delta K_2^2>$ and for $<\Delta b^2>$, how the extinction ratio varies with $\tau - K_2$, when L_C is taken as a parameter and $b = \pi$ Rad/m. It can be seen that the tolerances $\{<\Delta K_2^2>\}^{\frac{1}{2}}$ and $\{<\Delta\tau^2>\}^{\frac{1}{2}}$ increase as $(\tau - K_2)$ increases, while the tolerance $\{<\Delta b^2>\}^{\frac{1}{2}}$ increases rapidly only when $(\tau - K_2) L_C > \pi$. The right portions of the curves in Fig.3 are

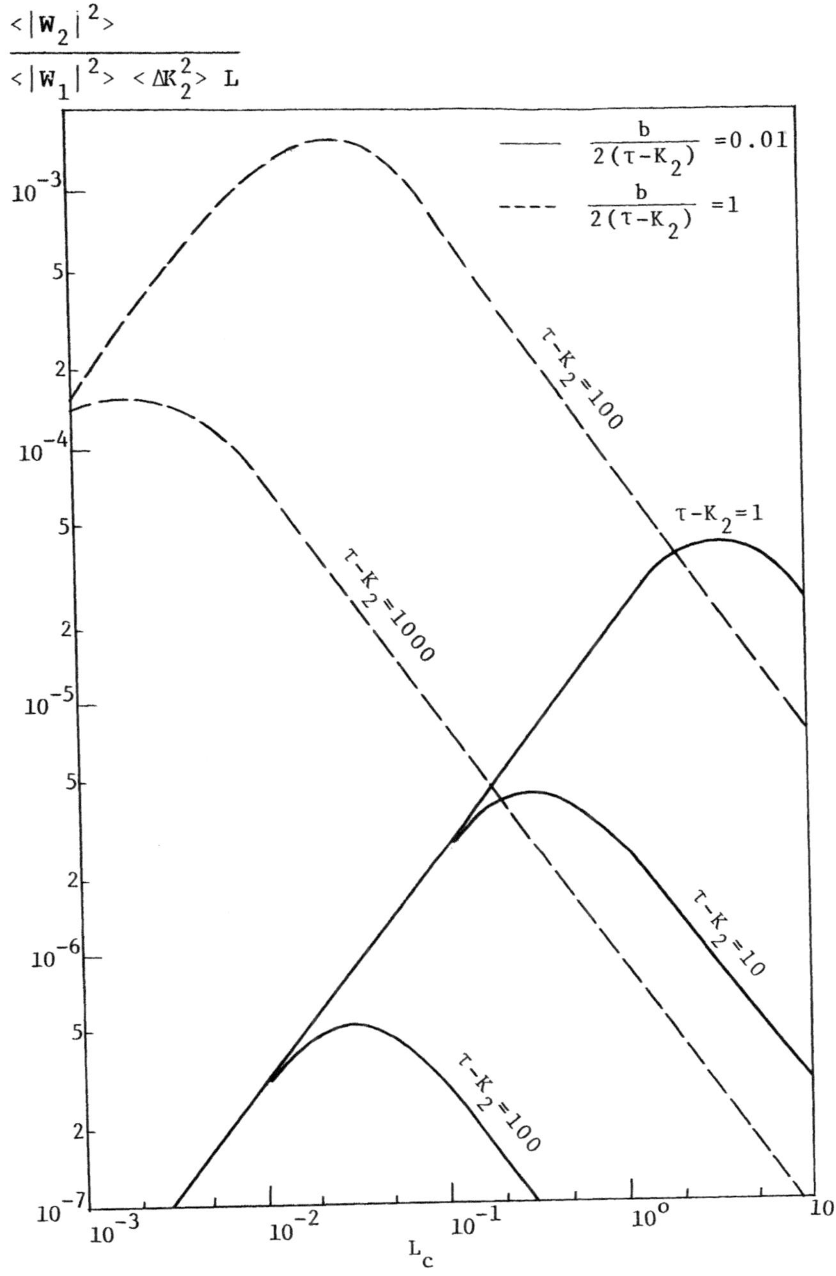

Fig.1 Normalized Extinction Ratio versus Correlation Distance

Fig.2 & 3. Normalized Extinction Ratio vs τ-K_2 (L_c as a parameter; $b/2=0.5\pi$ Rad./m)

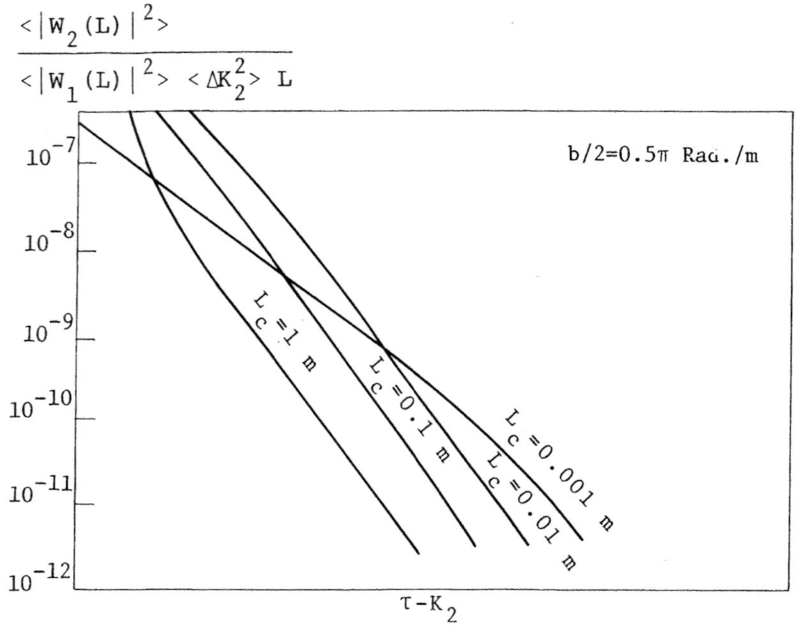

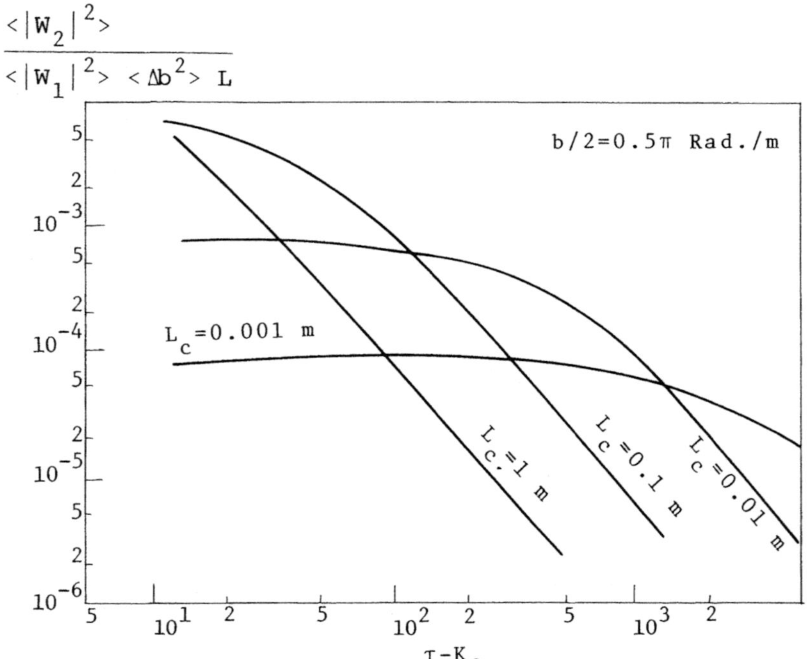

applicable to other values of b, so long as $(\tau-K_2)>>b/2$.

	L = 1 km		L = 10 km	
	L_C=0.01m	L_C=0.001m	L_C=0.01m	L_C=0.001m
$\Delta\tau/\tau$ (%)	25.23	36.39	7.979	11.507
$\Delta e/e$ (%)	50.47	72.78	15.96	23.01
R (mm)	53.74	44.75	95.50	79.58
p (g.w/cm)	2.2925	3.306	0.7249	1.045

Table 1. Tolerances of $\Delta e, \Delta\tau$, R and p
(b=rad/m, τ=200, K_2=0)

The result of the tolerance calculation is given in Table 1, for a specific example with n=1.45, 2a=5,5μm, 2d=120μm, $\Delta n/n$=2%, λ=1.32, v=2.4. The relations between Δb and Δe, p, R (p is transverse stress, R is the radius of curvature), are available in the literature.[(2)] The calculated values show that tolerances of Δe, R and p are essentially constant for different ellipticities, if $\tau>>b/2$. From this calculation, spun fiber as a nonconventional optical waveguide version is advantageous also from tolerance considerations.

CLOSING REMARKS

Although only a few specific examples are given in the paper for illustration of the theory, it is worth noting that the approach adopted herein, which makes use of super-local modes and their iterative solutions, is expected to be applicable to a wider class of problems relating to nonconventional optical fibers, as compared with previously published methods.

REFERENCES

1. Qian Jing-ren, On coefficients of coupled-wave equations,

Acta Electronica Sinica, No.2, 46 (1982)
2. J• I• Ichi and T. Kimura, Polarization behavior in multiply perturbed single-mode fibers, Trans. IEEE QE-18, No.1, 59 (1982)
3. R. Ulrich and A. Simon, Polarization optics of twisted single-mode fibers, Applied Optics, Vol.18, No.13, 2241 (1979)
4. Huang Hung-chia, On local normal modes in optical fiber and film waveguides, Scientia Sinica, Vol.22, No.10, 1147 (1979)
5. Huang Hung-chia, Weak coupling theory of optical fiber and film waveguides, Radio Science, Vol.16, No.4, 495 (1981)
6. Huang Hung-chia, COUPLED MODES AND NONIDEAL WAVEGUIDES (Collected Papers), Microwave Research Institute (MRI) of the Polytechnic Institute of New York, November 1981.
7. Huang Hung-chia, Generalized theory of coupled local normal modes in multiwave guides, Scientia Sinica, Vol.9, No.1, 142 (1960)
8. H. E. Rowe and W. D. Warters, Transmission in multi-mode waveguide with random imperfections, Bell System Technical Journal, Vol.41, No.3, 1031 (1962)

MONOMODE FIBRES FOR LONG HAUL TRANSMISSION SYSTEMS

J E MIDWINTER

1. FIBRE DESIGN

Monomode fibre systems using silica based fibres are designed to operate in the 1300nm and/or 1500nm transmission windows. In selecting a design, a large number of separate factors must be optimised, some which are inter related and others of which are independent. The starting point for most designs is to observe that to be single moded, the cut-off wavelength must be close to or shorter than the operating wavelength. This immediately defines a set of core diameter and index difference values that are allowable assuming a simple step index structure fibre and usng the V value relation with V less than 2.4. By examining the limiting cases we can see some general relations emerging.

Consider the small index difference and large core diameter situation. Such a fibre is weakly guiding and very susceptible to microbending loss on cabling. However, it has a low dopant concentration in the core and hence the fundamental attenuation of the fibre is low because additional dopant tend to increase the Rayleigh scattering and other attenuation mechanisms. In addition, the spot size will be relatively large leading to relatively low splice losses. The waveguide dispersion will be small so that the zero of dispersion will be close to that for pure silica in the 1300nm window.

At the other extreme, the high index difference fibre has a smaller core diameter, generally a smaller mode spot size and a more tightly guided field that is less susceptible to microbending attenuation. Higher dopant concentrations in the core material increase Rayleigh scattering and can give rise to additional absorption effects (Ref.1). The waveguide dispersion is larger, shifting the zero of resultant dispersion to longer wavelength, possibly into the 1500nm window. The smaller mode spot size will increase the fibres sensitivity to splicing loss through mechanical misalignment.

To summarise, a fibre having the minimum possible attenuation and index difference at 1300nm is difficult to cable. Using a larger index difference, in the region of 0.005, leads to a structure that is readily cabled. The attenuation will be very slightly higher but values in the range 0.4 to 0.5dB/km are readily obtained. The zero of dispersion will remain in the 1300nm window, while in the 1500nm window, lower attenuation

(0.2 to 0.3dB/km) and higher dispersion (15-20 ps/(km.nm)) are obtained. Using a higher index difference, say 0.015, leads to a smaller core fibre with a zero of dispersion near to 1500nm. However, such a fibre will normally show a higher minimum attenuation than that of a non-shifted fibre and early work suggested that it would have an attenuation value at the zero of dispersion very close to that of the unshifted fibre at its zero dispersion point (see Ref.1). From a system designers point of view, therefore, little if any advantage accrued from the use of such a fibre.

2. RESULTS ON EARLY CABLED FIBRE AND SYSTEMS

In the UK, the first monomode fibre systems will operate in the 1300nm window and will carry traffic at a data rate of 140Mbit/s. They may be upgraded to 560Mbit/s operation at a later date. A key design target has been the achievement of a 30km repeater spacing since this corresponds to the maximum distance between surface buildings with secure power feed on the UK Trunk Network. The fibres to be used will have a zero of dispersion in the 1300nm window and a core/cladding index difference of about 0.005. In early tests, fibres of this type have been used in a number of cable trials, the major results of which are summarised in Table I.

The fibre cables reported in Table I were all jointed by fusion splicing using a microprocessor controlled jig. The fibres were aligned visually under a twin axis microscope by their outer surfaces without transmitted-power monitoring. In the cases of all links, satisfactorily low splice losses were achieved, (see Table II). The BTRL link was made with fibres of varying design as was the GEC/BTRL link. Variations in the design parameters between fibres accounted for the higher splice losses compared to those obtained in the STC/BTRL link which used fibre manufactured to a uniform design. However, in each case it was concluded that the results indicated satisfactory control over the key fibre parameters of core concentricity, NA, core diameter etc.

The major results of transmission systems studies using these cables are then given in Table III. These results indicate two things rather clearly. High bit rate transmission is possible over long unrepeatered sections of monomode fibre at 1300nms and the 30km repeater spacing research target is entirely feasible. In addition, the systems studies showed that there was no dispersion penalty at this wavelength as expected.

These successful trials will lead to the first operational systems being ready for service in late 1983 or early 1984. They will be carefully evaluated during and after installation but it now seems likely that from the middle of this decade, monomode 1300nm systems will become the standard equipment for new cable installations operating at 140Mbit/s or higher in the UK. Worldwide interest in this work indicates that many other countries are now beginning to follow this same line of thinking.

3. EXPLOITATION OF 1500nm WAVELENGTH

The much lower attenuation obtainable at the longer 1500nm wavelength clearly makes use of that window potentially attractive. However, at 1500nm, the dispersion is in the region of 17ps/(nm.km). Thus a long link of, say, 100km used with a laser having a linewidth of 3nms (fairly typical of many of todays lasers) would show a pulse spreading or jitter of 5.1ns. Actually, since a multi longitudinal mode laser shows large random fluctuations in the output power between the various modes (mode partition noise, see Ref.2) this represents a randomly varying pulse shape, shifting within a 5ns time slot. Such an effect leads to disastrous jitter in a system.

Two options are open for overcoming this problem. Very recent work at BTRL (Ref.3) has shown that it is possible to combine both very low attenuation and zero dispersion in the 1500nm window, with a fibre design which combined a zero of dispersion at 1560nm and a minimum attenuation at that wavelength of 0.23dB/km. It is quite possible that such a fibre design will become the norm for future monomode systems. More complex design routes have been followed by other groups, notably the W fibre design, which minimise dispersion over a very wide spectral range. However, it is extremely difficult to make and it is not obvious that the performance it offers can actually be used, as we shall see.

Before we leave the subject of dispersion, it is important to note that another option exists. A narrow linewidth laser when modulated with 1GHz data bandwidth shows a spectral broadening of less than 0.01nms. Combining this number with 17ps/(nm.km) dispersion and 100kms, we find a mere 17ps broadening predicted. Since recent work at BTRL (Ref.4) has shown that 1500nm semiconductor lasers can be controlled so that their unmodulated linewidth is less than 10kHz and that this natural linewidth remains stable during modulation, we can state that the use of such a source allows the "dispersive" fibre design to be used in a high bit rate system. Thus, it is not really necessary to use the "zero dispersion" fibre, let alone the more complex fibre designs which minimise dispersion over a wide spectral range.

A number of experimental transmissions have been carried out at 1500nm wavelength on very long sections at BTRL. A jointed length of 102km of fibre on drum was used for studies of 140Mbit/s transmission (see Ref. 5) and for video transmission using Pulse Frequency Modulation (PFM) (see Ref.6). More recently, a long cabled fibre link was assembled using Cables 2 and 3 of Table I. Its length when spliced was 90.6kms (not the sum of the two separate routes since some link sections were removed). The 140Mbit/s transmision study was made at 1520nms with a margin of 2.3dB. These experiments point clearly to the future potential of this window for long haul transmission.

4. COHERENT SYSTEMS

Very recent developments seem to open up a wide range of possibilities for further system development. For some time it has been recognised that todays receivers using direct intensity detection, are about 20dB away from the quantum limit for detection of binary coded data at the same bit rate. One way of closing this gap would be to employ coherent receivers of the homodyne or heterodyne type but this has appeared insuperably difficult. The development of the injection locked semiconductor source at 1520nm based upon the 1520nm HeNe gas laser as frequency reference suddenly made experimental study of these predictions possible (Ref.4).

The results of these studies have shown clearly that very substantial gains in receiver sensitivity are possible (Refs.7,8). In Table IV, we list the receiver performances that have been obtained experimentally in our laboratories at 1520nm wavelength. Some of these involve measurements made over long fibre lengths installed in cable in duct, others are back to measurements made on the laboratory bench.

Field deployment of such technology would obviously require a source to act as Local Oscillator with exceptional stability and narrow spectral linewidth that did not rely upon the HeNe laser reference. Such a source has already been approximated in recent work on external cavity 1520nm semiconductor lasers at BTRL.(Ref.9).

The operation of a coherent system over long cable links raises the question of polarisation stability since such receivers are polarisation sensitive. Contrary to expectation, our own experience has shown that the polarisation stability on long cabled links is excellent and this will be discussed in a companion paper. (Ref.10).

Another aspect of importance for coherent systems is the effect of the narrow spectral linwidth transmitter on non-linear interactions in the fibre. Recent measurements have shown that such sources can stimulate Brillouin scattering with thresholds of a few milliwatts optical power (Ref.11). This would very seriously limit the transmitter power that could be exploited in such a fibre system. However, it has also been shown that the choice of a suitable modulation scheme such as Phase Shift Keying (PSK) suppresses this effect by at least 20dB. This will also be discussed in greater detail ina companion paper. (Ref.12).

Taking these observations together, we see that the use of coherent systems technology can improve receiver sensitivity by 15 to 20dB. Equally, it is possible to exploit much higher transmitter power provided that suitable modulation formats are employed, such as PSK. These effects all operate with advantage in the 1500nm window where fibre attenuations fall to 0.2dB/km or less. The result is a capability in principle of using silica fibre to span distances of several hundreds of kilometres without regenerators if that is desired.

5. SUMMARY

We have seen that 1300nm monomode technology has great attraction for land or undersea use in the immediate future. Very long repeater sections, perhaps of 30km, are possible in an engineering context with bit rates of 140 or 565Mbit/s today.

These figures are impressive in their own right but they do not reflect any fundamental upper limit to the performance of that monomode fibre. Advances in regenerator technology will extend that capability still further, opening up the possibility of higher bit rate, longer repeater separation or both. Increasingly, we find that the technology offers possibilities that are in advance of todays planning requirements, not because there are no applications but because there has not been time to think through the implications fully. It is clear that monomode optical systems will increasingly dominate long haul intercity high bit rate cable transmission as the decade progresses and it seems certain that there will be a shift in R&D towards very advanced regenerator development and away from more sophisticated fibre development.

6. ACKNOWLEDGEMENTS

The author wishes to thank the Director of Research of British Telecom Research Laboratories for permission to publish this paper.

7. REFERENCES

1. "Interplay of design parameters and fabrication conditions on the performance of monomode fibres made by MCVD", B J Ainslie,K J Beales,C R Day,J D Rush, IEEE J Quantum Electronics Vol QE-17, No 6,June 1981.

2. " Dispersion effects using a 1.52 micron laser", D A Frisch and I D Henning, Electronics Letters Vol.18, p129-130, 4 Feb.1982

3. "Monomode fibre with ultra low loss and minimum dispersion at 1.55 microns", B J Ainslie, K J Beales, D M Cooper, C R Day, and J D Rush. Electronics Letters Vol.18, Pages 842-844, 16 Sept. 1982.

4. "Megahertz linewidth from 1.5 micron semiconductor laser with HeNe laser injection", R Wyatt, D W Smith and K H Cameron, Electronics Letters Vol 18, p282-293, 1 April 1982.

5. "A 102km unrepeatered monomode fibre system experiment at 140Mbit/s with an injection locked 1.52 micron laser transmitter", D J Malyon, A P McDonna, Electronics Letters Vol 18, No 15, pp 445-447, 27 May 1982.

6. "Unrepeatered video transmission using pulse frequency modulation over 100 km of monomode optical fibre", D J T Heatley, Electronics Letters Vol.18,No.9,pp.369-371, 29 April 1982.

7. "Experimental assessment of a 140Mbit/s coherent optical receiver at 1.52 microns", T G Hodgkinson, R Wyatt and D W Smith, Electronics Letters Vol 18, No 12, pp 523-525, 10 June 1982.

8. "1.5 micron Optical Heterodyne System operating over 30km of monomode fibre. Electronics Letters to be published.

9. "10kHz linewidth 1.5 micron InGaAsP external cavity laser with 55nm tuning range", R Wyatt and W J Devlin, Electronics Letters, to be published 1983.

10. "Polarisation stability requirements for coherent optical fibre transmission systems", D W Smith. See this volume.

11. "Suppression of stimulated Brillouin Scattering during transmission of high power narrowband laser light in monomode fibre", D Cotter, Electronics Letters Vol 18, p638-640, 22 July 1982.

12. "Optical Nonlinearity in Fibres; a new factor in systems design", D Cotter. See this volume.

TABLE I. Results of fibre cable studies.

	Cable 1	Cable 2	Cable 3
Total cable length (kms)	7.5	7.6	15.5
Total fibre length in cable (kms)	31.6	31.5	62.0
Total number of fiber splices	18	30	37
Total attenuation in dB (incl.splices) at 1300nm.	17.5	21.7	30.0
at 1500nm	16.1	15.7	20.4

Note, for cable details see below.

TABLE II. Splice losses obtained in cabled links.

	Cable 1	Cable 2	Cable 3
Total number of splices	18	30	37
Mean splice attenuation (dB)			
at 1300nm	0.17	0.22	0.11
at 1500nm	0.11	-	0.08

Note:- For cable details see below

TABLE III. Results of fibre cable systems studies.

	Cable 1	Cable 2	Cable 3
Repeater section length (kms)	31.6	31.5	62.0
At a Bit Rate of 140Mbit/s Operating Margin in dB. (1300nm)	14.5	16.1	-
At a bit rate of 565Mbit/s Operating margin in dB. (1300nm)	5.0	3.4	1.0

NOTE:- Cable 1 made at BTRL with BTRL fibre and measured on cable drum; Cable 2 made by BEC/TCL using GEC and BTRL fibres and installed in duct running from BTRL to the local town of Woodbridge. Measurements quoted for cable jointed and installed in duct. Cable 3 made by STL/STC using their own fibre and installed in ducts running from BTRL to the nearby town of Ipswich.

TABLE IV. Coherent receiver sensitivities at BTRL.

Type of Rx.	Local Osc Power dBm	Modulation Mbit/s.type.	Sensitivity dBm (1:10^{8})
Homodyne	-32	140 ASK.	-59
Heterodyne	-27	8 ASK	-66 (-77)
Heterodyne	-22	34 ASK	-59 (-66)
Homodyne	-24	140 PSK	-64

TABLE II. Splice losses obtained in cabled links.

	Cable 1	Cable 2	Cable 3
Total number of splices	18	30	37
Mean splice attenuation (dB) at 1300nm	0.17	0.22	0.11
at 1500nm	0.11	-	0.08

Note:- For cable details see below

TABLE III. Results of fibre cable systems studies.

	Cable 1	Cable 2	Cable 3
Repeater section length (kms)	31.6	31.5	62.0
At a Bit Rate of 140Mbit/s Operating Margin in dB. (1300nm)	14.5	16.1	-
At a bit rate of 565Mbit/s Operating margin in dB. (1300nm)	5.0	3.4	1.0

NOTE:- Cable 1 made at BTRL with BTRL fibre and measured on cable drum; Cable 2 made by BEC/TCL using GEC and BTRL fibres and installed in duct running from BTRL to the local town of Woodbridge. Measurements quoted for cable jointed and installed in duct. Cable 3 made by STL/STC using their own fibre and installed in ducts running from BTRL to the nearby town of Ipswich.

TABLE IV. Coherent receiver sensitivities at BTRL.

Type of Rx.	Local Osc Power dBm	Modulation Mbit/s.type.	Sensitivity dBm (1:10)
Homodyne	-32	140 ASK.	-59
Heterodyne	-27	8 ASK	-66 (-77)
Heterodyne	-22	34 ASK	-59 (-66)
Homodyne	-24	140 PSK	-64
Direct Detection (for comparison).	-	140 ASK	-47

SINGLE-POLARIZATION SINGLE-MODE OPTICAL WAVEGUIDING SCHEMES

T. OKOSHI

DEPARTMENT OF ELECTRONIC ENGINEERING
UNIVERSITY OF TOKYO

1. INTRODUCTION

In ordinary axially symmetrical optical fibers, the polarization state of the propagated wave is subject to unstable fluctuation when ambient conditions change. In a multimode fiber, such instability usually causes little trouble except for its possible effect on modal noise. However, in a single-mode fiber, the following problems arise [1][2].

1) In an optical communication system, the received signal level fluctuates when the receiver is sensitive to the polarization. This situation takes place when an optical IC is used in the receiver, or in heterodyne-type optical communications in the future.

2) Even when a fiber is designed to be axially symmetrical, slight elliptical deformation may exist. This separates the propagation constants of two orthogonal HE_{11} modes which otherwise degenerate with each other, and causes the so-called polarization mode dispersion in the group delay.

3) In measurements using a single-mode fiber, such as magnetooptic current sensing or laser gyroscope, or in coherent optical communications such as those employing PCM-PSK (phase-shift keying) or PCM-FSK (frequency-shift keying), the polarization instability deteriorates the measurement error or the bit-error rate.

In all the above cases, use of single-polarization single-mode optical fiber (hereafter SPSM fiber for short) will be more desirable than an ordinary one if other characteristics (such as loss and bandwidth) are identical. This talk reviews the history, classification, principle, performance obtained so far, theoretical approaches, and related measurement techniques of the SPSM fibers.

2. HISTORY OF RESEARCH AND DEVELOPMENT

The short history of the research and development of SPSM fibers can conveniently be divided into the following four periods:

1976-1977

In this first period, the modal birefringence caused by an unintentional elliptical core deformation was the principal subject of the research. Intentional SPSM schemes were not investigated.

1978-1979

Various kinds of SPSM fiber schemes, such as dumbbell core fibers, stress-induced (elliptical cladding) SPSM fibers, were proposed, mainly by the investigators of Bell Laboratories. The side-pit fiber was proposed in Japan.

1980-1981

The center of research seems to have moved to Japan and Europe in this period. Investigators of CNET (France) proposed an entirely new scheme, the circularly-SPSM fiber, in the fall of 1980. Various novel linearly-SPSM fibers, such as elliptical jacket fiber, side-tunnel fiber, PANDA fiber and four-sector-core fiber were proposed and experimented with in Japan.

1982-1983

The development of various fibers continued. On the other hand, British Telecom people reported in the fall of 1982 that the polarization-state fluctuation was not so drastic when the fiber was cabled and enstalled. This report lead to a new discussion on the polarization-maintaining capability that will be required in practical communications systems.

3. CLASSIFICATION OF VARIOUS SPSM FIBERS

Various SPSM fiber schemes now being investigated can be classified as follows.

A. Linearly SPSM Fibers

Fibers belonging to this category are designed so that only one of the two linearly polarized modes [HE_{11x}(horizontal) and HE_{11y}(vertical) modes] can be propagated. These fibers can further be classified as A-1 and A-2:

A-1. Absolutely SPSM Fibers (Bileaky Fibers)

The fiber is designed so that the transmission losses for the HE_{11x} and HE_{11y} modes are largely different; hopefully so that one propagation mode is cut off. Such a scheme may also be called the "bileaky fiber" in contrast to the "birefringent fiber." The side-tunnel and side-pit fibers belong to this category.

A-2. Linearly Birefringent Fibers

The fiber is designed so that the propagation constants of HE_{11x} and HE_{11y} modes are different. Thus, these two polarization modes are "decoupled," and the launched one will remain travelling without power transfer to the other. This scheme can further be classified as

A-2-1. Stress-Induced Linearly Birefringent Fibers, in which the propagation-constant difference is produced by an axially nonsymmetrical internal stress; the elliptical cladding fiber, elliptical jacket fiber, PANDA fiber, and four-sector-core fiber are included in this category, and

A-2-2. Geometrically Birefringent Fibers, in which the propagation-constant difference is produced by an axially nonsymmetrical refractive index distribution. The elliptical and dumbbell core fibers belong to this category. However, we should note that the stress-induced birefringence also exists in such fibers.

B. Circularly SPSM Fibers

A round (axially symmetrical) fiber is twisted to produce a difference in the propagation constants of the clockwise and counter-clockwise circularly polarized HE_{11} modes. Thus, these two circular polarization modes are decoupled. These fibers, therefore, may also be called the "circularly birefringent" fibers.

In the following sections, the design principles and the performance obtained so far with the various approaches will be reviewed.

4. ABSOLUTELY SPSM FIBERS (BILEAKY FIBERS)

The difference between the absolutely SPSM fiber and the linearly birefringent fiber is depicted in Fig.1. This figure shows mode curves for HE_{11x} and HE_{11y} modes with the normalized frequency v in the abscissa and normalized propagation constant β/k in the ordinate. When the fiber has an axially nonsymmetrical structure, the curves for the two polarization modes will be separated. (However, in some cases the separation is very small and can hardly be noticed on a graph of this size.)

In an absolutely SPSM fiber [3], the two polarization modes must have different cutoff frequencies as shown by solid curves in Fig.1(a). However, in an ordinary optical fiber, no cutoff exists for HE_{11} modes as shown by broken curves in the same figure. A hint for realizing an absolutely SPSM fiber can be found in the following fact: in an axially symmetrical fiber having an "index valley" with appreciable width and depth around the core as shown in Fig.2, so that

$$\int\{n(r)-n_2\}2\pi r dr < 0 \qquad (1)$$

where n_2 denotes the cladding index, a cutoff region appears for the HE_{11} mode[3].

We may then predict that, in index distributions as shown in Fig.3 where, for example, $n_2=1.46$, $n_1=1.47$ and n_p(pit)=1.45, or that in Fig.4 where $n_2=1.46$, $n_1=1.47$ and n_t(tunnel)=1.00, the HE_{11x} and HE_{11y} modes will have different cutoff frequencies. The structures in Figs.3 and 4 are called side-pit fiber [3][4] and side-tunnel fiber[5], respectively.

Both the theoretical analysis[3] and experiment[4] show that the absolute SPSM region is not wide enough for the side-pit fiber; the theoretical relative bandwidth $K=|v_{cx}-v_{cy}|/v_{cy}$ is only 2.5% for $\Delta=(n_1-n_2)/n_2=1\%$ and $(n_2-n_p)/n_2=1\%$[3]. However, in a side-tunnel fiber, K can be 7% for $\Delta=1\%$ and 17% for $\Delta=5\%$. It has been confirmed experimentally that the transmission attenuation of a side-tunnel fiber is largely different for the vertical and horizontal polarization modes [5]. In other words, in such a fiber which has been designed as an absolutely single-polarization fiber, at least one polarization becomes strongly leaky, thus producing the "bi-leaky" property.

5. LINEARLY BIREFRINGENT FIBERS

In a linearly birefringent fiber, the both polarization modes can be propagated, but the propagation constant difference $\Delta\beta$ is made as large as possible. When a structural fluctuation is present along the axis, the power transfer between the two modes per unit length of the fiber is known to be proportional to $\Delta\beta^{-4} \sim \Delta\beta^{-8}$. Therefore, we can reduce the coupling between modes effectively by increasing $\Delta\beta$.

The modal birefringence can be generated by an axially nonsymmetrical refractive-index distribution, or an axially nonsymmetrical stress distribution. Usually, the latter produces much stronger birefringence.

One of the strongest birefringences has been reported by Katsuyama and others of Hitachi, Ltd., with a stress-induced birefringent fiber having a circular core, circular inner cladding, elliptical outer cladding, and circular jacketing[6]. A beat length between the two polarization modes (given as $L=2\pi/\Delta\beta$) shorter than 0.88mm has been obtained with his scheme.

There are many other types of the stressed-induced birefringent fibers, such as the PANDA fiber[7] in which the stress is produced by two boron-doped regions located on both sides of the core, but a little apart from it.

6. CIRCULARLY BIREFRINGENT FIBERS

The circularly birefringent fiber was proposed first by Jeunhomme and Monerie of CNET, France, in 1980[8]. In this scheme, a twist with rate of ϕ[rot/m] is given to an axially symmetrical fiber as shown in Fig.5. It is known both theoretically and experimentally that, if a linearly polarized light is launched onto such a fiber, a "drag" of the polarization plane takes place in the same direction as the twist, but with a rate $\alpha\phi$[rot/m] where $\alpha \simeq 0.07$. This means that a difference is generated between the propagation constants of the clockwise and counter-clockwise circular polarizations, which is geven as $\Delta\beta = 2\alpha\phi$.

The circular birefringence thus obtained is usually much weaker than that obtained with stressed-induced linearly birefringent fibers. Even if we assume $\phi = 100$[rot/m], $\Delta\beta \simeq 28\pi$ [rot/m] and $L = 2\pi/\Delta\beta \fallingdotseq 7$cm.

7. COMPARISON OF LINEARLY AND CIRCULARLY SPSM FIBERS

The features of the linearly and circularly SPSM fibers are compared in Table 1. It is dound that the latter has many attractive features. However, the maximum birefringence obtained so far is about two orders of magnitude smaller than in the former.

On the other hand, there is still a debate on what amount of birefringence (polarization-maintaining capability) is really needed in actual optical fiber communication channels. A recent paper from British Telecom Research Laboratories reported that the polarization state in a "cabled" fiber is not so unstable: the time constant of fluctuation is more than several hours[9]. If so, the use of the circularly SPSM fiber becomes attractive because it allows easy connection and splicing at the sacrifice of a weaker birefringence.

8. CONCLUSION

The technical background, history of research, classification, design principles and performance of various SPSM optical fiber schemes have been reviewed. For the time being we should continue the research on various schemes in parallel.

REFERENCES

[1] I.P. Kaminov, "Polarization in Optical Fibers," IEEE Jour. of Quantum Electron., Vol.QE-17, No.1, pp.15-22, Jan. 1981.

[2] T. Okoshi, "Single-polarization single-mode optical fibers," IEEE Jour. of Quantum Electron., Vol.QE-17, No.6, pp.879-884, June 1981.

[3] T. Okoshi and K. Oyamada, "Single-polarization single-mode optical fiber with refractive-index pits on both sides of the core," Electron. Lett., Vol.16, No.18, pp.712-713, Aug.28, 1980.
[4] T. Hosaka et al., "Single-mode fibers with asymmetrical refractive-index pits on both sides of core," Electron. Lett., Vol.17, No.15, pp.191-193, March 5, 1981.
[5] T. Okoshi et al., "Side-tunnel fiber: an approach to polarization-maintaining optical waveguiding scheme," Electron. Lett., Vol.18, No.19, pp.824-826, Sept.16, 1982.
[6] T. Katsuyama et al., "Low-loss single-polarization fibers," Electron. Lett., Vol.17, No.13, pp.473-474, June 25, 1981.
[7] T. Hosaka et al., "Low-loss single-polarization fibers with asymmetrical strain birefringence," Electron. Lett., Vol.17 No.15, pp.530-531, July 23, 1981.
[8] L. Jeunhomme and N. Monerie, "Polarization-maintaining single-mode fiber cable design," Electron. Lett., Vol.16, No.24, pp.921-922, Nov.20, 1980.
[9] T.G. Hodgkinson et al., "Experimental 1.5μm coherent optical fiber transmission system," 1982 ECOC, Cannes, Paper No. AXII-5.

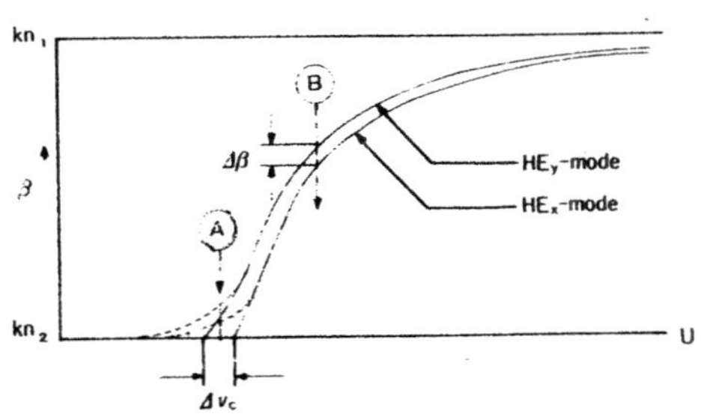

Fig.1 Comparison of the operating frequencies of an absolutely SPSM (bileaky) fiber (A) and a linearly birefringent fiber (B).

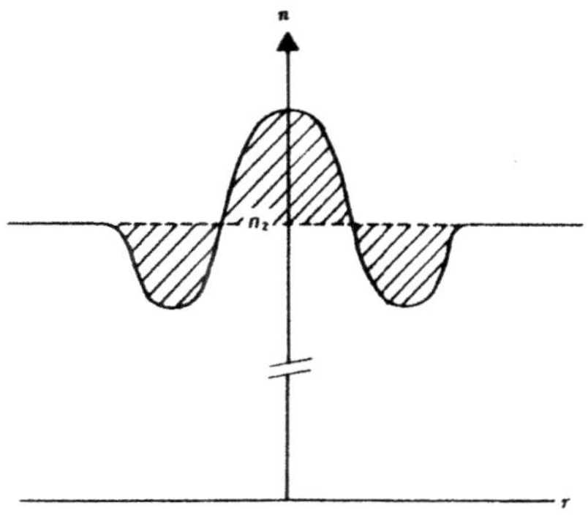

Fig.2 Condition for realizing a cutoff region in HE_{11} mode in an axially-symmetrical optical fiber.

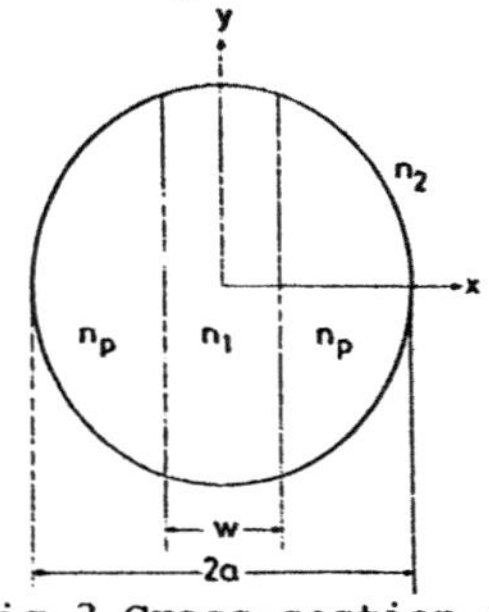

Fig.3 Cross-section of side-pit fiber.

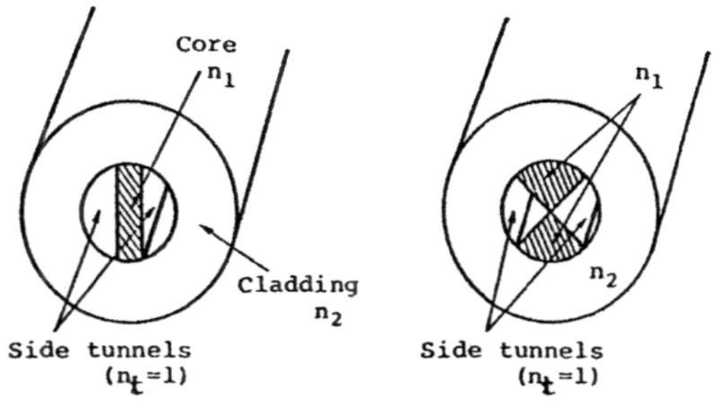

Fig.4 Two examples of the side-tunnel fiber.

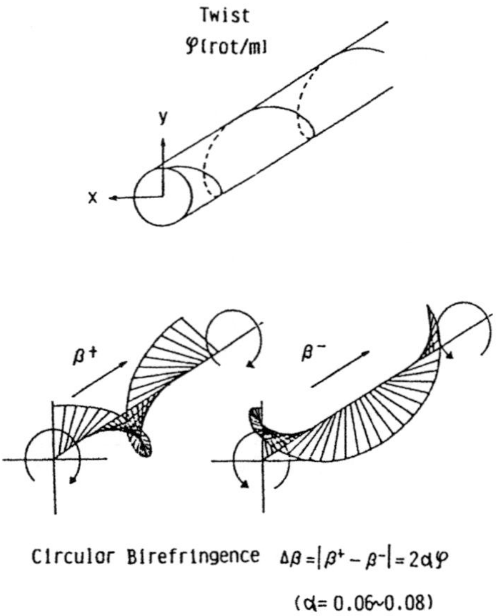

Fig.5 Principle of the circularly SPSM fiber.

	Lin. SPSM	Circ. SPSM
Fabrication	Difficult	Easy
Polarization stability	Probably better	Probably poorer
Connection	Difficult	Easy
Polarization dispersion	Large	Small
Polarization matching with LO injection	Necessary	Unnecessary

Table 1: Comparison of linearly and circularly SPSM fibers.

MEASUREMENT OF CHROMATIC DISPERSION IN SINGLE MODE FIBRES BY INCOHERENT SOURCES

B. COSTA, M. PULEO, E. VEZZONI

CSELT, Via Reiss Romoli 274 - Torino (Italy)

ABSTRACT

We have demonstrated for the first time that accurate chromatic dispersion measurements are feasible in single mode fibres, making use of sinusoidally modulated incoherent sources. This technique has been extended in the long wavelength range, thus allowing the accurate determination of the zero chromatic dispersion wavelength in an easy way and with simple instrumentation.

1. INTRODUCTION

The accurate knowledge of the chromatic dispersion properties of single mode fibres is a key factor which is needed both to verify the fabrication parameters of fibres and to allow the optimum design of transmission systems. Different methods have been reported till now, which allow the measurement of such properties. Most of them, although giving satisfactory results, make use of cumbersome techniques and expensive measuring set-ups like multi-laser sources [1, 2, 3] or non linear effects in crystals [4] or fibres [5] pumped by high power pulsed lasers (Raman effect). These set-ups, moreover, can hardly be operated outside a strict laboratory environment.

An alternative technique has been recently demonstrated, using sinusoidally modulated LED sources, easy to operate and giving highly accurate results [6, 7] for multimode optical fibres.

This paper shows for the first time that it is possible to measure chromatic dispersion properties using the same sinusoidal technique in single mode fibres, too.

2. EXPERIMENTAL SET-UP

The scheme of the measuring set-up, similar to the one reported in details in [7] for multimode fibres, is shown in Fig.1. The pigtailed LED, changed according to the spectral region of interest, is sinusoidally modulated at the highest useful frequency. The emitted optical power is coupled to the single mode fibre; at the output end, the wavelength is selected through a monochromator and the light is detected by a suitable APD coupled to a low noise amplifier; the phase variations vs. wavelength are measured by a vector voltmeter, with a resolution of 0.1 degree. A calibration measurement is carried out, at the output of the pigtail, in order to take into account wavelength dependent phase shifts other than those due to the fibre. An optical attenuator allows to set, during calibration, the same level that was achieved during measurement, at each wavelength; this avoids undesired phase shifts depending on the signal level at the APD.

All LED's are provided with standard 50/125 graded index pigtails, resulting in high coupling losses (about 15 dB) at the multimode-single mode interface. Despite this high loss, satisfactory S/N ratios have been achieved, at the modulation frequency of 32 MHz due to the excellent performance of the receiver.

3. RESULTS AND DISCUSSION

At first, measurements have been carried out on a single mode fibre manufactured by Corning Glass, 1.1 km long, (3.5 dB/km at 0.9 micron). LED's have been used with center wavelengths 0.83 (Northern Telecom), 0.9 (Plessey), 1.1 (Lasertron) micron. Measurements beyond this range have not been possible due to the high attenuation of the fibre above 1.2 micron.

The detector was a Si APD. type RCA C30954E, enhanced for 1.1 micron detection. The output slit of the monochromator was set to give a 6 nm wavelength resolution.

Fig.2 reports the measured time delay vs. wavelength, together with a fitting equation (continuous line). Assuming a minimum acceptable S/N of 30 dB (electrical), an almost total covering of the spectral range between 0.77 and 1.14 μm has been achieved. The gap around 0.97 micron has been overcome by extending the fitting curves from the two adjacent sub-ranges.

By differentiating the curve fitting the experimental delays the chromatic dispersion can be evaluated, ad shown in Fig.3. These values are in good agreement with the ones found in literature; let's notice that they keep into account both material and waveguide dispersion.

The measurement has been repeated in the 1.1 μm range using a Ge APD; as expected, a lower S/N ratio has been found due to the lower optimum gain of this detector. In spite of the higher noise (resulting in a narrower useful wavelength range) the chromatic delays measured around 1.1 micron were in good agreement with those achieved with the Si device. No measurement was possible at longer wavelengths due to the sharp increase of the fibre attenuation beyond 1.2 micron, therefore it was not possible to identify the zero chromatic dispersion wavelength.

A second measurement has been carried out on a single mode fibre, with low attenuation up to 1.6 micron, (1360 m. long, 0.5 dB/km attenuation at 1.3 micron, manufactured by Corning Glass) in order to verify the practicability of the technique in the long wavelength range, and mainly around the zero dispersion wavelength, where only Ge APDs are likely to be used. Fig.4 shows the relative delays measured between 1.25 and 1.38 micron by means of an edge emitting LED peaked at 1.34 micron. The continuous line is a fitting curve; its differentiation gives the chromatic dispersion curve reported in Fig.5, that crosses the zero dispersion axis around 1333 nm. A very good repeatability has been observed among different measurements.

For the first time, the wavelength of zero chromatic dispersion of a single mode fibre has been directly obtained with the sinusoidal technique.

Furthermore, a single LED source has been sufficient to this aim.

Although the minimum accepted S/N level was 30 dB (electrical), as established for the lower wavelengths when using Si APDs, the measured delays were somewhat noisier, with a root mean square deviation from the fitting curve of about 20 ps. This is probably due, at least partially, to the higher quantum noise of the Ge APD.

Let's notice that, due to the high coupling loss between the 62 micron standard pigtail and the single mode fibre, the mean optical power coupled into the single mode fibre with the present set up is very low, about 400 nW, resulting in a 60 dB S/N ratio (electrical) at the center wavelength of the LED.

So, as a first improvement, a direct coupling between edge emitting diodes and single mode fibres should allow a more efficient power coupling (up to 2 microwatts, with currently available devices), and should conseguently enhance the S/N ratio of a least 10-15 dB (electrical).

Then, on the other hand, the use of a higher modulation frequency (i.e. 90 MHz against the present 32 MHz) should result in a lower weight of the phase noise.

Furthermore, PIN-FET detectors with higher sensitivity could be used although, at present, they are built with a small size (to achieve a low capacitance) that can't be easily coupled to the strongly astigmatic output of the monochromator.

4. CONCLUSIONS

In conclusion, we have demonstrated for the first time that accurate chromatic dispersion measurements are feasible in single mode fibres, making use of sinusoidally modulated incoherent sources. Moreover, the accurate determination of the zero dispersion wavelength has been obtained, using a single LED. In view of some improvements that should enhance the S/N ratio, thus extending the dynamic range, this simple technique seems very promising, and could soon become a standard method for the characterization of chromatic dispersion in single mode optical fibres.

REFERENCES

1. D.Gloge, E.L.Chinnock and T.P.Lee: "Ga As twin laser set up to measure mode and material dispersion in optical fibres", Appl. Opt., Vol. 13, No. 2, pp.251.
2. B.Sordo, F.Esposto and B.Costa: "Experimental study of modal and material dispersion in spliced optical fibres", Fourth European Conference on Optical Communication, Genova, (Sept. 12-15, 1978), pp. 71.
3. C.Lin, A.R.Tynes, A.Tomita and P.L.Lin: "Pulse delay measurements in single mode fibers using picosecond In Ga As P injection lasers in the 1.3 μm spectral region", Proc. OFC '82, Phoenix, AZ, (Apr. 13-15, 1982), pp.28.
4. D.N.Payne, A.H.Hartog: "Determination of the wavelength of zero material dispersion in optical fibres by pulse-delay measurements", Electron. Lett., 13, 21, pp.627.
5. L.G.Cohen, C.Lin: "Pulse delay measurements in the zero material dispersion wavelength region for optical fibres", Appl. Opt. 16, 12, pp. 3136.
6. B.Costa, D.Mazzoni, M.Puleo and E.Vezzoni: "Highly accurate and simple technique for material dispersion measurements in optical fibres", Proc. 7th European Conference on Optical Communication, Copenhagen, (Sept. 8-11, 1981), pp. 5.3.
7. B.Costa, D.Mazzoni, M.Puleo and E.Vezzoni: "Phase shift technique for the measurement of chromatic dispersion in optical fibres using LED's", IEEE Journal of Q.E., Vol. QE-18, No. 10, pp. 1509-1514.

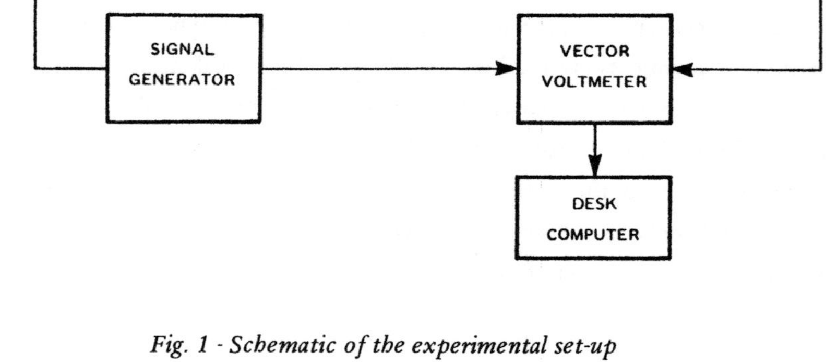

Fig. 1 - Schematic of the experimental set-up

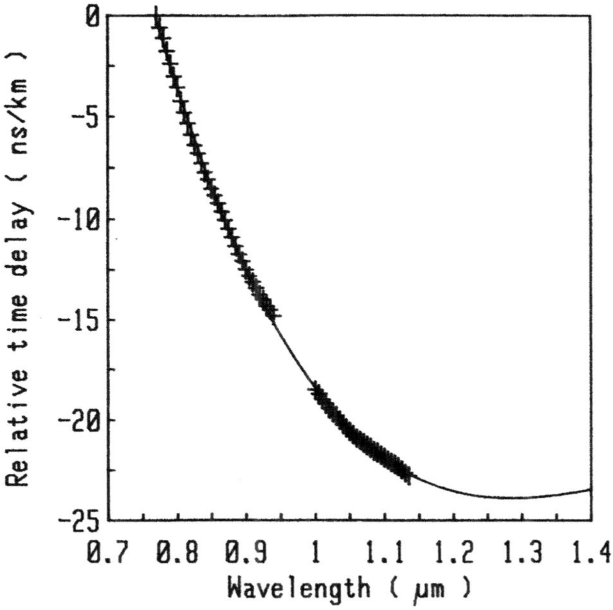

Fig. 2 - Relative time delay vs. wavelength: experimental points (crosses) and fitting equation (continuous line)

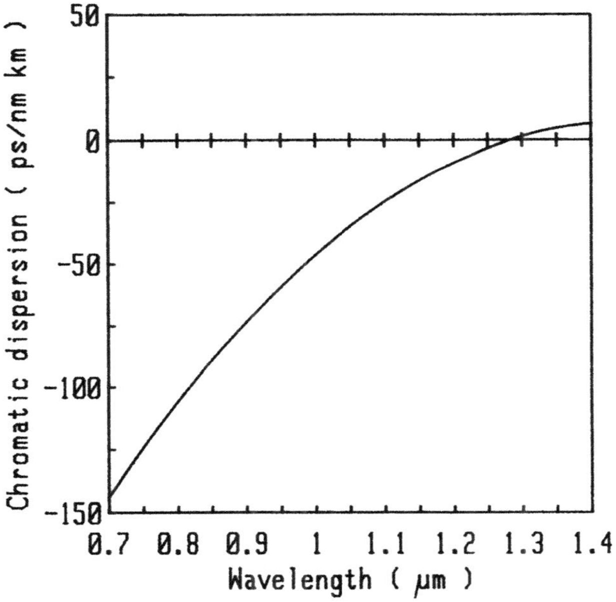

Fig. 3 - Chromatic dispersion of the single mode fibre obtained by differentiating the fitting curve of Fig. 2

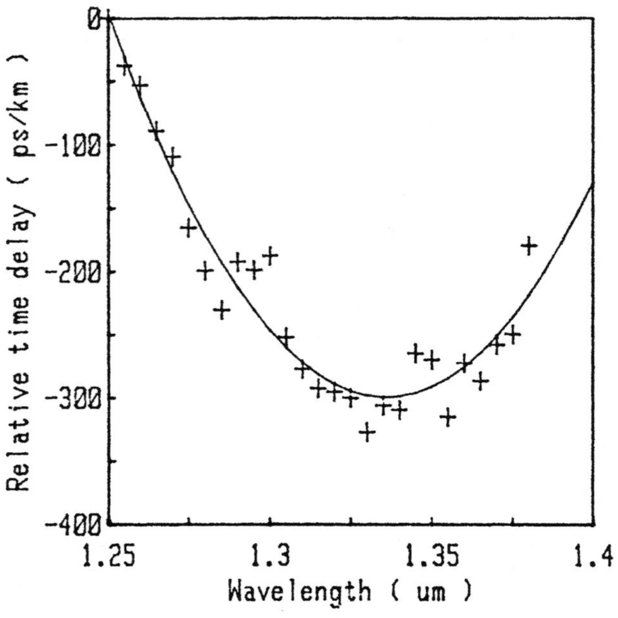

Fig. 4 - Relative time delay vs. wavelength around the zero dispersion wavelength of a low loss single mode fibre: experimental points (crosses) and fitting equation (continuous line)

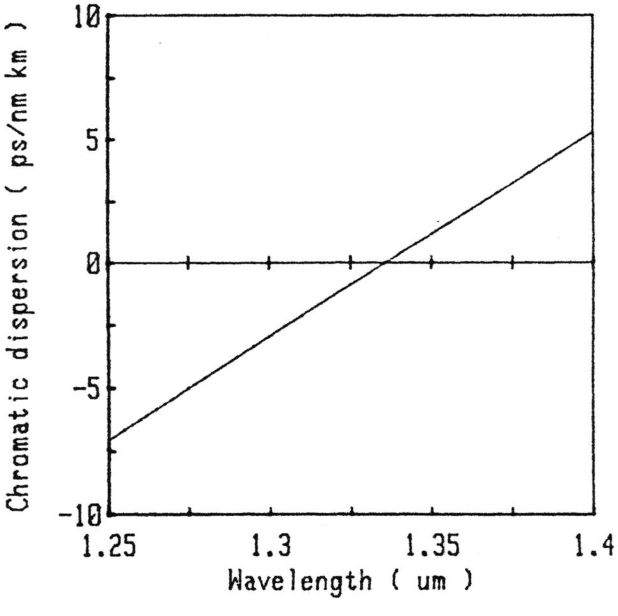

Fig. 5 - Chromatic dispersion of the single mode fibre obtained by differentiating the fitting curve of Fig. 4

EXTRACTION OF SINGLE-MODE FIBER ESI PARAMETERS BY USING A SIMPLE TECHNIQUE FOR THE MEASUREMENT OF BEND LOSS

A.B. SHARMA, A.-H. AL-ANI, AND S.J. HALME
HELSINKI UNIVERSITY OF TECHNOLOGY, COMMUNICATIONS LABORATORY, SF-02150 ESPOO 15, FINLAND

1. INTRODUCTION

It has been shown by a number of authors /1-3/ that for a practical range of V values, the field profile of the fundamental mode can be closely approximated by a Gaussian function more or less independent of the exact shape of the fiber's refractive-index profile. This property has naturally led to the idea of replacing (for calculations) the complex profiles typical to MCVD fibers, by equivalent-step-index (ESI) models /2-4/ that yield reasonably accurate estimates of parameters such as constant bend, transition (microbend), and coupling losses /3/. The prediction of dispersion behaviour from ESI parameters is presently somewhat uncertain /5/.

In practice, a number of experimental techniques have been developed for ascertaining ESI parameters /6-9/. The basic approach is to measure some parameter of interest, such as the fiber-to-fiber splicing loss (as a function of offset, tilt, or end-face separation), and to search for the ideal step-index fiber with the same property. Of course, this raises the question of differences amongst the ESI parameters obtained via the different methods, and indeed an international comparison experiment is currently in progress in Europe under the auspices of COST 208.

Here, we present the preliminary results of another technique for the determination of ESI parameters based on matching measured constant-bend loss as a function of the bend radius. Experimentally, our technique essentially eliminates the influence of externally produced transition losses, so that comparison with published constant-bend loss formulae /10-13/ becomes possible. The technique does not depend on fiber-end quality, and is based on a loss measurement set-up that is conventional except for the processing electronics /14/. So far, we have only carried out repeatability experiments on a single monomode fiber, and although many problems still remain, our preliminary results appear encouraging. Beside the inherent match to the measured bend loss, we also find the predicted cut-off wavelength to be in close agreement with that specified by the manufacturer.

2. PRINCIPLE OF THE METHOD

The theoretical loss of an ideal step-index fiber can be written in form /10-13/:

$$2\alpha = \frac{A}{\sqrt{R}} \exp(-BR), \tag{1}$$

where α is the power attenuation in nepers per unit length, R is the radius of curvature of the fiber axis, and A and B are given by:

$$A = \tfrac{1}{2}\sqrt{[\pi/(a\,W^3)]}\cdot[U/(V\,K_1(W))]^2 , \tag{2}$$

and

$$B = 4\Delta W^3/(3aV^2). \tag{3}$$

In the above, a is the core radius, Δ is the relative index difference of core and cladding, K_1 is the modified Bessel function, while U, V, and W are the usual parameters given by:

$$V^2 = U^2 + W^2 , \tag{4}$$

$$U^2 = a^2[k_0^2\,n_1^2 - \beta^2] , \tag{5}$$

$$W^2 = a^2[\beta^2 - k_0^2 n_2^2] , \tag{6}$$

where $k_o = 2\pi/\lambda$, λ is the wavelength, n_1 is the core index, n_2 is the cladding index, and β is the axial propagation constant. A good approximation of β can be obtained from the eigenvalue equation /15, 16/ below:

$$U = acos(U/V) + \pi/4 . \tag{7}$$

From the above, it is clear that A and B are functions of a, n_1, n_2, and λ. Generally, n_2 is very close to the refractive index of fused quartz; its absolute value is of less interest than the relative index difference Δ, so that we may reasonably set $n_2 = 1.459$. Then, for a given λ, A and B are functions of the two unknowns a and n_1, which are also the ESI parameters of interest. Thus, in principle, if the loss is measured for two different radii of curvature (at a known λ), then the values of A and B for the equivalent step-index fiber can be determined, thus yielding two equations (in n_1 and a) that can be solved by the Newton-Raphson method. Note that every pair of roots at the beginning of an iteration loop uniquely defines U, V, and W via eqns. 4 to 7.

3. EXPERIMENTAL TECHNIQUE

The experimental set-up consists of a monochromator as the optical source providing a chopped (190 Hz) light beam that is focused onto the fiber end. The output from the far end is detected by using a germanium detector whose output is synchronously sampled and integrated as described in /14/. The fiber used in the experiments was specified to have a core diameter of 5.5 μm and a cut-off wavelength of 1.15 μm. Cladding-mode strippers were used at the input and output, although these are probably unnecessary for the long fiber used.

A few meters beyond the input, the fiber is "folded back" and used to form a bifilar winding as shown in Fig. 1. The beginning and end of the straight section of the loop is taped as shown in the figure and the fiber gently wound on to a polished cylinder and secured using tape 2. Tape 1 is used to fasten the fiber onto a table, while tape 3 is used to secure the bend created by looping back. The cylinder is raised off the table by supports (not shown) on its end, and the fiber section between tape 1 and the cylinder is kept straight according to a line drawn on the table and a mechacal "cursor" butted to the cylinder supports.

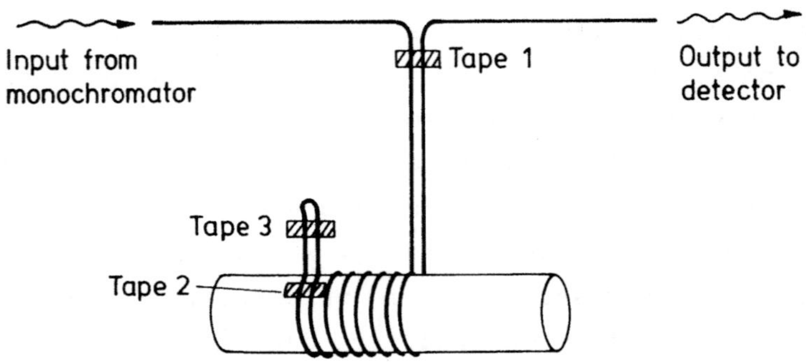

FIGURE 1. Schematic diagram of the bend-loss apparatus

The measurement consists of measuring the power reaching the detector by unwrapping one turn from the cylinder at a time. One turn is always left on the cylinder to prevent disturbance form tape 2, while the maximum number of turns is limited such as to prevent any significant transition at the edge of tape 1. Under these conditions the loss behaviour between tapes 1 and 2 may be considered to be homogeneous, and the power in the bends to decay esponentially as:

$$P(z) = P(0) \exp[-2\alpha(2\pi\phi)n] \quad , \qquad (8)$$

where P(z) is the detected power. P(0) is the power entering the bend, ϕ is the diameter of the bend (including fiber thickness), n is the number of turns on the cylinder, while the additional factor of 2 accounts for the lenghts in any one turn. Here, we have assumed that the contribution of other loss mechanisms including random microbends is negligible, which is reasonable for short lengths of low-loss fiber. In fact, the fiber used in the experiment was secondary coated, further reducing any potential problems due to cylinder roughness. The validity of these assumptions is supported by the typical plots in Fig. 2, showing reasonable straight line fits to measured data at several wavelengths. This data was obtained for a primary-coated fiber that we used during the earlier stages of apparatus development.

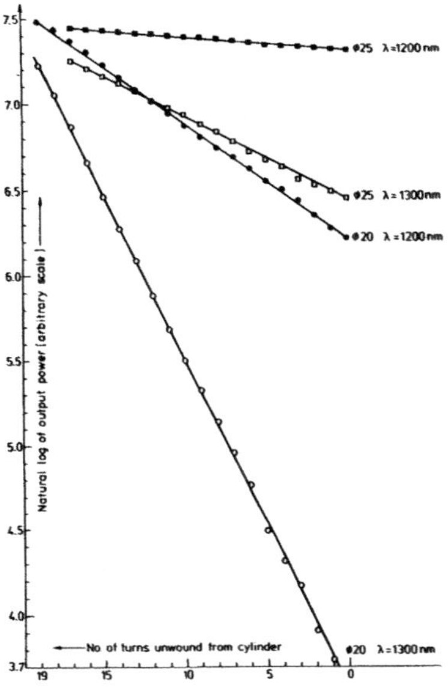

FIGURE 2. Some typical bend-loss results for a primary-coated fiber for rod diameters of 20 and 25 mm and at operating wavelengths of 1200 nm and 1300 nm

Clearly, a logarithmic plot of the detected power against the number of turns can be used to extract the value of α for the particular value of ϕ. In practice, we obtain α via linear regression to the natural logarithm of eqn. (8).

We repeat the measurements for several values of the radius of curvature R $(=\phi/2)$, and obtain the constants A and B of eqn. (1) via the curve fitting procedure given in the next section.

4. COMPUTATION PROCEDURE

A direct least-squares estimation of A and B from (1) involves problems because of the necessity of iteration of a numerically difficult function. A procedure that seems to work fairly well is to use the natural logarithm of (1). As usual, we then form the residuals:

$$V_i = \ln A - \tfrac{1}{2} \ln R_i - BR_i - \ln \alpha_i \quad , \tag{9}$$

and minimize the quantity

$$S = \sum_i V_i^2 \tag{10}$$

with respect to A and B. The following results are obtained:

$$A = \exp\left\{\frac{1}{2N}\sum_i \ln(\alpha_i^2\, R_i) + \frac{B}{N}\sum_i R_i\right\} \tag{11}$$

$$B = \frac{-\sum_i (R_i - \sum_k R_k/N)\,\ln(\alpha_i^2 R_i)}{2\sum_i (R_i - \sum_k R_k/N)^2} \tag{12}$$

where N is the total number of data points. Thus, A and B can be easily evaluated numerically. The results are then used in (2) and (3) to iterate for n_1 and a using the Newton-Raphson method extended to two variables (see e.g. /17/). Our initial roots are selected by looping through all combinations within the normal range of n_1 and a, until a pair sufficiently close to the true roots is found, i.e. a pair that leads to convergence.

5. RESULTS

We have found that the bend loss consistently tends to be lower than expected, and yields ESI core radii that are too low. However, it has been proposed that such an effect would occur due to stress-induced refractive index changes, and could be taken into account by specifying an effective bend radius which for fused quartz would be given by /18/:

$$R_e = 1.27\ R\ . \tag{13}$$

We have used this correction factor with success, and to date have obtained the following results for our fiber:

$$\left.\begin{aligned} a_e &= 4.30\ \mu m \pm 0.07\ \mu m \text{ (one standard deviation)} \\ \Delta_e &= 2.71 \times 10^{-3} \pm 0.03 \times 10^{-3} \text{ (one standard deviation)} \end{aligned}\right\} \tag{14}$$

$$\lambda_{co} = 1.21\ \mu m \pm 0.01\ \mu m \text{ (one standard deviation)} \tag{15}$$

The cut-off wavelength in (15) has been computed from the result in (14) and is seen to be in good agreement with the specified value of 1.15 μm.

We would next like to comment on the scatter associated with the measured data. The results of several bend loss measurements is shown in Fig. 3.

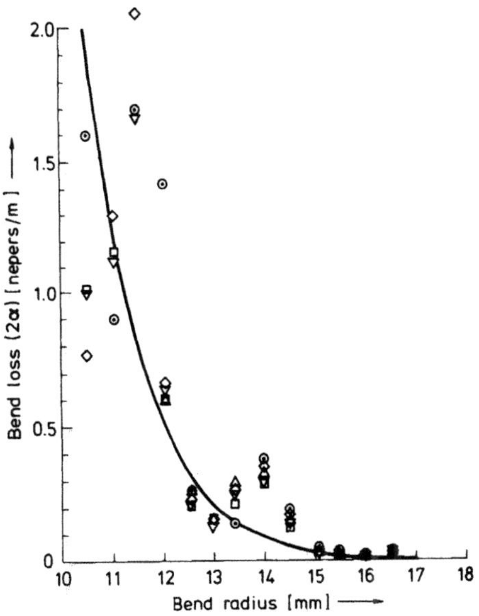

FIGURE 3. Bend loss as a function of bend radius (from several independent trials for the same fiber) showing deviations from the theoretically predicted behaviour of the ESI fiber (solid curve)

It can be seen that a fair amount of scatter exists at the higher values of loss, and that a hump appears consistently at a radius of about 14 mm. The solid line curve is a theoretical plot for the ESI fiber corrected for the bend radius as dictated by (13). We have no satisfactory explanation for the hump, but we are in the process of introducing more careful experimental control to reduce the scatter. The results we have presented here have been obtained by the rejection of points that clearly deviate from the almost straight line when the loss is plotted on a logarithmic scale. A further justification is provided by Fig. 4 which shows that for most fibers of interest today, the ratio of A to B remains greater than unity.

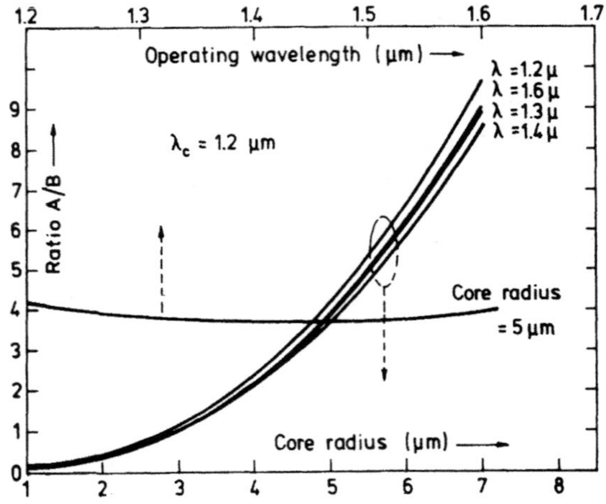

FIGURE 4. Behaviour of the ratio of parameters A and B for a cut-off wavelength of 1200 nm

If the "out of range" points are not rejected the curve fitting procedure rapidly leads to ratios of A and B that cannot be explained in terms of the theory.

6. CONCLUSIONS

We have presented a preliminary report on a new method for the measurement of constant bend loss and our approach for the extraction of parameters of a step-index fiber with the same bend-loss behaviour. The experimental technique is simple and the influence of transition losses seems to be negligible. The ESI parameters appear to be reasonably repeatable and yield a cut-off wavelength in close agreement with the specified (measured) value. The random and systematic errors associated with some of the data points are excessive and imply the necessity for improvement of experimental procedure. However, the method appears to be viable, has the advantage of being applicable to bare, as well as primary and secondary coated fibers, and is not sensitive to fiber-end quality or positioning.

REFERENCES

1. Marcuse D. 1978. "Gaussian approximation of the fundamental modes of graded index fibers", J. Opt. Soc. Am. 68, 103.
2. Snyder A.W., Sammut R.A. 1979. "Fundamental modes of graded optical fibers", J. Opt. Soc. Am. 69, 1663.
3. Matsumura H, Suganuma T. 1980. "Normalization of single mode fibers having an arbitary index profile", Appl. Opt. 19, 3151.
4. Brinkmeyer E. 1979. "Spot size of graded index single mode fibers: profile independent representation and new determination method", Appl. Opt. 18, 932.
5. Bhagavatula V.A. 1982. "Estimation of single mode waveguide dispersion using an equivalent-step-index approach", Electron. Lett. 18, 319.
6. Millar C.A. 1981. "Direct method for determining equivalent-step-index profiles for monomode fibers", Electron. Lett. 17, 458.
7. Alard F, Jeunhomme L, Sansonetti P. 1981. "Fundamental mode spot-size measurement in single-mode optical fibers", Electron. Lett. 17, 958.
8. Masuda S, Iwama T, Daido Y. 1981. "Non-destructive measurement of core radius numerical aperture, and cut-off wavelength for single-mode fibers", Appl. Opt. 20, 4035.
9. Gambling W.A, Payne D.N, Matsumura H, Dyott R.B. 1976. "Determination of core diameter and refractive-index difference of single mode fibres by observation of the far field pattern", IEE J. Microw. Opt. Acoust. 1, 13.
10. Lewin L. 1974. "Radiation from curved dielectric slabs and fibers", IEEE Trans. MTT22, 718.
11. Kuester E.F, Chang D.C. 1975. Surface-wave radiation loss from curved dielectric slabs and fibers", IEE J. Quant. Electron. QE11, 903.
12. Marcuse D. 1976. "Curvature loss formula for optical fibers", J. Opt. Soc. Am. 66, 216.
13. Snyder A.W, White I, Mitchell D.J. 1975. "Radiation from bent optical waveguides", Electron. Lett. 11, 332.
14. Sharma A.B, Halme S.J, Butusov M.M. 1981. "Optical Fiber Systems and their Components", pp 157-159 (Springer).
15. Hartog A.H, Adams M.J. 1977. "on the accuracy of the WKB approximation in optical di-electric waveguides", Opt. Quant. Electron. 9, 223.
16. Sharma A.B, Saijonmaa J, Halme S.J. 1980. "Comparison of pulse responses of step-index fibers", Electron. Lett. 16, 557.
17. Sokolnikoff I.S, Redheffer R.M. 1966. "Mathematics of Physics and Modern Engineering" (McGraw-Hill).
18. Nagano K, Kawakami S, Nishida S. 1978. "Change of the refractive index in an optical fiber due to external forces" Appl. Opt. 17, 2080.

REFERENCES

1. [illegible] D., 1970. "[illegible] [illegible]", [illegible] member, [illegible] [illegible] [illegible] [illegible] [illegible]. [illegible] pp. 1[illegible].
2. [illegible] M., [illegible] and [illegible], 1978. "[illegible] [illegible] [illegible] spatial [illegible]", [illegible] [illegible] [illegible].
3. [illegible] [illegible] H., [illegible] 1980. "[illegible] [illegible] [illegible] [illegible] [illegible] [illegible] [illegible] [illegible]", [illegible] [illegible].
4. [illegible] [illegible] S., 197[illegible]. "[illegible] [illegible] [illegible] [illegible] [illegible] [illegible] [illegible] [illegible]", [illegible].
5. [illegible] [illegible] [illegible], 1981. "[illegible] [illegible] [illegible] [illegible] [illegible] [illegible] [illegible] [illegible] [illegible]", Electrophoresis, [illegible].
6. [illegible] [illegible] A., 1981. "[illegible] method [illegible] [illegible] [illegible] [illegible] [illegible] [illegible] [illegible]", [illegible].
7. [illegible] [illegible] [illegible], 1981. "[illegible] [illegible] [illegible] [illegible] [illegible]", Electrophoresis, [illegible] 1981.
8. [illegible] [illegible] [illegible] [illegible] [illegible] [illegible] [illegible] [illegible] [illegible] [illegible].
9. [illegible] [illegible] [illegible] [illegible] [illegible] [illegible] [illegible] [illegible] [illegible] [illegible].
10. [illegible] [illegible], 1974. "[illegible] [illegible] [illegible] [illegible]", [illegible].
11. [illegible] [illegible] [illegible] [illegible] [illegible], Electrophoresis, 1981 [illegible].
12. [illegible] [illegible] [illegible] [illegible] [illegible] [illegible].
13. [illegible] [illegible] [illegible] [illegible] [illegible] [illegible].
14. [illegible] [illegible] [illegible] [illegible] [illegible] [illegible] [illegible].
15. [illegible] [illegible] [illegible] [illegible] [illegible] [illegible] [illegible].
16. [illegible] [illegible] [illegible] [illegible] [illegible] [illegible] [illegible].
17. [illegible] [illegible] [illegible] [illegible] [illegible] [illegible].
18. [illegible] [illegible] [illegible] [illegible] [illegible] [illegible] [illegible].

DETERMINATION OF SINGLE-MODE FIBER REFRACTIVE INDEX PROFILES BY A PROPAGATION MODE NEAR-FIELD SCANNING TECHNIQUE

KATSUMI MORISHITA

Osaka Electro-Communication University
Neyagawa-shi, Osaka, 572 Japan

1. INTRODUCTION

The characteristics of an optical fiber, such as spot size, single-mode condition, and dispersion characteristics, are principally determined by its refractive index profile. It is important to establish an efficient and accurate method for measuring the refractive index profile. For a single-mode fiber, however, the measurement of the profile is very difficult because of its small core diameter and low refractive index difference. Several methods were proposed for the measurement of the refractive index profile of single-mode fibers, such as the far-field radiation pattern technique [1],[2], the scattering pattern method [3], the reflection method [4], the focusing method [5], the refracted near-field technique [6], and the interferometric slab method [7].

This paper presents a new method of measuring the refractive index profile of a single-mode fiber. The profile is determined from the measurement of the near-field pattern of a propagation mode, i.e., the light intensity distribution across the output face of a propagation mode, by a new equation derived from the scalar wave equation. A result of computer simulation shows that this method is useful for refractive index profile measurement systems from the practical point of view.

2. THEORY

It is assumed that the refractive index of the specific fiber depends only upon the distance r from the center axis, and the permeability of the medium is equal to that of vacuum. The scalar approximation technique is applied to optical fiber analysis. The field distribution of a propagation mode in the fiber to be measured, ϕ_r, is obtained by scanning the near-field pattern. The real refractive index profile of the fiber is expressed by n_r, which is an unknown function. These quantities satisfy the following scalar equation:

$$\frac{1}{r}\frac{d}{dr}(r\frac{d\phi_r}{dr}) + (k^2n_r^2 - \beta_r^2 - \frac{m^2}{r^2})\phi_r = 0. \tag{1}$$

where k and m are the free space wave number and an azimuthal quantum number, respectively, and β_r is an unknown propagation constant.

First, we assume that the refractive index profile of the specific fiber is n insted of n_r. Then the field distribution ϕ and the propagation constant β can be obtained by solving the following equation:

$$\frac{1}{r}\frac{d}{dr}(r\frac{d\phi}{dr}) + (k^2n^2 - \beta^2 - \frac{m^2}{r^2})\phi = 0. \tag{2}$$

Second, to estimate the deviations in the field distribution, the propagation constant, and the refractive index profile from those for the fiber to be measured, we let

$$\begin{aligned} \phi &= \phi_r + \Delta\phi \\ \beta^2 &= \beta_r^2 + \Delta\beta^2 \\ n^2 &= n_r^2 + \Delta n^2. \end{aligned} \tag{3}$$

Fig.1 shows a schematic illustration for the refractive index profiles of the real fiber to be measured and the test fiber whose index profile is assumed. We derive an expression for Δn^2, i.e., n_r, by means of ϕ_r and $\Delta\phi$. The deviation of the refractive index profile is written by using (1)-(3) as

$$\Delta n^2 = \frac{\Delta\beta^2}{k^2} - \frac{1}{k^2\phi_r}\left(\frac{1}{r}\frac{d}{dr}\{r\frac{d(\Delta\phi)}{dr}\} + (k^2n^2 - \beta^2 - \frac{m^2}{r^2})\Delta\phi\right). \tag{4}$$

The first term on the right-hand side of (4) is an unknown constant, but the remainder terms are the known functions. Since the constant term does not affect the mode pattern, n_r^2 obtained from this mode pattern has an uncertainty of constant. However, this disadvantage is removed as follows, measuring the refractive index of the cladding, which contains information on that constant factor.

$$\Delta n^2 = -\frac{1}{k^2\phi_r}\left(\frac{1}{r}\frac{d}{dr}\{r\frac{d(\Delta\phi)}{dr}\} + (k^2n^2 - \beta^2 - \frac{m^2}{r^2})\Delta\phi\right). \tag{5}$$

The unknown constant of (4) does not affect the difference distribution of the squared refractive index from that of the cladding. Therefore, even if the refractive index of the cladding is unknown, we can get the difference distribution of the squared refractive index by using (5).

3. COMPUTER SIMULATION

In this paper, the measurement of the refractive index difference distribution is simulated considering noise in the near-field pattern of a propagation mode. The scalar multilayer approximation method [8] is used to calculate the field distribution of a propagation mode for arbitrary refractive index profiles. In this numerical simulation, the index profile of the core is divided into 50 layers, and is computed by means of the power intensity distribution of the fundamental mode LP_{01}, which is easily excited in a single-mode fiber. The power intensity distribution is given by 121 numerical data, and the assumed index profile n is taken to be the step-index profile with index difference 4.5×10^{-3}.

Fig.2 shows results of computer simulation. The real refractive index profile to be measured, which is a step-index fiber with dip, is drawn by a solid curve in Fig.2(a). The power intensity distribution of the LP_{01} mode for the fiber with dip is shown by a solid curve in Fig.2(b). It is assumed that noise in the measured power intensity distribution is a gaussian random variable with mean zero and standard deviation 10^{-3}, as shown by dots in Fig.2(b). The power intensity distribution with noise is smoothed by a local least squares method using a curve of third order, and the refractive index profile is obtained from the smoothed intensity distribution (shown by a broken curve) and (5). The dot-dash curve in Fig.2(b) denotes the power intensity distribution of the LP_{01} mode for the calculated refractive index profile in Fig.2(a). The intensity error in Fig.2(b) indicates the difference from the real power intensity of the LP_{01} mode (solid curve). Fig.3 shows results of computer simulation for a different set of random numbers with standard deviation 5×10^{-3}. In this measurement, the error in the computed profile is sensitive to the accuracy of the measured intensity distribution and smoothing. If the intensity distribution is measured more precisely, the more accurate index profile can be obtained by little smoothing.

4. CONCLUSIONS

Based on the propagation mode near-field scanning technique, a new technique is proposed for measuring the refractive index profile of single-mode fibers. It can be concluded that this technique is useful for refractive index distribution measurement from the practical point of view.

This method requires simple sample preparation and relatively simple equipment for its instrumentation. However, it takes time consuming computer calculation. For the observation of single-mode fibers, a microscope is required to magnify the near-field pattern in the fiber end face, and a diode detector, a diode array detector or a vidicon would be used for scanning purposes.

REFERENCES

[1]W.A.Gambling, D.N.Payne, H.Matsumura and R.B.Dyott, IEE J. Microwaves Opt. Acoust., vol.1, no.13, pp.13-17, 1976.

[2]K.Hotate and T.Okoshi, Appl. Opt., vol.18, no.19, pp.3265-3271, 1979.

[3]T.Okoshi and K.Hotate, Appl. Opt., vol.15, no.11, pp.2756-2764, 1976.

[4]M.Tateda, Appl. Opt., vol.17, no.3, pp.475-478, 1978.

[5]H.M.Presby, D.Marcuse and W.G.French, Appl. Opt., vol.18, no.23, pp.4006-4011, 1979.

[6]K.I.White, Opt. Quantum Electron., vol.11, pp.185-196, 1979.

[7]G.W.Tasker, W.G.French, J.R.Simpson, P.Kaiser and H.M.Presby, Appl. Opt., vol.17, no.11, pp.1836-1842, 1978.

[8]K.Morishita, IEEE Trans. Microwave Theory Tech., vol.MTT-29, no.4, pp.348-352, 1981.

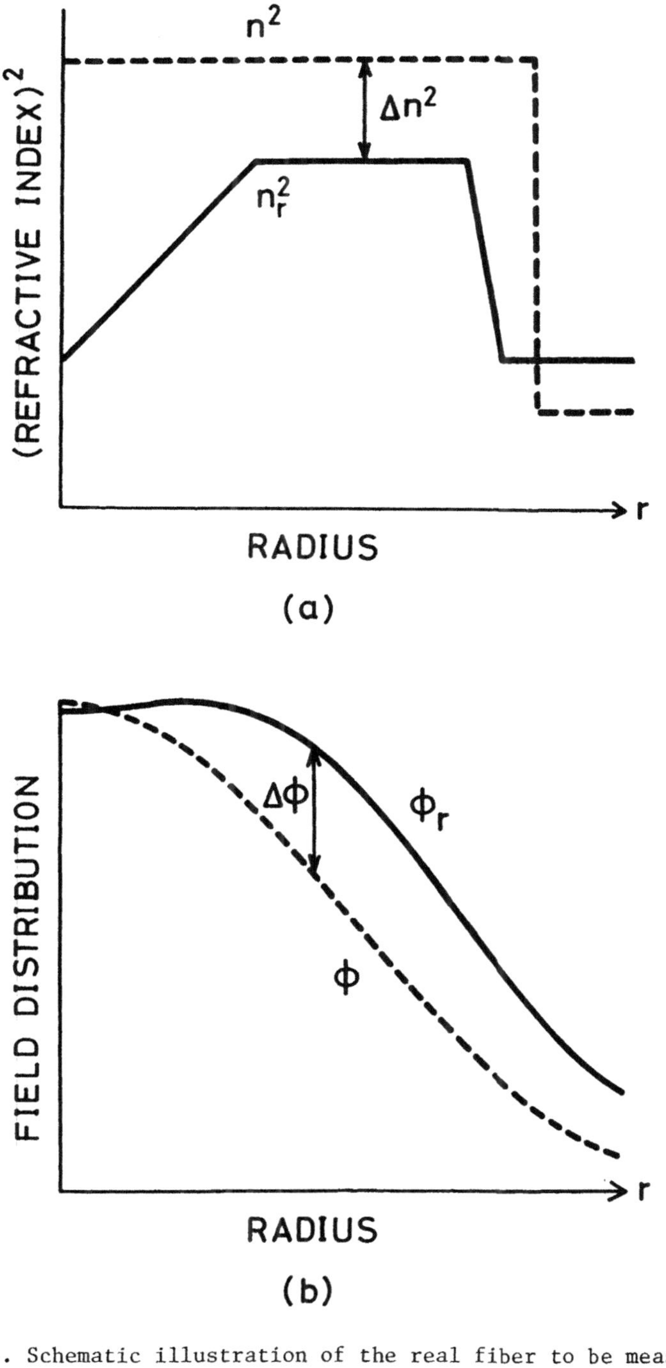

FIGURE 1. Schematic illustration of the real fiber to be measured and the test fiber whose index profile is assumed: (a)the real and the assumed refractive index profiles, n_r and n; (b)field distributions of propagation modes for n_r and n, ϕ_r and ϕ.

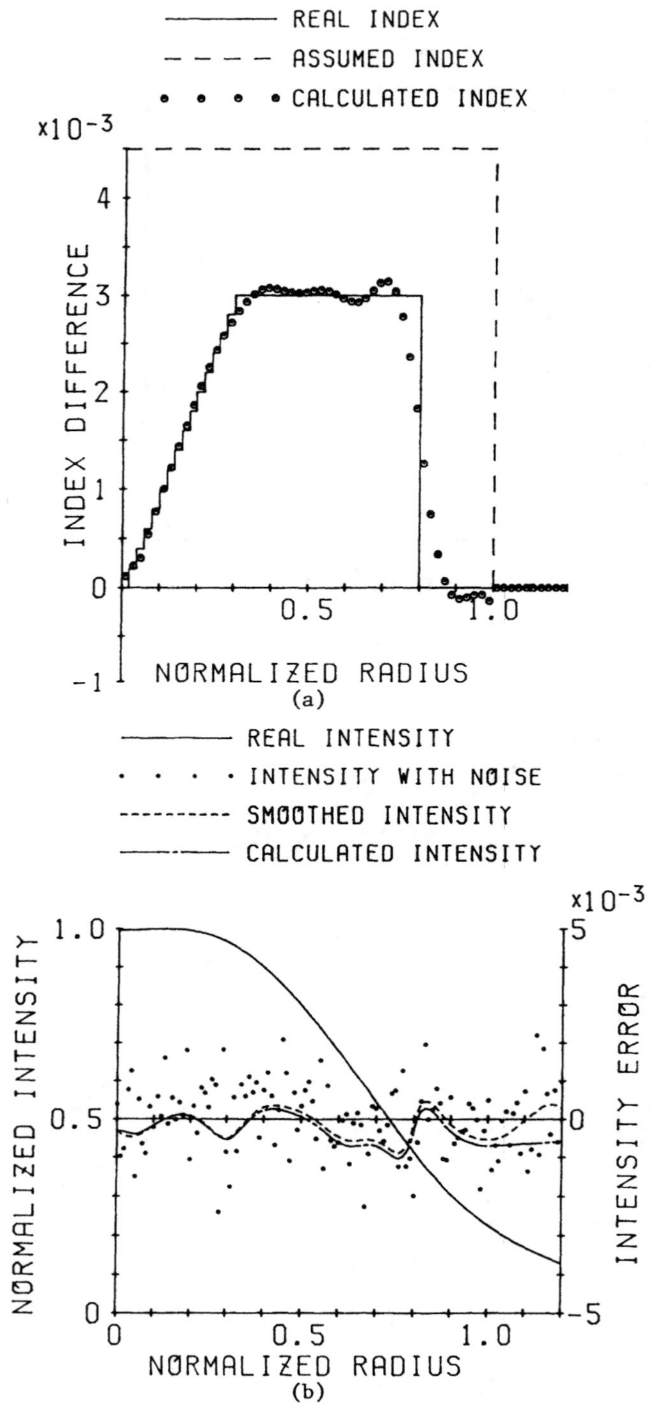

FIGURE 2. Computer simulation for random numbers with standard deviation 10^{-3}. (a)Refractive index profiles. (b)The power intensity distribution of the LP_{01} mode and the error in the perturved intensity distribution.

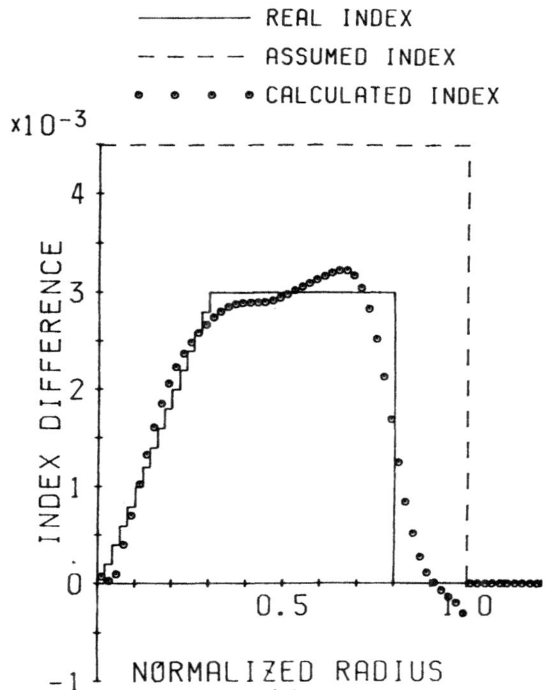

(a)

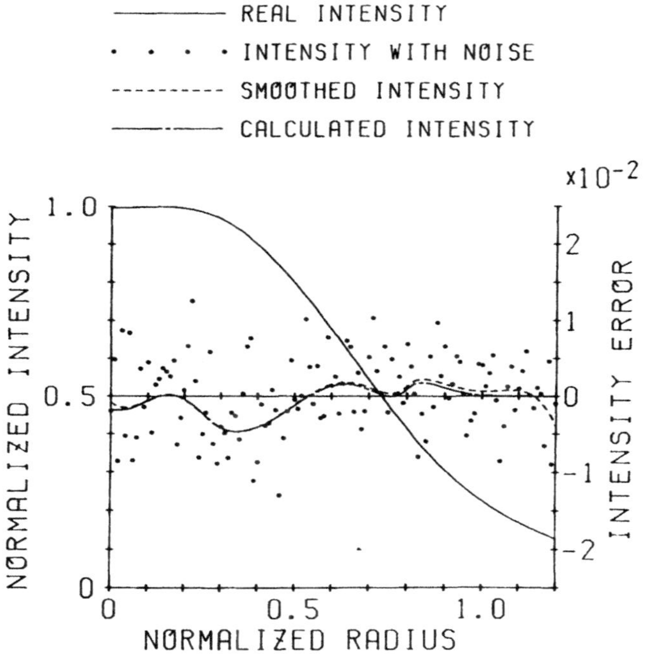

(b)

FIGURE 3. Same as Fig.2 for a different set of random numbers with standard deviation $5x10^{-3}$.

THE EFFECTS OF BIREFRINGENT FLUCTUATIONS ON A FIBER OPTIC GYRO

RONALD L. PHILLIPS and LARRY C. ANDREWS

1. INTRODUCTION

Fluctuations of the polarization of an optical wave in a Sagnac fiber optic gyroscope cause random variations in the output signal of the sensor which are indistinguishable from variations of the rotation rate. The random variations of the polarization of the optical wave are caused by random environmental changes in the fiber. These environmental fluctuations can induce a random coupling between the two propagating eigen-modes within the fiber. If a very large birefringence is generated within the fiber, either by internal strain or by shaping the core into an elliptical configuration, the environmentally induced cross-coupling can be neglected. The fiber is then polarization preserving if the input optical wave is launched exactly parallel to one of the two birefringent axes. If it is not, and both eigen-modes are excited, the net polarization at the output may fluctuate due to differences in the environmental effects on the indices of the two birefringent axes. It is the nearness of alignment of the input wave polarization that defines the limitations of the polarization preserving fibers to reduce erroneous signal fluctuations due to environmental variations.

The configuration for the fiber optic gyro rate sensor under consideration is shown in Fig. 1. It is convenient in the following analysis to describe the propagation of the laser beam through the system by using Jones matrices to represent the optical components of the system as they act upon the clockwise (CW) and counterclockwise (CCW) rotating optical wave.

2. INTENSITY OF THE OUTPUT

At the output of the acoustic optic modulator the two beams are first reflected by a mirror, which is simply an identity transformation since there is no loss of energy. The next component is a mirror divider, whose transmissive and reflective properties are characterized by

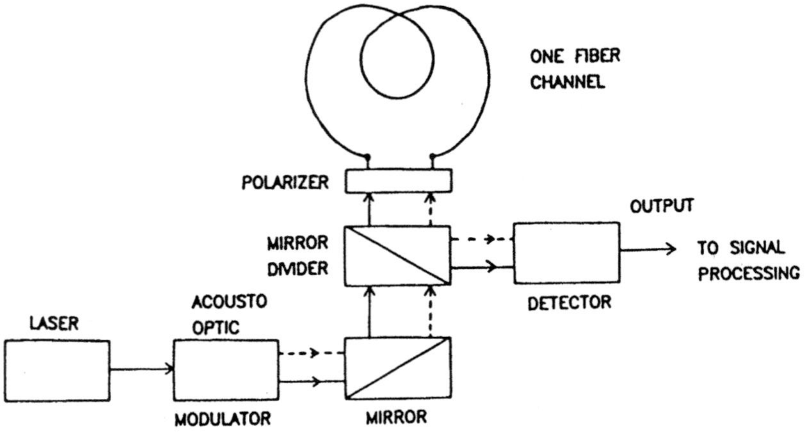

Fig. 1 Fiber optic rate sensor.

$$T = \begin{pmatrix} \tau_x & 0 \\ 0 & \tau_y \end{pmatrix}, \qquad R = \begin{pmatrix} \rho_x & 0 \\ 0 & \rho_y \end{pmatrix}. \tag{1}$$

Before entering the fiber, the beams are passed through a polarizer which we assume is perfect. Hence, we write

$$P = \begin{pmatrix} 1 & 0 \\ 0 & 0 \end{pmatrix}. \tag{2}$$

Both beams are then passed through the fiber — one in the CW direction and the other in the CCW direction. While in the fiber, both beams undergo transformations due to rotational effects and birefringence in the fiber. Let us assume the accumulative phase shift between the alignment of the CW and CCW beams due to birefringence in the fiber is ψ (which may be the sum of several phase shifts). If the input polarization makes an angle θ with respect to the birefringence axis, the transformations that take place within the fiber are characterized in the CW direction by

$$J_{12} = e^{i\phi/2} \begin{pmatrix} \cos\theta & \sin\theta \\ -\sin\theta & \cos\theta \end{pmatrix} \begin{pmatrix} e^{i\psi/2} & 0 \\ 0 & e^{-i\psi/2} \end{pmatrix} \begin{pmatrix} \cos\theta & -\sin\theta \\ \sin\theta & \cos\theta \end{pmatrix} \tag{3}$$

and by J_{21} in the CCW direction, which is the same as J_{12} except for the lead factor $e^{-i\phi/2}$. Here ϕ denotes the desired Sagnac phase angle.

The recombined CW and CCW beams at the detector can thus be represented by the sequence of transformations

$$\bar{U} = \underbrace{R\,P\,J_{12}\,P\,T\,\bar{E}_1}_{CW} + \underbrace{R\,P\,J_{21}\,P\,T\,\bar{E}_2}_{CCW}, \tag{4}$$

where $\bar{E}_1$ and $\bar{E}_2$ denote the fields of the two beams at the output of the acoustic optic modulator. Performing the matrix operations indicated in (4) and calculating the magnitude squared, we find that the intensity of the field is given by

$$I = |\bar{U}|^2$$

$$= K(1 + \cos^2 2\theta + \cos\psi \sin^2 2\theta)(1 + \cos\phi), \tag{5}$$

where K is a constant depending on the components of T, R, and the fields $\bar{E}_1$ and $\bar{E}_2$. The exact value of K does not affect the following analysis.

2.1 Special Cases

When birefringence is not present in the fiber, i.e., when $\psi = 0$, Eq. (5) reduces to

$$I = 2K(1 + \cos\phi), \tag{6}$$

which is considered the ideal case. That is, demodulation of (6) will lead to the correct Sagnac phase angle ϕ. From (5), we also note that if the angle θ between the polarizer transmission axis and the birefringence axis is zero, then (5) again reduces to the ideal case (6). Thus, either rotation or birefringence by themselves do not affect the value of the Sagnac angle ϕ. Rather, it is the combined effect of input rotation θ and birefringence ψ that contributes to an error in the measurement of ϕ.

3. THE APPARENT PHASE ANGLE

The combined effect of birefringence and the input rotation angle is to induce an apparent phase angle ϕ_a at the output of the system. That is, while the actual intensity of the system is that described by (5), the system assumes the intensity is that given by the ideal case (6). Hence, there is an apparent phase shift $|\phi - \phi_a|$ at the first threshold crossing such that (see Fig. 2)

$$(1 + \cos^2 2\theta + \cos\psi \sin^2 2\theta)\left[1 + \cos(\tfrac{1}{2}\pi + \phi - \phi_a)\right] = 2. \tag{7}$$

Writing $\cos(\frac{1}{2}\pi + \phi - \phi_a) = \sin(\phi_a - \phi)$ and solving (7) for ϕ_a, we find

$$\phi_a = \phi + \sin^{-1}\left(\frac{1 + c}{1 + c\cos\psi} - 1\right), \tag{8}$$

where

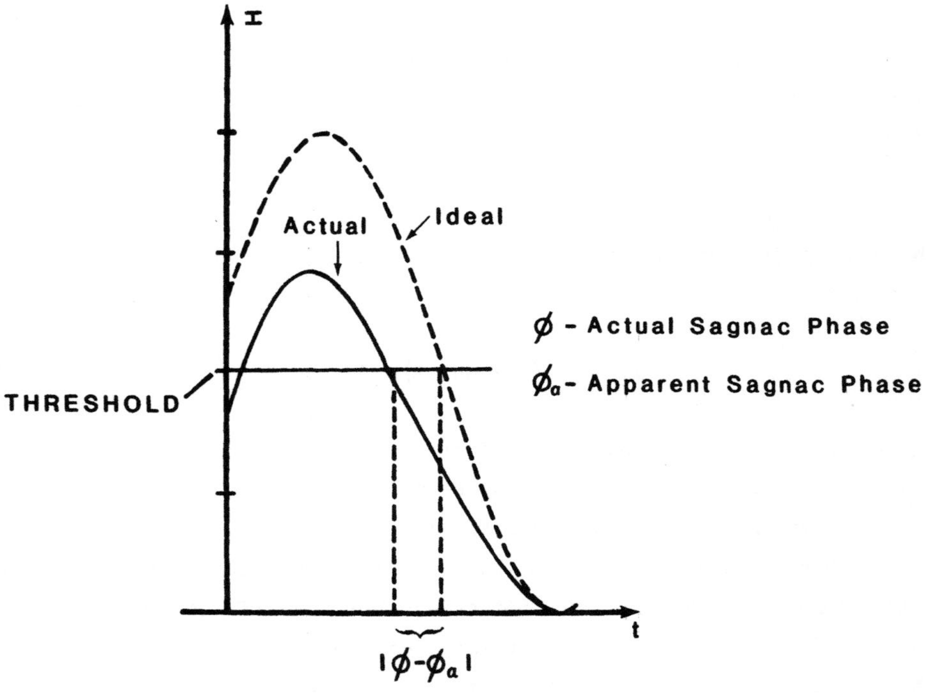

Fig. 2 Apparent phase angle.

$$c = \frac{\sin^2 2\theta}{1 + \cos^2 2\theta}. \tag{9}$$

For small values of the input rotation angle θ, the parameter c is also small and (8) can then be closely approximated by the simpler expression

$$\phi_a = \phi + c(1 - \cos\psi), \quad c \ll 1. \tag{10}$$

3.1 Statistics of the apparent phase angle

Because of the large fluctuations of the eigen-mode phase difference ψ that normally occur in the system, a reasonable assumption for ψ is that it is a random variable with uniform probability density function, i.e.,

$$p(\psi) = \frac{1}{2\pi}, \quad -\pi \leqq \psi \leqq \pi. \tag{11}$$

The mean value of ϕ_a is

$$\langle\phi_a\rangle = \phi + c - c\langle\cos\psi\rangle, \tag{12}$$

which for ψ uniformly distributed reduces to

$$\langle\phi_a\rangle = \phi + c, \quad c \ll 1. \tag{13}$$

Hence, c represents the average error in our measurement of the Sagnac phase angle, which can be subtracted out if we have knowledge of the input polarization angle θ. The variance of ϕ_a, defined by

$$\sigma_a^2 = \langle\phi_a^2\rangle - \langle\phi_a\rangle^2, \tag{14}$$

leads to the result

$$\sigma_a^2 = \frac{2\sin^4\theta}{\cos^3 2\theta}. \tag{15}$$

4. DISCUSSION

The analysis provided here shows that there is an apparent phase angle ϕ_a that the system detects, due to the combined effect of a nonzero input polarization angle θ and the random fluctuations of the phase difference ψ associated with the birefringence in the fiber. Under the assumption that the angle θ is small and that ψ is uniformly distributed, we have calculated the mean and variance of the apparent phase angle. If we now specify that the standard deviation of the output phase angle be no more than 1 deg/hr, then solving (15) for θ reveals that the input polarization angle θ must be aligned to within an angle of approximately 2.3 degrees of the birefringence axis in order to meet this requirement.

BIREFRINGENCE IN WEAKLY GUIDING HIGHLY ELLIPTICAL CORE FIBERS

Arun Kumar and R. K. Varshney
Department of Physics, Indian Institute of Technology
Hauz Khas, New Delhi-110 016 INDIA

ABSTRACT

A simple first order perturbation approach to study the birefringence of highly elliptical core weakly guiding optical waveguides is presented. The index profile is taken to be a rectangular waveguide for which exact scalar wave modes are known. Results for the birefringence obtained by the present analysis agree very well with the one obtained by exact numerical calculations.

INTRODUCTION

Single mode fibers which can preserve the state of polarization of the incident beam over large distances are known as single-mode single polarization (SMSP) fibers. These fibers have a large number of applications such as in coherent optical communication [1], fiber optics sensors [2] and in coupling with polarization sensitive integrated optics [3]. Three basic types of SMSP fibers reported in the literature are elliptical core fibers, side pit fibers and stress induced birefringent fibers [4]. Elliptical core fibers were not considered of much use initially because for good polarization holding one requires highly elliptical core single-mode fibers which are hard to fabricate as one of the core dimensions becomes impossibly small. However, the shift to longer wavelengths for long-distance communications has generated a great interest in elliptical core fibers in the recent past [5-13]. Dyott et al. [6] have reported an elliptical core fiber with GeO_2/SiO_2 core having $\Delta n=0.065$ and major to minor axis ratio $a'/b'=2.5$ with silica cladding. Simultaneous transmission in two opposite directions using an elliptical core fiber with $a'/b'=2.5$ has also been recently reported [14].

While a large number of papers were published in the recent

past studying the propagation characteristics of slighly elliptical core fibers, not much work has been reported on fibers with high ellipticity which are of much practical use. Although, the propagation characteristics of elliptical core fibers with any ellipticity can be obtained exactly using series of Mathieu Functions [15,16], it is quite combersome to obtain propagation constants upto the desired accuracy because eigen value equation involves infinite determinants. Dyott et al. [6] have generated a computer programme for the analysis given by Yeh [15] and have reported the results for birefringence for various elliptical core fibers. Recently Okamoto et al. [12] have used point matching method to study the propagation characteristics of such fibers; however, this method also requires combersome computer oriented calculations.

In the present paper we have developed a perturbation approach to study the propagation characteristics of highly elliptical core waveguides. We have considered the elliptical core waveguides as a perturbed form of appropriate rectangular waveguides for which exact scalar wave modes are known [17]. The results for birefringence obtained by the present simple approach agree very well with the ones obtained by Dyott et al. [6] using exact analysis of Yeh [15].

ANALYSIS

We consider an step index elliptical-core fiber with dielectric constant profile given by,

$$n^2(x,y) = n_1^2 \qquad \text{for } \frac{x^2}{a'^2} + \frac{y^2}{b'^2} \leq \frac{1}{4}$$
$$= n_2^2 \qquad \text{otherwise} \tag{1}$$

as a perturbed form of a rectangular waveguide, characterised by the following dielectric constant profile:

$$n^2(x,y) = n'^2(x) + n''^2(y)$$

where $$n'^2(x) = n_1^2/2 \qquad |x| < a/2$$
$$= n_2^2/2 \qquad |x| > a/2 \tag{2}$$

and $$n''^2(y) = n_1^2/2 \qquad |y| < b/2$$

$$= n_2^2/2 \qquad |y| > b/2$$

The dimensions a and b are choosen in such a way so that the core area and the aspect ratio of perturbed and unperturbed waveguides are equal (i.e. $a/a'=b/b'=\sqrt{\pi}/2$). The scalar wave modes of the above mentioned rectangular waveguide are given as follows [17]

$$\psi(x,y) = X(x)\ Y(y)$$

where
$$X(x) = A_1\cos\mu_1 x \quad \text{(symmetric)} \qquad |x| \leq a/2$$
$$A_1\sin\mu_1 x \quad \text{(Antisymmetric)} \qquad (3)$$
$$= A_2 e^{-\mu_2|x|} \qquad |x| \geq a/2$$

and
$$Y(y) = B_1\cos\nu_1 y \quad \text{(symmetric)} \qquad (4)$$
$$B_1\sin\nu_1 y \quad \text{(Antisymmetric)} \qquad |y| \leq b/2$$
$$= B_2 e^{-\nu_2|y|} \qquad |y| \geq b/2$$

where

$$\mu_1 = [R_o^2\frac{n_1^2}{2} - \beta_1^2]^{\frac{1}{2}}, \quad \mu_2 = \frac{2}{a}[v_1^2-\mu_1^2\frac{a^2}{4}]^{\frac{1}{2}}$$

$$\nu_1 = [R_o^2\frac{n_1^2}{2} - \beta_2^2]^{\frac{1}{2}}, \quad \nu_2 = \frac{2}{b}[v_2^2-\nu_1^2\frac{b^2}{4}]^{\frac{1}{2}} \qquad (5)$$

$$v_1 = R_o\frac{a}{2}[n_1^2-n_2^2]^{\frac{1}{2}}, \quad v_2 = R_o\frac{b}{2}[n_1^2-n_2^2]^{\frac{1}{2}}$$

R_o is the free space wave vector and β_1, β_2 are constants such that $[\beta_1^2+\beta_2^2]^{\frac{1}{2}}=\beta$ is the propagation constant of the mode under consideration. Boundary conditions on ψ which give the eigenvalue equations for β_1, β_2 and relationships between A_1, A_2 and B_1, B_2 are given as follows:

The E_{pq}^x mode (polarized predominantly along the x-direction) would approximately correspond to a TM mode in the x-direction and a TE mode in the y-direction. We assume ψ to correspond to E_x and make E_x and $\partial E_x/\partial y$ continuous at $y=\pm b/2$ and $n^z E_x$ and $\partial E_x/\partial_x$ continuous at $x=\pm a/2$. This gives us the following eigenvalue equations for symmetric E_{pq}^x modes,

$$\eta\tan\eta = \sqrt{v_2^2-\eta^2}$$
$$\xi\tan\xi = \frac{n_1^2}{n_2^2}\sqrt{v_1^2-\xi^2} \qquad (6)$$

and relationship between A_1 and A_2, B_1 and B_2 as

$$\frac{A_1}{A_2} = \frac{n_2^2}{n_1^2}\,\frac{\bar{e}^{\,\mu_2 a/2}}{\cos\xi}$$
$$\frac{B_1}{B_2} = \frac{\bar{e}^{\,\nu_2 a/2}}{\cos\eta} \qquad (7)$$

where $\xi=\mu_1 a/2$ and $\eta=\nu_1 b/2$.

Similarly for symmetric E_{pq}^{x} mode (polarized predominantly along y - axis) one obtains the following equations determining the eigenvalue:

$$\eta\tan\eta = \frac{n_1^2}{n_2^2}\sqrt{v_2^2-\eta^2} \qquad (8)$$
$$\xi\tan\xi = \sqrt{v_1^2-\xi^2}$$

and relationship between A_1 and A_2, B_1 and B_2 as

$$\frac{A_1}{A_2} = \frac{\bar{e}^{\,\mu_2 a/2}}{\cos\xi}$$
$$\frac{B_1}{B_2} = \frac{n_2^2}{n_1^2}\,\frac{\bar{e}^{\,\nu_2 b/2}}{\cos\eta} \qquad (9)$$

We define the normalized propagation constants from the following relation

$$P^2 = \frac{\beta^2-R_o^2 n_2^2}{R_o^2(n_1^2-n_2^2)} \qquad (10)$$

For the unperturbed waveguide P^2 is given by

$$P_o^2 = \frac{\beta_o^2 - R_o^2 n_2^2}{R_o^2(n_1^2 - n_2^2)} \tag{11}$$

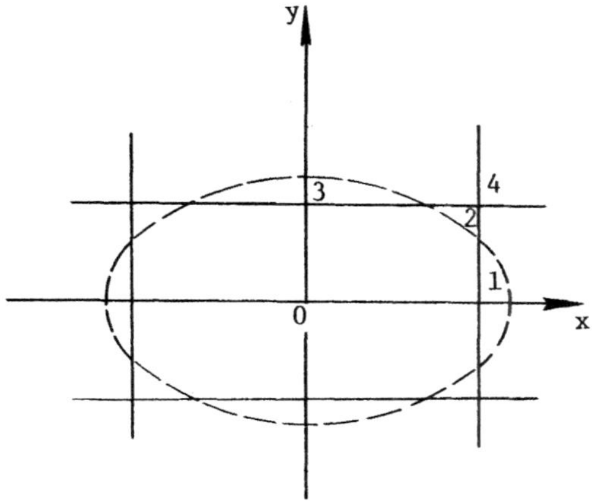

Fig.1 Solid lines form the unperturbed rectangular waveguide. Dashed curve shows the core cross-section of the elliptical waveguide.

where ξ and η are determined from equations (6) for E_{pq}^{x} modes and from equations (8) for E_{pq}^{y} modes. The propagation characteristics of the elliptical core waveguide can now be obtained by perturbing the dielectric constant in the regions marked 1 to 4 in Fig.1 by an amount δn^2 given by,

$$\begin{aligned} \delta n^2 &= (n_1^2 - n_2^2) && \text{in regions 1,3 \& 4} \\ &= (n_2^2 - n_1^2) && \text{in region 2} \\ &= 0 && \text{otherwise} \end{aligned} \tag{12}$$

For weakly guiding fibers, since $n_1 \cong n_2$, δn^2 would be a small quantity and the effect of δn^2 can be taken into account by the first order perturbation theory which gives the following expression for P^2:

$$P^2 = P_o^2 + P'^2 \tag{13}$$

where

$$P'^2 = \frac{1}{n_1^2 - n_2^2} \frac{\int_{-\infty}^{\infty}\int_{-\infty}^{\infty} |\psi|^2 \delta n^2 dxdy}{\int_{-\infty}^{\infty}\int_{-\infty}^{\infty} |\psi|^2 dxdy}$$

After putting ψ and δn^2 from equations (3), (4) and (12) in the above equation, one obtains

$$p'^2 = \frac{I_1 - I_2 + I_3 + I_4}{\int_{-\infty}^{\infty}\int_{-\infty}^{\infty} |\psi|^2 dxdy} \tag{14}$$

where I_1, I_2, I_3 and I_4 are the contributions to the perturbation from regions 1,2,3 and 4 respectively (see Fig.1). Thus

$$I_1 = 4 \int_0^{b\alpha/2} dy B_1^2 \cos^2 \nu_1 y \int_{a/2}^{a'\sqrt{\frac{1}{4} - y^2/b'^2}} A_2^2 \bar{e}^{2\mu_2 x}\, dx$$

$$I_2 = 4 \int_{a\alpha/2}^{a/2} dx A_1^2 \cos^2 \mu_1 x \int_{b'\sqrt{\frac{1}{4} - \frac{x^2}{a'^2}}}^{b/2} B_1^2 \cos^2 \nu_1 y dy$$

$$I_3 = 4 \int_0^{a\alpha/2} dx A_1^2 \cos^2 \mu_1 x \int_{b/2}^{b'\sqrt{\frac{1}{4} - \frac{x^2}{a'^2}}} B_2 \bar{e}^{2\nu_2 y} dy$$

$$I_4 = 4 \int_{a/2}^{\infty} dx A_2^2 \bar{e}^{2\mu_2 x} \int_{b/2}^{\infty} dy B_2^2 \bar{e}^{2\nu_2 y}$$

where $\alpha = [\frac{4}{\pi} - 1]^{\frac{1}{2}}$. After simplification above expressions reduce to

$$I_1 = \frac{A_2^2 B_1^2 \bar{e}^{\mu_2 a}}{2\mu_2 \nu_1} [2\alpha\eta + \sin 2\alpha\eta - 4\eta e^{\mu_2 a} \int_0^{\alpha} \cos^2 \eta y \bar{e}^{\mu_2 a\sqrt{\frac{4}{\pi} - y^2}} dy] \tag{15}$$

$$I_2 = \frac{A_1^2 B_1^2}{4\mu_1 \nu_1} [(2\eta + \sin 2\eta)\{2\xi(1-\alpha) + \sin 2\xi - \sin 2\alpha\xi\}$$

$$-4\xi \int_{\alpha}^{1} \cos^2 \xi x \{2\eta\sqrt{\frac{4}{\pi} - x^2} + \sin 2\eta\sqrt{\frac{4}{\pi} - x^2}\} dx] \tag{16}$$

$$I_3 = \frac{A_1^2 B_2^2 \bar{e}^{\nu_2 b}}{2\nu_2 \mu_1} [2\alpha\xi + \sin 2\alpha\xi - 4\xi e^{\nu_2 b} \int_0^{\alpha} \cos^2 \xi y \bar{e}^{\nu_2 y\sqrt{\frac{4}{\pi} - y^2}} dy] \tag{17}$$

$$I_4 = \frac{A_2^2 B_2^2}{\mu_2 \nu_2} e^{-\mu_2 a} e^{-\nu_2 b} \quad (18)$$

Normalization constant appearing in the denominator of equation (14) is also simplified and is given by

$$\int_{-\infty}^{\infty}\int_{-\infty}^{\infty} |\psi|^2 dxdy = \frac{A_2^2 B_2^2}{\mu_2 \nu_2} e^{-\mu_2 a} e^{-\nu_2 b} \left[1+\frac{A_1^2}{2A_2^2}\frac{\mu_2}{\mu_1} e^{\mu_2 a} x(2\xi+\sin 2\xi)\right]$$

$$x \left[1+\frac{B_1^2}{2B_2^2}\frac{\nu_2}{\nu_1} e^{\nu_2 b} (2\eta+\sin 2\eta)\right] \quad (19)$$

Normalized propagation constants p_x^2 or p_y^2 corresponding to E_{pq}^x or E_{pq}^y modes can be calculated using equations (13) - (19) after putting the values of A_1/A_2, B_1/B_2. Corresponding to the respective modes. The birefringence would then be given by

$$\Delta\beta = [p_x^2(n_1^2-n_2^2)+n_2^2]^{\frac{1}{2}} - [p_y^2(n_1^2-n_2^2)+n_2^2]^{\frac{1}{2}} \quad (20)$$

RESULTS AND DISCUSSION

Numerical calculations were carried out to calculate the birefringence of fibers with core and cladding refractive indices as n_1=1.48 and n_2=1.47 and with various ellipticities. For fibers with a'/b'=2.5, the results obtained by the present analysis agree within 8% with that of Dyott et al. [6]. It should be mentioned here that the accuracy of the present analysis increases with the increase in ellipticity of the fiber, e.g. for a'/b'=3.5 the solid and dashed curves agree within 5% as compared to 8% for a'/b'=2.5.

SUMMARY

In summary we have developed a simple first order perturbation approach to study the propagation characteristics of weakly guiding highly elliptical core fibers. The unperturbed waveguide is taken to be a rectangular waveguide with the same core area and aspect ratio as that of the elliptical waveguide and for which the scalar wave modes are known exactly. It has been shown that the results for the birefringence obtained by the perturbation approach agree very well with the numerical calculations reported by Dyott et al. [6].

ACKNOWLEDGEMENT

This work was partially supported by the Department of Science and Technology, India. The authors wish to thank Professor A.K. Ghatak for guidance and encouragement.

REFERENCES

1. F. Favre, L. Jeunhomme, I. Joindot, M. Monerie and J.C. Simon "Progress towards heterodyne-type single-mode fiber communication systems" IEEE J. Quantum Electron., Vol.QE-17 pp.897-906, June 1981.
2. R. Ulrich and M. Johnson, "Fiber-ring interferometer; Polarization analysis," Opt. Lett., Vol.4, No.5 pp.152-154, 1979.
3. R.A. Steinberg and T.G. Giallorenzi, "Performance limitation imposed on optical waveguide switches and modulator by polarization," Appl. Opt., Vol.15, pp.2440-2453, 1976.
4. D.N. Payne, A.J. Barlow and J.J.R. Hansen, "Development of low and high birefringence optical fibers" IEEE J. Quantum Electron, Vol.QE-18, No.4 pp.477-488, 1982.
5. C. Yeh, K. Ha, B. Dong and W.P. Brown, "Single-mode optical waveguides" Appl. Opt., Vol.18, No.10, pp.1490-1504, 1979.
6. R.B. Dyott, J.R. Cozens and D.G. Morris, "Preservation of polarization in elliptical fiber waveguides with elliptical cores" Electronics Lett., Vol.15, pp.380-382, 1979.
7. J.D. Love, R.A. Sammut and A.W. Snyder, "Birefringence in elliptically deformed optical fibers" Electron. Lett., Vol.15, pp.615-616, 1979.
8. R.A. Sammut, "Birefringence in Graded-index monomode fibers with elliptical cross-section" Elect. Lett., Vol.16, pp.156-157, 1980.
9. R.A. Sammut, "Birefringence in slighly elliptical optical fibers" Elect. Lett., Vol.16, pp.728-29, 1980.
10. N. Imoto, N. Yoshizawa, J. Sakai and H. Tsuchiya, "Birefringence in single-mode optical fiber due to elliptical core deformation and stress anisotropy" IEEE J. Quantum Electron., Vol.QE-16, pp.1267-1271, 1980.
11. R.A. Sammut, C.D. Hussey, J.D. Love and A.W. Snyder, "Modal analysis of polarization effects in weakly guiding fibers" IEE Proc., Vol.128, pp.173-187, 1981.
12. K. Okamoto, T. Hosaka and Y. Sasaki, "Linearly single polarization fibers with zero polarisation mode dispersion" IEEE J. Quantum Electron., Vol.QE-18, pp.496-503, 1982.
13. D.N. Payne, A.J. Barlow and J.J.R. Hansen, "Development of low and high birefringence optical fibers" IEEE J. Quantum Electron., Vol.QE-18, pp.477-488, 1982.
14. News item in Laser Focus, p.109, May 1982.
15. C. Yeh, "Elliptically dielectric waveguides," J. Appl. Phys. Vol.33, pp.3235-43, 1962.

16. C. Yeh, "Modes in weakly guiding elliptical optical fibers" optical and Quantum Electronics, Vol.8, pp.43-47, 1976.
17. A. Kumar, K. Thyagarajan and A.K. Ghatak, "Analysis of rectangular core dielectric waveguides; An accurate perturbation approach" Optics Letters (To be published).

PART THREE

OPTICAL WAVEGUIDE SYSTEMS

RECENT PROGRESS IN HETERODYNE/COHERENT OPTICAL FIBER COMMUNICATIONS

T. OKOSHI

DEPARTMENT OF ELECTRONIC ENGINEERING
UNIVERSITY OF TOKYO

1. INTRODUCTION

The present optical fiber communications are in a sense as primitive as the radio communications prior to 1930. The reason is that neither of these make use of the phase information of the carrier; in other words, both of these are non-coherent communications.

The modulation/demodulation scheme being employed in the present standard optical fiber communications is often called the intensity-modulation/direct-detection (hereafter IM/DD) scheme. The term IM stems from two facts: (1) that the light intensity (not the amplitude) is modulated linearly with respect to the input signal voltage; and (2) that basically no attention is paid to the phase of the carrier. (The original spectral spread of the optical carrier is usually much wider than the spread due to modulation.) The term DD stems from that the signal is detected directly at the optical stage of the receiver; neither the frequency conversion (heterodyne scheme) nor sophisticated signal processing at lower frequencies are performed.

On the other hand, in the history of radio communications, the heterodyne scheme became common since 1930, and is now widely used even in pocket radio sets. Sophisticated coherent modulations such as FM, PM, frequency-shift keying (FSK), and phase-shift keying (PSK) are also widely used in broadcasting and communications.

Here a question arises: will the IM/DD system continue to be predominating in optical communications for at least ten years from now? Or will it gradually retire to yield its present position to more sophisticated heterodyne and/or coherent systems?

2. ADVANTAGE OF HETERODYNE/COHERENT SCHEMES

The IM/DD system has a great advantage in system simplicity and low cost. On the other hand, some applications of the optical fiber communications exist in which a long repeater separation is our primary concern; an example is the optical fiber communications between islands. In such a case, the improvement of the equivalent receiver sensitivity by a heterodyne-type receiving technique, or by coherent modulation/demodulation scheme such as PCM-FSK or PCM-PSK may become advantageous, even at the sacrifice of simplicity and low cost.

Especially noteworthy is that the sensitivity improvement is particularly dramatic (10∿25 dB) at the 1.5∿1.6μm wavelength region, where the silica-fiber loss becomes minimum (∿0.2 dB/km) whereas good photodetectors (for the DD scheme) are not available. In this wavelength region, only a Ge-APD or quarternary-compound APD could be used which has rather poor noise performance.

Figure 1 shows the result of a computer analysis of the improvement in the required receiving level for obtaining a prescribed bit-error rate, in this case 10^{-9}, for the case when a Ge-APD is used. An excess noise factor x=1 and a dark current $I_d=10^{-9}$ Ampere are typical with Ge APDs. It is found in Fig.1 that the sensitivity improvement from the IM-baseband direct-detection scheme to the ASK heterodyne scheme is about 20dB. The sensitivity improvement of 20dB will result in a repeater-separation elongation of 100 km when the fiber loss is 0.2 dB/km.

The technical significance of the heterodyne scheme lies also in that it is a premise for the coherent optical communication, such as PCM-FSK or PCM-PSK, as is seen in the lower half of Fig.2.

As compared with the ASK heterodyne, 3dB further improvement can be expected with an FSK, frequency-shift keying, heterodyne scheme, and still further 3dB imptovement with a PSK, phase-shift keying, heterodyne system. Further 3dB improvement could be obtained if we could realize an ASK homodyne or PSK homodyne systems, but for the time being these seem technically rather difficult.

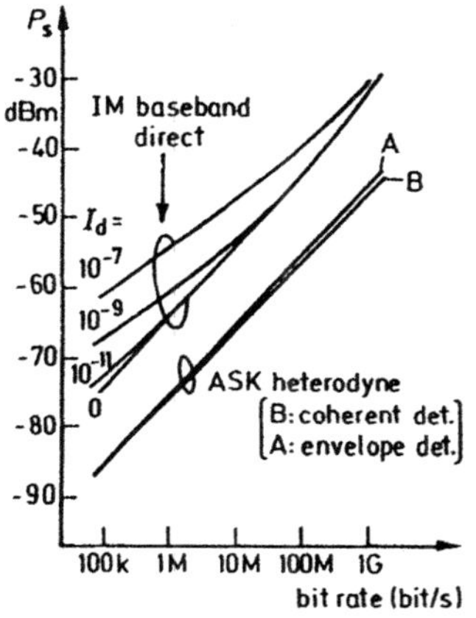

Fig.1 Comparison of the minimum receiving signal level for the IM/DD and ASK/heterodyne schemes in the long-wavelength region (BER=10^{-9}, x=1.0). The values of I_d, photodetector dark current, are given in amperes.

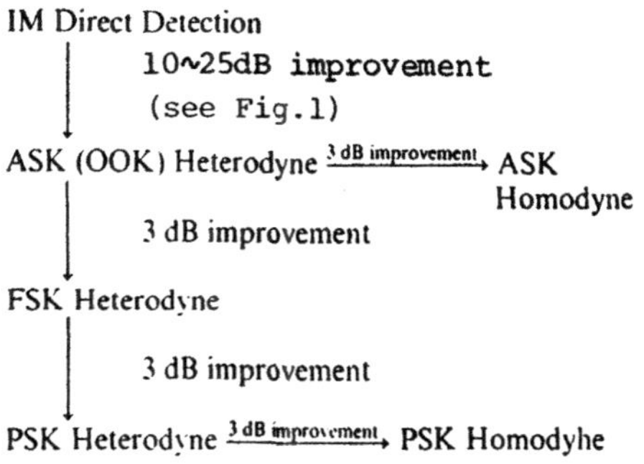

Fig.2 Improvement of the minimum receiving signal level by various heterodyne/coherent communication schemes.

In addition to the reduction of the minimum signal level for achieving a prescribed bit-error rate as has been described above, there are two other advantages in a heterodyne system.

The second advantage is that the FDM, the frequency-division multiplexing, with very fine carrier separation becomes possible because of the good frequency selectivity of the IF amplifier which is sharper than that of an optical branching filter. This allows, for example, an efficient use of the minimum dispersion wavelength region, that is, usually 1.3μm region of an optical fiber.

The third advantage is that the direct light amplification in the receiver becomes possible, as the noise outside signal bandwidth can effectively be rejected in a heterodyne system.

3. HISTORY OF THE RESEARCH

The above expectation on the receiver-sensitivity improvement is the principal motivation underlying the present effort toward the heterodyne/coherent optical fiber communications. On the other hand, it is accepted by all the specialists that the IM/DD systems will never retire, because the heterodyne/coherent systems are, and will continue to be, rather expensive.

In reviewing the history of this topic, we must first distinguish optical communications and optical fiber communications. The heterodyne-type optical communication is not a new topic. The optical communication systems conceived by Bell Laboratories investigators in the late 1960's were heterodyne systems imitating millimeter-wave communications. However, those systems were assumed to use lens waveguides and gas lasers, and look far from practical at present. After the advent of practical fibers and semiconductor lasers in 1970, the IM/DD optical fiber communications became predominant; the heterodyne scheme had been given up until 1978 mainly because of the poor spectral purity of semiconductor lasers.

The second phase of the research started in 1978. Our group in the University of Tokyo started it in the summer of 1978, and after that Musashino ECL in Tokyo, CNET in Lannion, France, and British Telecom Res. Labs. in Ipswich, England, also started the research. At present, the research toward the heterodyne/coherent system is performed in more than ten laboratories in the world.

4. RELEVANT TECHNICAL TASKS

Several technical problems must be overcome before a practical heterodyne system is realized in future.

First, the actual improvement in the receiving level is affected by many parameters, and detailed system analysis is required to know how much improvement in the receiving level can actually be expected.

The second, and probably the biggest problem is how to stabilize the frequency of semiconductor lasers to be used both as the transmitter and the local oscillator. Since a typical IF frequency would be 2GHz, which is about 10^{-5} times the signal frequency which is typically 200THz, the requirement for the frequency stability is very severe.

The third problem is the spectral purity of the semiconductor lasers. If the carrier is noisy, the frequency and phase fluctuation will greatly deteriorate the bit-error rate, thus increasing the minimum receiving level.

The fourth problem is the development of practical, simple FSK or PSK modulation/demodulation techniques.

The fifth problem is how to prevent the random fluctuation of the mixer efficiency due to the fluctuation of the polarization state of the received light signal.

The sixth task is to develop practical laser amplifiers.

Finally, combining some of, or all of the above techniques, experimental systems must be constructed to demonstrate the afore-mentioned advantages in a practical scale.

5. BIT-ERROR RATE MEASUREMENT

In the present talk the technical advantages and relevant technical tasks of heterodyne and/or coherent optical fiber communications are described, and the recent progress in the research and development toward those are reviewed. The topics are the frequency stabilization of semiconductor lasers, the improvement of its spectral purity, the polarization-state stabilization along the optical fiber waveguide, and the realization of practical systems and the measurement of their performance.

In the following, some results of the bit-error rate (BER) measurement of a heterodyne-type optical communication system will be presented.

Figure 3 shows examples of the result of the measurement.[(1)] Small circles, crosses and dots show the measured BER for three different local oscillator power levels. The experimental BER curves show fairly good agreement with the theoretical minimum curve, that is, the shot-noise limited curve, when the local power is sufficient. The difference is about 1dB, and is probably due to the intensity noise of the local oscillator.

The BER measurement of a system using two independent, frequency stabilized lasers has also been performed. It has been found that the BER is somewhat deteriorated due to the oscillator noise. The details of the data will be reported elsewhere.

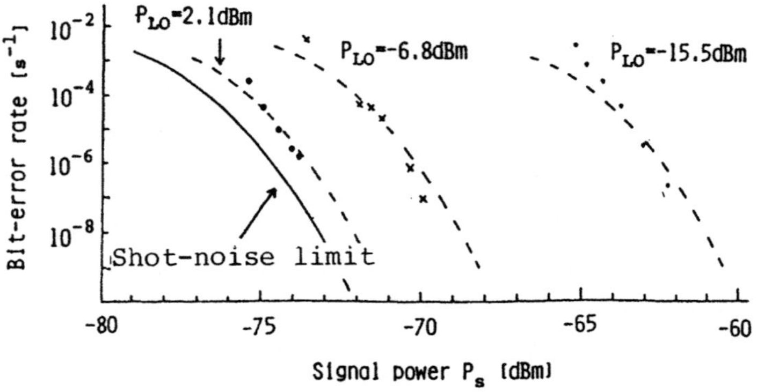

Fig.3 Measured and theoretical bit-error rate of a PCM-ASK heterodyne-type optical communication system

REFERENCE

(1) T. Okoshi, "Heterodyne and coherent optical fiber communications: Recent progress (Invited)," IEEE Trans. on M.T.T., Vol.MTT-30, No.8, pp.1138-1149, Aug. 1982.

POLARISATION STABILITY REQUIREMENTS FOR COHERENT OPTICAL FIBRE TRANSMISSION SYSTEMS

D W Smith, R A Harmon, T G Hodgkinson

British Telecom Research Laboratories, Martlesham Heath, Ipswich IP5 7RE
United Kingdom

1. INTRODUCTION

Heterodyne (1) and homodyne (2) detection schemes are presently being developed as an alternative to direct detection in optical fibre communications. These coherent systems can offer better receiver sensitivity and will allow the available bandwidth of monomode fibre to be used efficiently. However, their performance will be dependent on matching the polarisation state (S.O.P) at the output of the transmission fibre to that of the local oscillator, as Fig 1 illustrates. This paper discusses techniques that can be used by the system designer to overcome the problems of polarisation matching, and it presents experimental results obtained from a coherent transmission system operating over conventional circular core fibre.

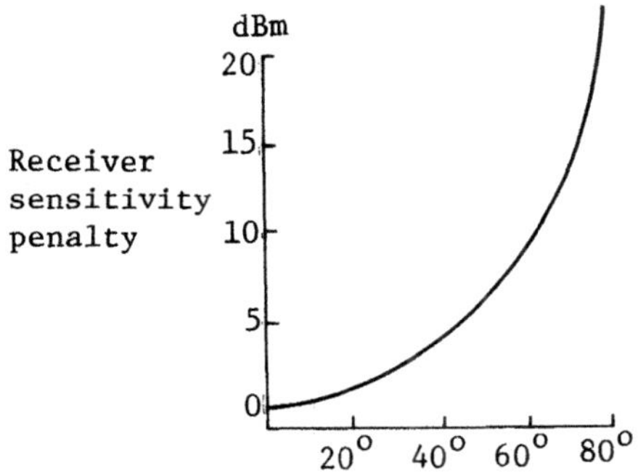

Dependence of coherence receiver sensitivity on polarization alignment

Fig.1 Polarization Misalignment Angle

2. POLARISATION HOLDING FIBRE

Ideally, polarisation holding fibre would be used for coherent transmission. This is designed to have either strong linear birefringence (3) or circular birefringence (4). Linear polarisation holding fibres are designed with sufficient built in birefringence to swamp external bending and pressure induced effects. However, unless special fibre designs using both an elliptical core and stress applying regions (5) are employed, the polarisation mode dispersion will be large (30 ps/km typical). Therefore when mode coupling occurs from either the occasional axially misaligned joint, or from twists, there will be an additional dispersed component in the received signal which will produce intersymbol interference. An alternative is to design a fibre with circular birefringence, usually by giving the fibre a regular twist (4). This avoids problems of axial joint alignment and is of course a simpler structure; but to achieve full immunity from the effects of external bending the required twist rate could damage the fibre (11).

3. CONVENTIONAL FIBRE

Conventional fibre can be used for coherent transmission systems if either a dual polarisation receiver is used, or an optical compensator is used to match the input signal with the local oscillator. The former approach will generally result in a 3 dB receiver sensitivity penalty, furthermore the receiver is more complex. The latter approach requires a feedback control system to operate the compensators (6). A single feedback loop can be used for controlling either the polarisation's orientation or ellipicity, and for this scheme a 3 dB penalty would also be expected. Alternatively a dual feedback loop would allow control over both parameters (7) and thus eliminate the penalty. The necessary compensators can be placed in either the local oscillator or received signal path; the former is preferable if the devices introduce significant attenuation. A range of compensation techniques are currently available; the choice will depend on the likely rate-of-change of the received S.O.P.

To determine the requirements for the receiver polarisation control system we have been measuring the stability (8) of the output S.O.P from long monomode fibre links. These measurements, for which a stable single frequency source at 1.52 microns was used, have been carried out for time periods up to 96 hours. We have monitored the output S.O.P both for fibres within the laboratory environment and cabled fibres which were installed in underground ducts adjacent to busy main roads. Two types of fibre cable have been tested, the first is a loose tube package cable containing BTRL fibre (9) and the second a tight extruded package cable containing STL fibre, both fibre lengths are approximately 30 Km. Results for both types of installed fibre have shown polarisation variations occuring only very slowly; less than 1 dB power change in the required polarisation state during a 72 hour period, Fig 2 shows a typical result. Furthermore, operation of an experimental coherent system over the installed cables indicated that there was no noticable microphonic effects on the output S.O.P; in fact identical sensitivity was measured for the heterodyne receiver when operating over both a 1 metre fibre path and the installed links, as is shown in Fig 3 by the bit-error-rate measurements. For this measurement only occasional manual polarisation adjustment of the receiver local oscillator was necessary. Further experiments have now demonstrated stable operation of a heterodyne system operating over 123 Km of fibre at a data rate of 34 Mbit/s. For this system the fibre path was made up of 30 Km of installed cabled fibre and 93 Km of fibre wound on reels.

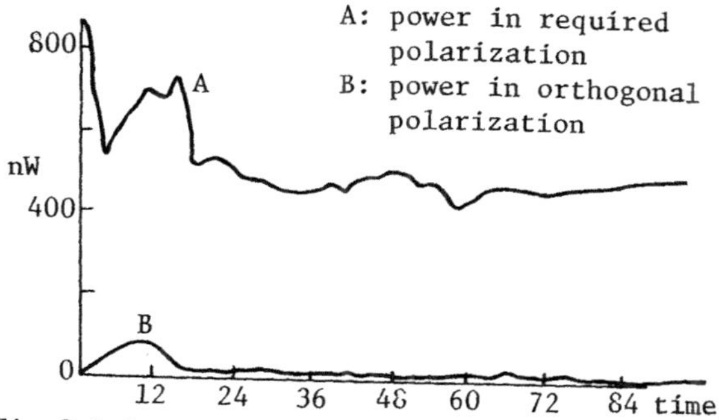

Fig.2 Polarization Stability for 30km Installed STL Cable over a 96 Hour Period

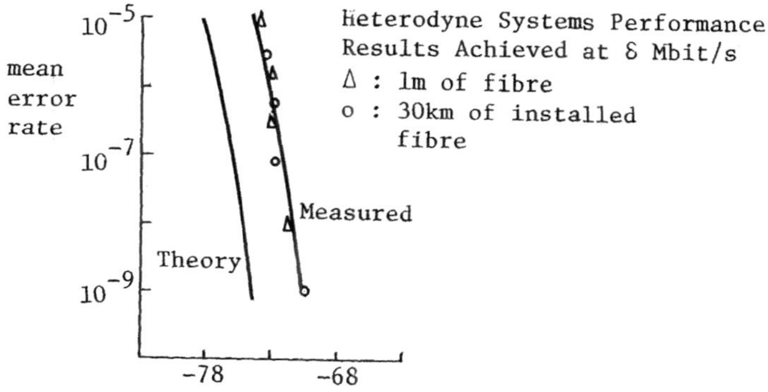

Fig.3 Mean Received Signal Power, dBm

Because of residual birefringence producing polarisation mode dispersion, the output S.O.P will generally depend on the transmitted optical frequency, and hence the applied modulation signal; the only exception is when there is coincidence between the launched S.O.P and one of the two elliptical orthogonal eigenstates of the fibre (12). Therefore measurements have been made of output S.O.P against source frequency for a range of fibre lengths and launch polarisation states. Fig 4 shows the result of a measurement made over a laboratory link containing 102 Km of BTRL fibre using a tunable C.W. single longitudinal mode semiconductor laser at 1.5 microns. These measurements indicate that polarisation mode dispersion would not be a significant problem for bit-rates less than 1 Gbit/s, over this particular fibre. To reduce polarisation dispersion still further, for higher bit-rate systems, the spun preform technique looks promising (10).

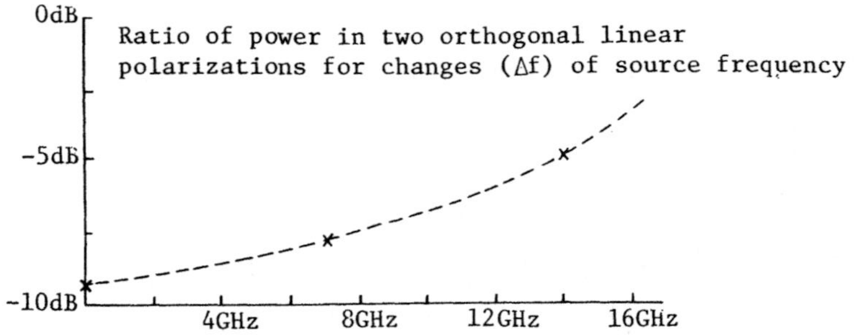

Fig.4 Mearsured Effect of Source Frequency on Output Polarization

4. CONCLUSION

The demonstration of heterodyne detection over long lengths (123 Km) of conventional monomode fibre introduces new possibilities for increasing the repeater separation and transmission capacity of optical fibre links. This could allow existing transmission routes to be upgraded by making changes only to the terminal equipment, leaving the transmission media undisturbed.

We would like to acknowledge colleagues in the fibre technology section at BTRL for the supply of fibre and the director of research of British Telecom for permission to publish this work.

REFERENCES

1. Hodgkinson TG.et al. Electron. Letters, p929, 1982.
2. Hodgkinson TG.et al. Electron. Letters, p523, 1982.
3. Okoshi T. IEEE JQE, p879, 1981.
4. Jeunhomme L. et al. Electron. Letters, p921, 1980.
5. Okamoto K. et al. IEEE JQE, p496, 1982.
6. Giese E. et al. Optics Letters, p337, 1982.
7. Ulrich R. App. Physics Letters, p840, 1979.
8. Harmon RA. Electron. Letters, p1058, 1982.
9. Ainslie BJ. et al. IEEE JQE, p854, 1981.
10. Barlow AJ. et al. Electron, Letters, p726, 1981.
11. Barlow AJ. et al. Electron. Letters, p389, 1981.
12. Bazarov EN. et al. Sov. Tech. phys. p396, 1981.

FEASIBILITY AND REQUIREMENTS FOR DISPERSION COMPENSATION IN COHERENT FOC

C. S. IH
University of Delaware
Department of Electrical Engineering
Newark, Delaware 19711

SUMMARY

Despite many difficulties, a coherent Fiber Optical Communication (FOC) system is very attractive for long distance transmission because of its potentially enormous information bandwidth. Over long distance, fiber dispersion remains an important factor in limiting the information bandwidth. We discuss in this paper the feasibility of dispersion compensation using a novel optical modulation technique by which either one of the sidebands or the carrier can be suppressed.

We recently described and demonstrated the feasibility of a new modulation method termed as TDHM, (Time-Domain-Holography Modulation) [1-5]. A variation of the TDHM, termed as DBM (Double-Beam Modulation), which is suitable for digital transmission has also been suggested [6].

Basically TDHM is achieved with an arrangement similar to a Mach-Zehnder interferometer excepting that an AO (Acousto-Optical) or SAW (Surface-Acoustic-Wave, for Integrated-Optics implementation) modulator is included in one of the arms. The Mach-Zehnder waveguide interferometer was proposed as a modulator by Martin [7]. However, by incorporating an AO or SAW modulator in one of the arms, the modulated laser beam is automatically put on a "(sub)carrier" because the photodiode is a square-law detector. The laser beam can then be effectively Amplitude, Frequency, Phase or SSB modulated (with respect to the subcarrier) through the input of the AOM [2,4, 5]. The basic arrangement for the TDHM is shown in Fig. 1.

In the DBM [6], the laser beam is modulated directly either in an analog or digital form. The beam is then passing through a SWAOM (Standing-Wave Acousto-Optical Modulator) at the Bragg angle. The diffracted beam is also put on a "(sub) carrier" which is twice the driving frequency of the SWAOM, F_c. The diffracted beam contains two superimposed frequency components, with one frequency up-shifted by the F_c and the other down-shifted by the F_c. They are equivalent to the upper and lower sideband of a conventional AM. The DBM is therefore identical to the DSB (Double-Side-Band) modulation with the carrier being suppressed. A basic arrangement of DBM is shown in Fig. 2.

The proposed novel modulation techniques offer an advantage for being capable of compensating the fiber dispersion in a coherent transmission system. In a conventional AM (or IM), if the propagation constant, β, is linear with respect to the frequency, the two sidebands maintain their proper phase relationship with respect to each other and to the carrier. The signals will be transmitted without distortion and the bandwidth is infinite. In a nonlinear system as is usual for practical fibers, the β can be expanded into a Taylor series. The first derivative of β, $\dot{\beta}$, can be identified as the group delay and the second derivative a (major) dispersion term.

In a non-linear system, the two sidebands can no longer maintain their proper phase relationship. The result is a high frequency attenuation (bandwidth reduction or pulse broadening) and waveform distortion.

Since TDHM has only one sideband, its transmission characteristics are different [1]. All higher derivatives of β can be identified as phase terms. Its transfer function has an infinite response (amplitude) even for non-linear fibers but with a larger quadratic phase shift.

Since the relative positions of the Fourier spectrum of a signal are reversed in the two sidebands, the dispersion terms for the two sidebands have opposite signs. Therefore if we use the U-TDHM (upper-sideband) and L-TDHM (lower-sideband) alternately, the dispersion term can be effectively equalized or compensated. This equalization applies to all higher even orders.

The third order dispersion can be compensated by designing the optical fiber in such a way that the third derivative of beta is zero at the transmitting wavelength. This can be achieved by introducing an impurity absorption at a certain wavelength [8]. If matching the fiber and laser diode transmitting wavelength is too difficult or too costly, a technique similar or an alternative to that described by Lin,etal [9] can be used.

In DBM, there are two sidebands but no carrier. We expected that the waveform distortions for the two sidebands were similar but with a different delay in arriving at the receiving end. The delay between the two sidebands can be compensated by pre-delaying one of the sidebands. One arrangement using an optical delay is shown in Fig. 5. Another using a driving current delay is shown in Fig. 6.

In conclusion, we have shown that, unlike the conventional AM, the proposed TDHM and DBM offer the possibility for dispersion equalization. The DBM system is most suitable for digital transmission in FDM [6]. The DBM-FDM system will be in reach for operation to 4.5 GHz with five channels at 2.5, 3.0, 3.5, 4.0, and 4.5 GHz and at 200 mbit/sec per channel. Both the TDHM and DBM can be implemented in the Integrated-Optics (IO) form [5,6] using techniques developed for IO spectrum analyzers [10,11]. The maximum operating frequency of the SWAOM or SWSAWM is only 2.25 GHz. Sapphire acousto-optical cells have been reported to operate up to 13 GHz [12]. Acousto-optical cells of good efficiency can be expected at 2.25 GHz using lithium niobate crystals. The capacity loading of the photodetector at very high frequency can be minimized by using an optoelectronic downconverter described by MacDonald and Hill [13]. When all the related developments are employed, the DBM-FDM scheme in conjunction with a conventional WDM (Wavelength-Division-Multiplexing) could provide a FOC system whose information bandwidth is practically unlimited.

References

1. C.S. Ih and Bing-Yan Chen, Characteristics of and Requirements for Subnanometer-Wavelength Multiplexing, Tech. Digest of the 1981 International Conference on Integrated Optics and Optical Fiber Communication (IOOC '81), April 27-29, 1981, San Francisco, CA.
2. C.S. Ih, Bing-Yan Chen and A. Mallya, A Direct RF Compatible Optical Fiber Transmission System, Proceedings of the International Conference on LASER '80, December 15-19, 1980, New Orleans, LA.
3. C.S. Ih and Bing-Yan Chen, Subnanometer-Wavelength Multiplexing for Fiber-Optical Communication, JOSA, 70, 1569 (Dec. 1980).
4. C.S. Ih, JOSA, 68, 1384 (Oct., 1978).
5. U.S. patent No. 4,210,803, July 1, 1980.
6. C.S. Ih, M. P. O'Boyle and Bing-Yan Chen, A Novel FDM System for Digital FOC, submitted to the Fourth International Conference on Integrated Optics and Optical Fiber Communication, June 27-30, 1983, Tokyo, Japan.
7. W.E. Martin, A New Waveguide Switch/Modulator for Integrated Optics, Appl. Phys. Lett., 26, 526 (May 1975).
8. A. Sugimura, K. Daikoku, N. Imoto and T. Miya, "Wavelength Dispersion Characteristics of Single-Mode Fibers in Low-loss Region", IEEE Journal of Quantum Electronics, QE-16, 215 (Feb. 1980).
9. C. Lin, H. Kogelnik and L.G. Cohen, "Optical-pulse Equalization of Low-Dispersion Transmission in Single-Mode Fibers in the 1.3-1.7 micrometer Spectral Region", Optics Letters, 5, 476 (November 1980).
10. D. Mergerian and E.C. Malarkey, Integrated-Optic Spectrum Analyzer: Current Status, p. 114, Technical Digest of the Third International Conference on Integrated Optics and Optical Fiber Communication (IOOC '81), April 27-29, 1981, and references cited therein.
11. T.R. Ranganath, Integrated-Optic Spectrum Analyzer: A first Demonstration, p. 114, Technical Digest of the Third International Conference on Integrated Optics and Optical Fiber Communication (IOOC '81), April 27-29, 1981, San Francisco, CA.
12. J.T. Flynn, R. Kaul, and T.M. Reeder, Optical Heterodyne Detection of Microwave Signals from Acousto-optical Interactions, Proceedings of the IEEE, 388 (March 1976).
13. R.I. MacDonald and K.O. Hill, Avalanche Optoelectronic Downconverter, Optics Letters, 7, 83 (Feb. 1982).

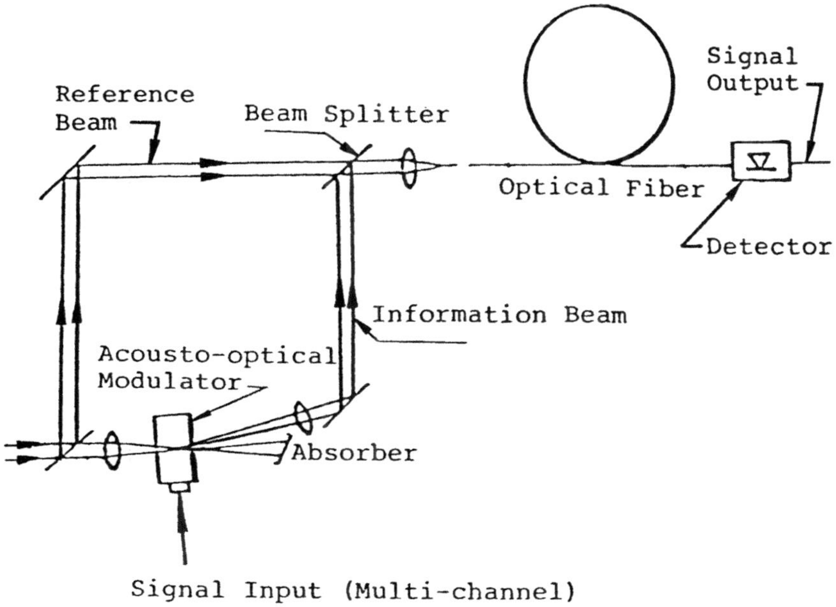

Fig. 1. Basic TCHM Arrangement

Fig. 2. DBM Modulator

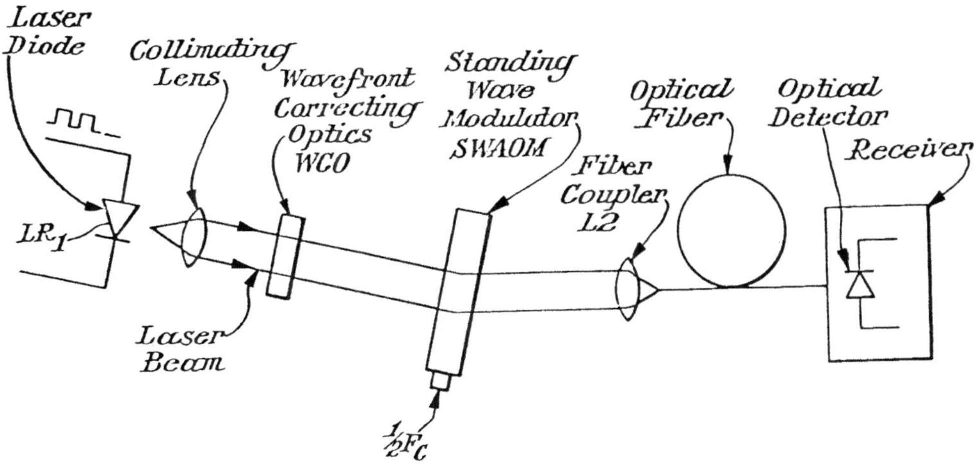

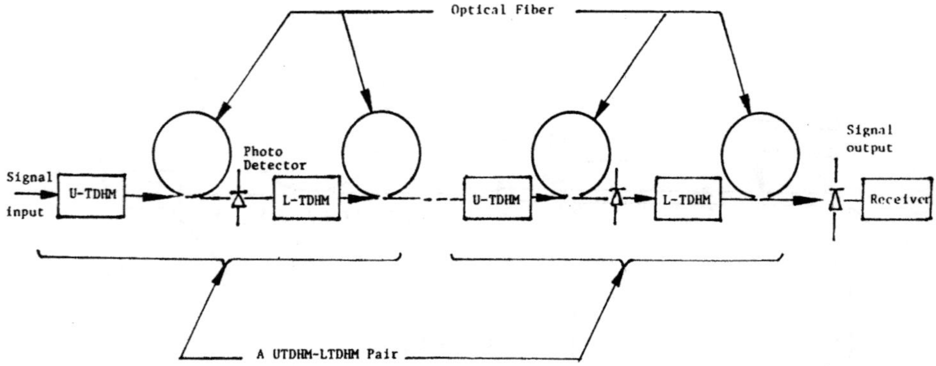

Fig. 3. Dispersion Equalization Using UTDHM - LTDHM Pairs

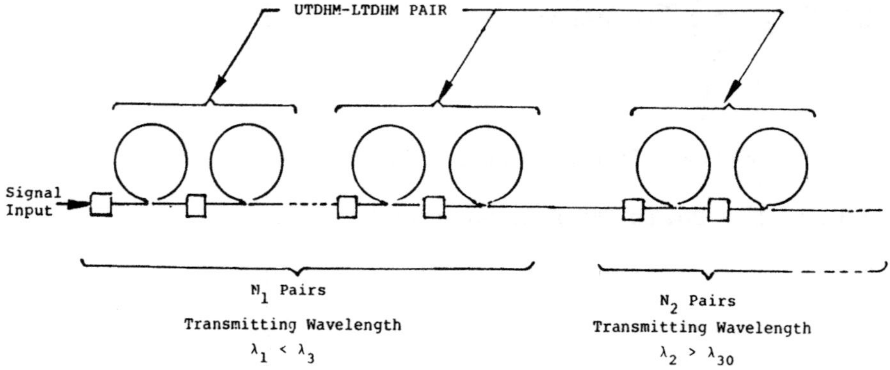

Fig. 4. Dispersion Equalization Using Difference Transmitting Wavelength (λ_1 and λ_2) and UTDHM - LTDHM Pairs.

Fig.5.

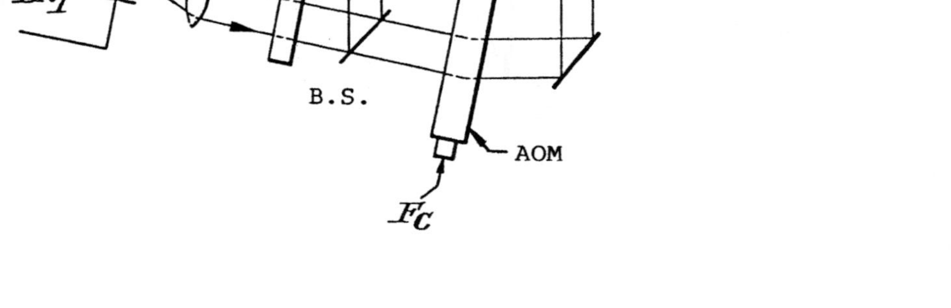

Fig. 6.

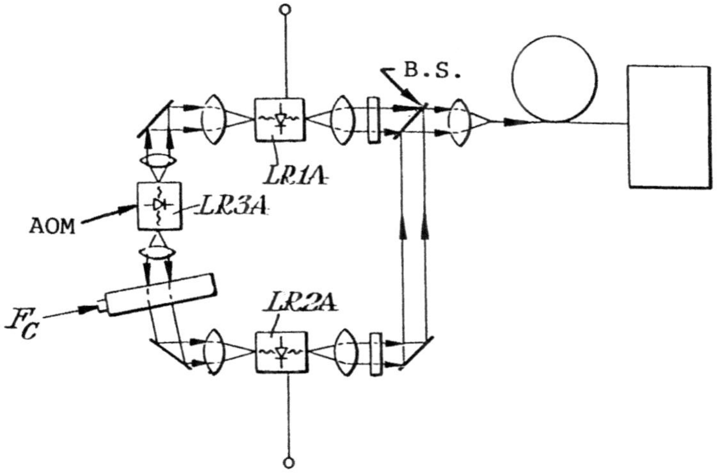
B.S.
LR1A
AOM
LR3A
F_C
LR2A

Fig. 6A.

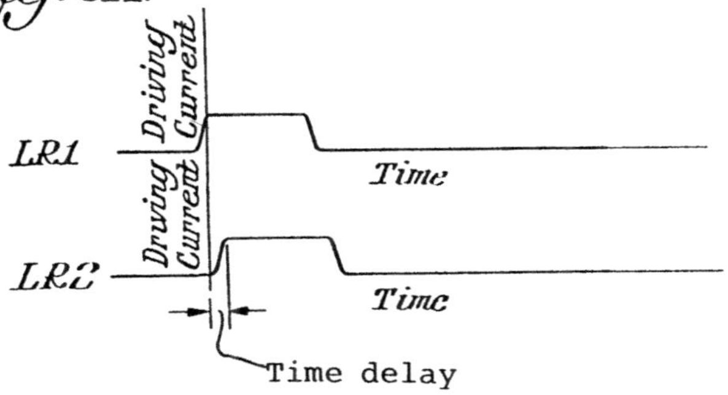
Driving Current
LR1
Time
Driving Current
LR2
Time
Time delay

OPTICAL NONLINEARITY IN FIBRES: A NEW FACTOR IN SYSTEMS DESIGN

D COTTER
British Telecom Research Laboratories
Martlesham Heath, Ipswich, Suffolk IP5 7RE, UK

1. INTRODUCTION

It was realised about ten years ago that optical nonlinearity could place an ultimate practical limitation on the power levels used in fibre transmission systems [1,2]. Over the past decade a wide range of interesting nonlinear effects have been studied extensively [3-5] and have found useful application in optical frequency convertors, tunable sources [6] and pulse compressors [7]. Nevertheless nonlinear effects in fibres have not until now had any significant influence on the operation of optical communications systems. Optical nonlinearity has been thus far, a negligible factor in the design of these systems.

This situation is no longer true. For the first time nonlinear effects are beginning to determine the operating limits for some advanced optical systems, and in some cases these effects can influence the whole design approach.

Many of the advanced optical communications systems currently under development use narrow-linewidth single-frequency lasers. Such lasers, having linewidths of about 100 MHz or less, are being employed to eliminate spectral mode-hopping [8], to overcome bit rate limitations due to chromatic dispersion in the fibre, and in coherent transmission systems [9]. The latter systems require laser linewidths of less than one megahertz and are likely to be employed in the 1.5-1.6 μm wavelength range using extremely low-loss silica fibre. These conditions (a narrow excitation linewidth and long interaction length by virtue of the low attenuation coefficient) are precisely those which favour the nonlinear process of stimulated Brillouin scattering (SBS). If the optical power launched into the fibre exceeds some critical threshold level, then SBS can occur. In this process a significant proportion of the optical power travelling through the fibre may be converted into a second light wave, shifted in frequency, travelling <u>backwards</u> towards the transmitter. SBS can be detrimental in an optical communications system in a number of ways: by introducing severe additional signal attenuation, by causing multiple frequency shifts in some cases, and by introducing a high intensity backward coupling into the transmission optics. It will therefore be necessary to operate at power levels below the threshold for SBS, and this could place a severe limitation on the launch power and thus also the repeater spacing.

The losses in silica fibres have now been reduced to levels close to the theoretical limit set by Rayleigh scattering and IR absorption. At the same time new systems techniques, such as coherent heterodyne and homodyne detection, can increase the receiver sensitivity to a level approaching the quantum noise limit. Therefore it appears that in the next few years the major advances in increasing the optical transmission distance (for example, to allow submarine inter-continental communications without submerged repeaters) are likely to be made by increasing the transmitter power. However, the very low fibre loss, the narrow laser linewidths required, and higher power levels together mean that nonlinear optical effects, principally SBS, could pose the major systems limitation.

In this paper we review the current status of research on SBS in monomode fibres. It is found that in low-loss silica fibres SBS can occur and be severely detrimental at power levels of only a few milliwatts, which is well within the envisaged operating range for future optical transmission systems. The important dependence of the SBS gain and threshold on the spectrum and modulation characteristics of the optical transmitter is under investigation. Some earlier misconceptions concerning the effect of the modulation bandwidth on the SBS threshold have been identified and clarified; this indicates that SBS may be a greater problem than was previously realised. However some techniques for suppressing SBS by using various forms of phase modulation of the optical carrier appear promising, and are the subject of current research.

2. STIMULATED BRILLOUIN SCATTERING

The first observation of SBS in an optical fibre was by Ippen and Stolen in 1972 [10] using a pulsed laser. Continuous-wave SBS in a fibre has been generated using different types of fibre resonators [11]. However it is only recently [12,13] that continuous-wave SBS has been observed in a single pass through multi-kilometer lengths of low-loss silica fibre. We observed SBS at power levels as low as 5 mW when pumping a 13 km length of

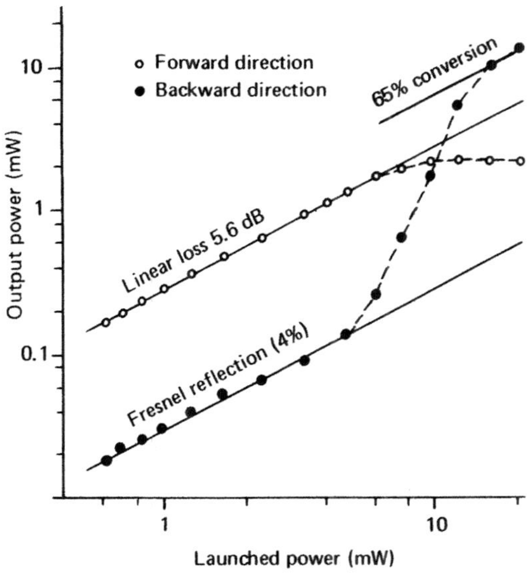

FIGURE 1: Power emitted from each end of a 13.6 km monomode fibre as a function of launched power at 1.32 μm

low-loss fibre (0.4 dB/km) at 1.32 μm using a single-longitudinal mode Nd:YAG laser with a linewidth less than 1 MHz. This is the lowest power level at which any nonlinear scattering process has been observed, and this results from the unique properties of optical fibres: tight lateral confinement ($\sim$ 5 μm), multi-kilometer interaction length and low attenuation coefficient. Figure 1 is a typical result, showing the output power from each end of the fibre as a function of the launched power. At low input power the output power monitored in the backward direction was due only to the Fresnel reflection from the cleaved end face of the fibre. However, at input powers exceeding 5 mW, the output power in the backward direction was seen to increase rapidly in a nonlinear fashion, and the efficiency of conversion to the backscattered wave reached 65%. At low input powers the power emitted from the distant end of the fibre was related linearly to the input power, determined by the linear loss of the fibre. However, at input powers exceeding about 6 mW, the output power became nonlinear. At input powers greater than 10 mW the output power in the forward direction reached a saturated maximum of about 2 mW.

We find that there is good agreement between the observed threshold powers and the theoretical values calculated using the bulk parameters of fused silica [3,14]. For a fibre having 0.2 dB/km attenuation and length exceeding 15 km, the SBS threshold is expected to be around 2.4 mW. A fibre with the same specification and having the ability to maintain polarisation would exhibit a SBS threshold of about 1.2 mW. These power levels are already well within the range envisaged for future transmission systems using silica fibres. Moreover it has been proposed recently that fibres manufactured from materials, such as fluoride glasses, having an infrared absorption edge at a wavelength longer than that of silica would have a very low intrinsic loss, perhaps as low as 0.001 dB/km. The elasto-optic coefficients of these materials are unlikely to be smaller than that of silica. As such fibres would be employed in very long links, we predict that the threshold powers for SBS could be in the microwatt region.

It is not immediately apparent how the observations of SBS with narrow-band cw lasers relate to the case where the transmitted optical wave is modulated to convey data. To determine the operating limits of optical communication systems we are currently investigating the dependence of the SBS threshold on parameters such as types of modulation, modulation indices and bit rate. Preliminary theoretical work [15], being confirmed by laboratory experiments now in progress, indicates that in terms of the average optical power, the SBS threshold in an on-off amplitude-keyed (ASK) system will be the same or only about a factor of 2 higher than for a cw unmodulated transmitter. Moreover, for bit rates greater than about 100 Mbit/s, the SBS threshold is found to be independent of bit rate in ASK. At first glance this result may appear surprising. It certainly contradicts a widely-held view that in a digital transmission system operating at bit rates greater than the Brillouin linewidth (16 MHz at 1.55 μm) the average optical power at which SBS would occur should be very much greater than the corresponding narrowband cw threshold because the pulse bandwidth would be significantly greater than the Brillouin linewidth. However a careful frequency-domain analysis shows this to be incorrect; the modulated signal comprises an optical carrier together with a spectrum of upper and lower sidebands whose amplitude and spectral distribution depend on the bit rate and coding statistics. It is the carrier, of constant amplitude and containing one half of the total average power, which can give rise to SBS [14].

Quite different conclusions are reached for systems using phase-shift (PSK) or frequency-shift (FSK) modulation. The time-dependent analysis [15] shows that SBS could be suppressed in a binary PSK system by using a 180° keyed phase shift. The physical explanation for this is that for significant SBS to occur, the pump field must be able to build up by electrostriction a strong coherent acoustic wave. In the case of ASK with a narrowband laser and external modulation, the optical pulses have a fixed phase relationship and are thus additive in their effect of generating a coherent acoustic wave. However in PSK, provided the appropriate phase shift is chosen, the net acoustic excitation produced by the optical field can be minimised. This is because the acoustic excitation produced by the field in the time periods representing the binary digit zero is counteracted by the phase-shifted field in the periods representing the digit one. Similarly it is found that for FSK operation, provided the frequency shift is sufficiently large, the continual phase changes of the optical field will result in only a small net acoustic excitation and thus the SBS gain can be reduced.

As the present time the use of PSK appears to be the most attractive method of suppressing SBS in coherent optical transmission systems. In fact homodyne PSK is also attractive because it offers the best advantage in terms of improved signal-to-noise ratio [16]. Further work is underway in our laboratory to develop phase modulation techniques to suppress SBS, and to exploit these in high-power coherent transmission.

ACKNOWLEDGEMENTS

I gratefully acknowledge helpful discussion with D W Smith, R Wyatt, K I White and C J Todd. Permission of the Director of BTRL to publish is acknowledged.

REFERENCES

1. Smith RG: Appl Optics 11 (1972) 11, 2489-2494.
2. Stolen RH: Proc IEEE 68 (1980) 10, 1232-1236.
3. Stolen RH: in <u>Optical Fiber Telecommunications</u>, eds S E Miller, A G Chynoweth, Academic Press, New York (1979).
4. Hill KO et al: in <u>Fiber Optics - Advances in Research and Development</u>, eds B Bendow, S J Mitra, Plenum, New York (1979).
5. Mollenauer LF et al: Phys Rev Lett 45 (1980) 13, 1095-1098.
6. Cohen LG, Lin C: IEEE J Quant Elec QE-14 (1978) 11, 855-859.
7. Shank CV et al: Appl Phys Lett 40 (1982) 9, 761-763.
8. Malyon DJ, McDonna AP: Elec Lett 18 (1982) 11, 445-447.
9. Hodgkinson TG et al: Elec Lett 18 (1982) 12, 523-525.
10. Ippen EP, Stolen RH: Appl Phys Lett 21 (1972) 11, 539-541.
11. Hill KO et al: Appl Phys Lett 28 (1976) 10, 608-609.
12. Uesugi N et al: Elec Lett 17 (1981) 11, 379-380.
13. Cotter D: Elec Lett 18 (1982) 12, 495-496.
14. Cotter D: J Opt Comm (in press)
15. Cotter D: Elec Lett 18 (1982) 12, 504-506.
16. Yamomoto Y, Kimura T: IEEE J Quant Elec QE-17 (1981) 6, 919-935.

PART FOUR

OPTICAL WAVEGUIDE THEORY AND PRACTICE

INVERSION TRANSFORMATIONS IN GEOMETRICAL OPTICS AND ALLIED SUBJECTS

S. CORNBLEET

1. INTRODUCTION

Of the many transformations that exist in theoretical physics, that of inversion is the least understood or applied. Geometrically, the inversion of a space in a sphere, is such that the position co-ordinate $\underset{\sim}{r}$ in one space, referred to the centre of the sphere as origin, is transformed into a co-ordinate $\underset{\sim}{r}'$ in the inverse space, where $\underset{\sim}{r}$ and $\underset{\sim}{r}'$ are in the same straight line, and $rr' = k^2$, the radius of the sphere. k is termed the radius of inversion.

Recently, in the field of perfectly imaging optics (microwave optics) the application of this transform to two particular problems has been studied, one the basic reflection/refraction process (1) and another to imaging in a spherically symmetrical non-uniform medium (2). Both have practical applications which will be demonstrated. Each subject has an analogue in another branch of physics, the refraction process in electromagnetic wave theory, and propagation in a non-uniform medium in the dynamics of a particle in a variable potential field. If one studies the inversion transformations that apply in those areas, the results and the conditions are not identical with the geometrical optics inversions. These anomalies have not yet been resolved. In themselves, they raise several questions of interest, and the answers as yet are only tentative. The indications are that the transformations agree if they are performed on generalized co-ordinate parameters in a representation space appropriate to each problem. What precisely these are can only be indicated in an article of this kind, but the solution holds great promise for the unification of various aspects of physics as well as practically, for the transformation of a known solved problem into the solution of a similar problem in the inverse space. This is akin to the method of conformal mapping in the theories of electrostatics and hydrostatics.

2. DAMIEN'S THEOREM AND THE INVERSION OF REFRACTING SURFACES

The first inversion involves the concept of the "zero distance phase front" (3). This is defined to be the surface formed by the refraction of rays from a source S in a surface g (Fig. 1a), where each ray is continued backward to a point P where $\eta_1 SM = - \eta_2 MP$ ... (1
the negative sign indicating the reversal of direction.

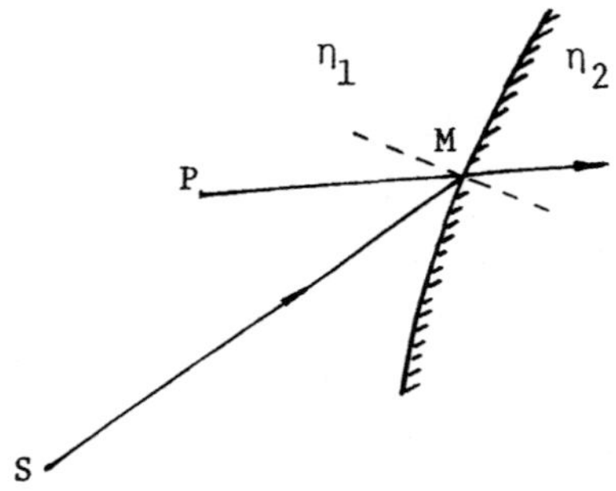

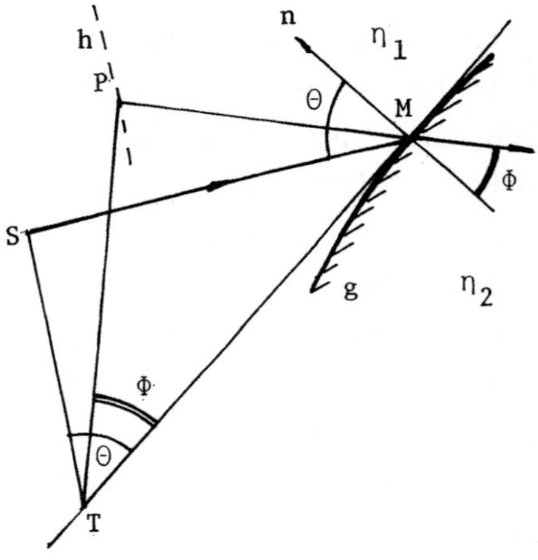

FIGURE 1. a) Construction of the zero distance phase front   b) Geometry of the zero distance phase front

The resultant surface h would then be the equivalent source for the ray complex after the refraction as if the entire space were filled with the second medium. Hence the term zero phase. Geometrically it can be shown that the perpendiculars PT at P and ST at S meet at T which lies on the tangent to g at M (Fig. 1b). Hence h can be obtained from algebraic geometry alone and not by ray by ray tracing as is usual in optics design. It is a whole-surface interaction appropriate to optics-in-the-large. For a reflection the same construction applies but $\eta_1 = 1$, and $\eta_2 = -1$ in equation 1, thus P will be on the far side of the reflector. For example, for a source S at the focus of a parabolic reflector, P would lie on the directrix. Damien (4) now gives a most unusual property of this refraction construction. Stated as a theorem it is:- "If h is the zero distance phase front of a source S and a surface g, the inversion of g in a circle centred on S is g', and the inversion on h in the circle is h', then g' is the zero distance phase front of h' for the same refractive index". (Figure 2). The proof of this theorem too is purely geometrical and it is essential to note that a surface is transformed into a phase front and vice versa. In practice the theorem can be used to design offset and multiple beam reflectors, and to obtain the final surface of a system that will bring all the rays to a perfect focus (if the system is orthotomic).

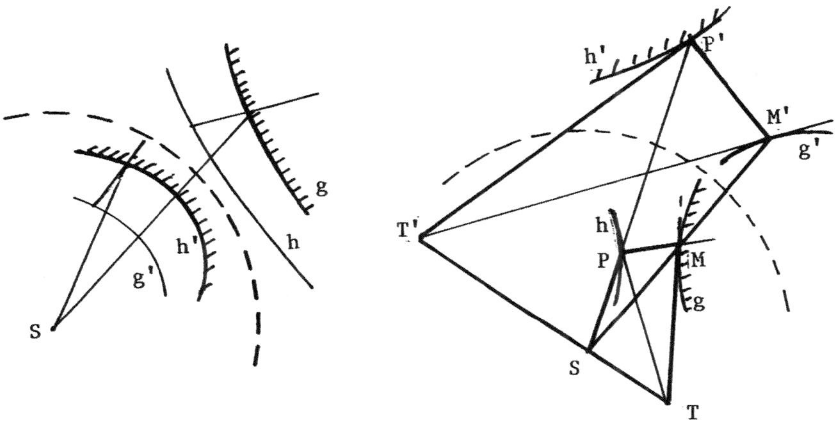

FIGURE 2. Damien's theorem and illustrative proof.

The studies of the applications of inversion to electromagnetic problems goes back as far as to Maxwell himself. One of his papers (5) gives the proof that the only conformal mapping of a three space onto itself is, apart from the elementary translation, rotation or magnification, is the inversion, a fact later established in differential geometry by Forsyth (6). This latter analysis can (with considerable algebra) be extended to a Euclidean space of four dimensions. Inversions of the complex space-time of four dimensions was performed by Bateman (7), whose criteria were the invariance of the wave and eikonal equations (to a scale factor). Bateman applied the inversion:-

$$X = \frac{x}{r^2-c^2t^2}; \quad Y = \frac{y}{r^2-c^2t^2}; \quad Z = \frac{z}{r^2-c^2t^2}; \quad T = \frac{t}{r^2-c^2t^2} \quad \ldots(2$$

$$X = \frac{x}{z-ct}; \quad Y = \frac{y}{z-ct}; \quad Z = \frac{r^2-1}{2(z-ct)}; \quad T = \frac{r^2+1}{2(z-ct)} \quad \ldots(3$$

$$r^2 = x^2 + y^2 + z^2$$

to the problem of obtaining a new lens from a known lens of previous design. Here however we find that refracting surfaces transform directly into refracting surfaces and not into phase fronts. He was followed by Cunningham (8), who gave the complete transformation of the electromagnetic field under inversion. He showed that, when the inversion was accompanied by a contraction or expansion, the constituent relations transform exactly as for the Lorentz transformation of special relativity. In the Lorentz transformation, the wave and eikonal equations are covariant, but time cannot be made simultaneously zero in the two spaces. In the inversion, with no expansion or contraction, this can be the case and the inversion then becomes the static geometrical inversion of space. Cunningham's inversion in full is:-

$$\underset{\sim}{E}' = d^{-4}(r^2-c^2t^2)\{(r^2-c^2t^2)\underset{\sim}{E} + 2\underset{\sim}{r}\times(\underset{\sim}{r}\times\underset{\sim}{E})-2ct(\underset{\sim}{r}\times\underset{\sim}{H})\}$$

$$\ldots(4$$

$$\underset{\sim}{H}' = d^{-4}(r^2-c^2t^2)\{-(r^2-c^2t^2)\underset{\sim}{H}-2\underset{\sim}{r}\times(\underset{\sim}{r}\times\underset{\sim}{H})-2ct(\underset{\sim}{r}\times\underset{\sim}{E})\}$$

from which it can be seen that with $t = 0$, $E\times H.\underset{\sim}{r} = E'\times H'.\underset{\sim}{r}'$ and $\underset{\sim}{r} = \underset{\sim}{r}'$ under inversion. That is the Poynting vector in one space inverts into the Poynting direction in the transformed space. The Poynting vector can be considered to be the ray direction, and hence rays will convert into rays as will be seen to occur in another context.. However, we have again that the surface of a reflector transforms into another reflector in contradiction to the theorem of Damien.

If one takes Bateman's transformation, equation 2, putting $T = t = 0$, one again has the ordinary inversion of three space. There is no way that one can imply that $t = 0$ in one space refers to a zero distance phase front and $T = 0$ in the other to a refracting surface.

One other correlation between rays and the electromagnetic field is of interest. Stavroudis (9) shows that if the variable refractive index function $\eta(x,y,z)$ is included in the space co-ordinates, then the combination of the Frenet-Seret formulae for curves in space and the ray equations give a triad of orthogonal vectors $\eta\hat{n}$, $\eta\hat{b}$ and $\eta\hat{t}$ which obey Maxwell like equations, $\hat{t}$ being the tangent vector, $\hat{n}$ the principal normal and $\hat{b}$ the binormal at any point on the ray. $\hat{t} = \hat{n} \times \hat{b}$ plays the part of the Poynting vector. This concept of refractive co-ordinate space becomes of great relevance in the next section.

3. THE NON-UNIFORM SPHERICALLY SYMMETRICAL MEDIUM

The study of rays in a non-uniform isotropic medium also originated with Maxwell, but most early work can be attributed to biologists who obtained curious focusing effects in the eyes of marine and other animals. The archetype spherical lens is the Maxwell fish-eye. This focuses all the rays from a point into another point in the complete refractive space (q.v.), and is proved by Luneburg (10) to be the only ideal lens capable of doing this (apart from the simple plane reflector). In the following, refractive index laws are to be derived which can give values of the refractive index which are positive and less than unity, negative, zero and infinite but not complex. This range of values allows the description of a "complete refractive space". It can then be shown that the rays will form complete closed paths, and, with the inclusion of the circle at infinity, these will include curves such as parabolas and hyperbolas, as closed curves in the projective plane. The rays in Maxwell's fish-eye then form the complete coaxal circle system shown in Figure 3. The refractive index law is $\eta(r) = \frac{2}{1+r^2}$

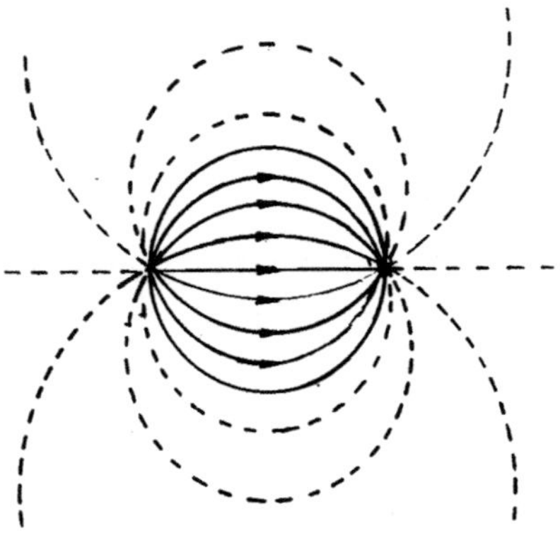

FIGURE 3. Rays in the refractive space of Maxwell's fish-eye

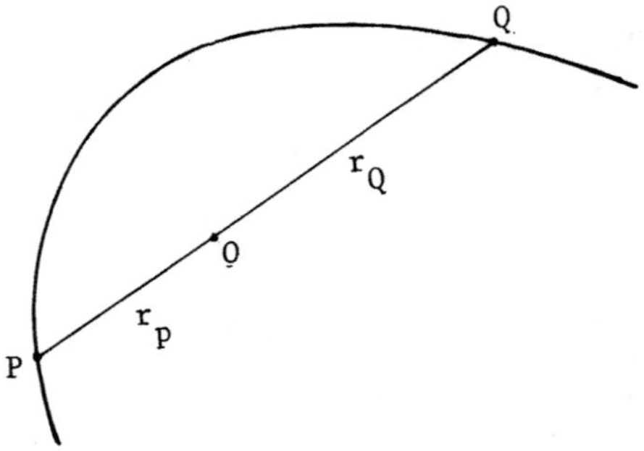

FIGURE 4. Closed ray path in a spherically symmetrical medium

The general solution for a spherical medium has the form

$$r_P r_Q = a \; ; \qquad r_P/r_Q = b \qquad \text{...(5a, 5b}$$

where P and Q are image points in the medium (Fig. 4).
The solution gives:-

$$r\eta(r) = f\left[\phi\left(\frac{r}{a}\right) \oplus \phi\left(\frac{a}{r}\right)\right] \qquad \text{...(6}$$

where f and ϕ are arbitrary functions of their arguments, and $\oplus$ is an associative law of combination.

The first relation in equation 5 gives the fish-eye and its generalisations. For example,

$$f(x) = 1/x \; ; \quad \phi(r) = r^p \text{ gives } \eta = \frac{2r^{p-1}}{1+r^{2p}}$$

The second relation gives the Luneburg lens with refractive index law $\eta(r) = (2 - r^2)^{\frac{1}{2}}$. The rays are then ellipses going through the points (1, -1) (Figure 5a).

One other lens needs to be noted, the Eaton lens. It can be derived from the Maxwell fish-eye by the transformation

$\eta \to r$, $r \to \eta$ and thus has the law

$$\eta(r) = \left(\frac{2}{r} - 1\right)^{\frac{1}{2}}.$$

The rays are all ellipses but this time with the centre of the medium at one focus, with major axes all of the same length lying along the horizontal axis (Figure 5b).

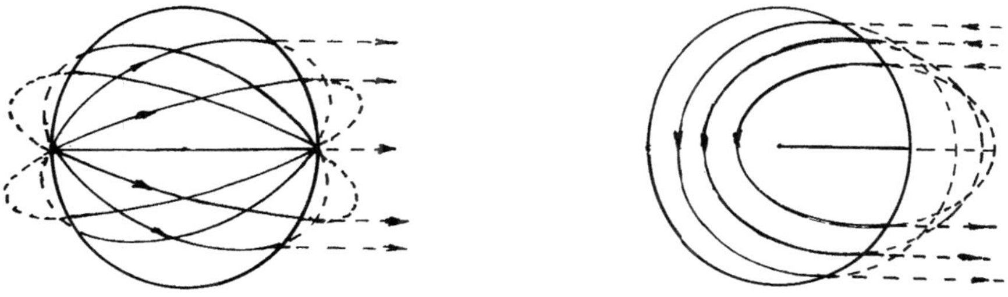

FIGURE 5. a) Rays in a Luneburg lens. b) Rays in the Eaton lens.

As a lens inside the unit circle the rays are retroreflected after one circuit of the origin at which there is a simple pole. Under the same transformation the Luneburg lens is invariant.

There are a number of other spherical lenses with specified focusing properties, but we can use the three above to illustrate the transformation to be discussed. All of them have a geodesic analogue on a surface of revolution. (Figure 6). The general law for this surface is

$$\sigma = A\rho + B\sin^{-1}C\rho \qquad \ldots(7a$$

(11). From this we can derive the most general law of refractive index

$$Cr^{\frac{2}{BC}} - 2r^{\frac{1}{BC}}(\eta r)^{\frac{A}{BC} - 1} + C(\eta r)^{\frac{2A}{BC}} = 0 \qquad \ldots(7b$$

The transformation $\eta \leftrightarrow r$ is then (12) simply $A \to A + 1$. The geodesic form allows the design of integrated optical components shaped indent-

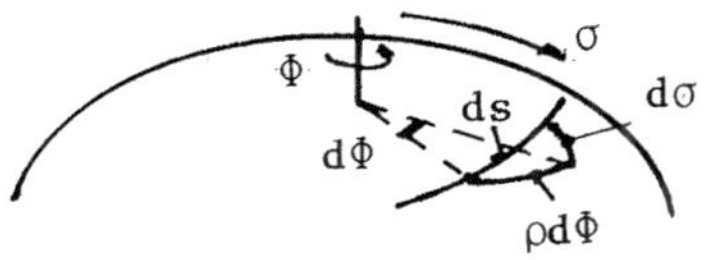

Comparing geodesics

$$d\sigma^2 + \rho^2 d\phi^2 = \eta^2 dr^2 + \eta^2 r^2 d\theta^2$$

whence $\rho = r\,\eta(r)$; $d\sigma = dr\,\eta(r)$

FIGURE 6. Geodesics on a surface of revolution

ations (or elevations) in the guiding film over a substrate. There are other forms which can be used effectively for this purpose.

4. TRANSFORMATION OF THE SPHERICAL MEDIUM

One such transformation $\eta \to r$ $r \to \eta$ has already been given. Consider now two spaces the first in which the medium has refractive index $\eta(x,y,z)$ and phase front surfaces $S(x,y,z)$ = const., and the second with index $n(\lambda,\mu,\nu)$ and phase fronts $T(\lambda,\mu,\nu)$. Then as in Luneburg (10 p.173), if these are connected by a Legendre transformation

$$\frac{\partial S}{\partial x} = \lambda, \ \frac{\partial S}{\partial y} = \mu, \ \frac{\partial S}{\partial z} = \nu$$

$$\frac{\partial T}{\partial \lambda} = x, \ \frac{\partial T}{\partial \mu} = y, \ \frac{\partial T}{\partial \nu} = z \qquad \ldots(8$$

or $S + T = x\lambda + y\mu + z\nu$

then the eikonal in one medium is transformed into the eikonal in the other, for

$$\left(\frac{\partial S}{\partial x}\right)^2 + \left(\frac{\partial S}{\partial y}\right)^2 + \left(\frac{\partial S}{\partial z}\right)^2 = \eta^2; \ \left(\frac{\partial T}{\partial \lambda}\right)^2 + \left(\frac{\partial T}{\partial \mu}\right)^2 + \left(\frac{\partial T}{\partial \nu}\right)^2 = n^2 \quad \ldots(9$$

Applied to the fish-eye $= \dfrac{2}{1+r^2} = \dfrac{2}{1+x^2+y^2+z^2} =$

$$= \left\{\left(\frac{\partial S}{\partial x}\right)^2 + \left(\frac{\partial S}{\partial y}\right)^2 + \left(\frac{\partial S}{\partial z}\right)^2\right\}^{\frac{1}{2}}$$

transforms to $\{\lambda^2 + \mu^2 + \nu^2\}^{\frac{1}{2}} = \dfrac{2}{1+\left\{\left(\frac{\partial T}{\partial \lambda}\right)^2 + \left(\frac{\partial T}{\partial \mu}\right)^2 \left(\frac{\partial T}{\partial \nu}\right)^2\right\}^{\frac{1}{2}}}$

or $n^2 = 2/r - 1$ and we have the Eaton lens. This is therefore equivalent to the transformation $\eta \leftrightarrow r$.

It is obvious that no construction is going to give a zero distance phase front in a continuously varying medium. However we can find an approriate inversion theorem applicable to the spherical medium. This is found not to apply directly to the co-ordinate space but to the refractive space defined by $f(r) = r\eta(r)$. The new space is obtained from $\bar{m}(r) = f(1/r)$.

Then it can be shown that the rays in the $\bar{\eta}(r)$ medium are the direct geometrical inversion of the rays in the $\eta(r)$ medium (2). That is so since

$$\eta^2 ds^2 = \eta^2 r^2 \left\{ \frac{dr^2}{r^2} + d\theta^2 + \sin^2\theta d\phi^2 \right\} \quad \text{with } \eta r \rightarrow \bar{\eta} r' \text{ and } r \rightarrow \frac{1}{r'} \quad \ldots(10$$

$$= \bar{\eta}^2 r'^2 \left\{ \frac{dr'^2}{r'^2} + d\theta^2 + \sin^2\theta d\phi^2 \right\} = \bar{\eta}\bar{d}s^2$$

and which generalizes to give the following theorem:-
" If a spherically symmetrical medium defined by the refractive index law $r\eta(r) = f(r)$ is transformed to a new medium by $r\bar{\eta}(r) = f(1/r^n)$ then rays in the first medium with equation $F(r,\theta) = 0$ transform into rays in the new medium with equation $F(1/r^n, n\theta) = 0$ ".
With general values of n the theorem applies only to an equatorial plane, and the optical line element is scaled by a factor n. When $n = 1$ the transformation is a pure geometrical inversion and in this case we find that it is Maxwell's fish-eye that is the invariant. For general n it transforms the fish-eye to its generalized form as follows:-

$$r\eta = \frac{2r}{1+r^2} \quad \therefore \quad r\bar{\eta} = \frac{2/r^n}{1+\frac{1}{r^{2n}}} \quad \text{or} \quad \bar{\eta} = \frac{2r^{n-1}}{r^{2n}+1}$$

This, as was shown, derives from the law of equation 6. It was also found by Luneburg as the result of a conformal mapping (10 p.179). It should be noted that the usual ray path integral solution can also be derived as the geodesics of the refractive space defined by

$$\eta^2 ds^2 = \eta^2 dr^2 + \eta^2 r^2 d\theta^2 + \eta^2 r^2 \sin^2\theta d\phi^2 \quad \ldots(11$$

If applied to the general refractive index law, equation 7b, it becomes the transformation $B \rightarrow B^{-n}$, but the arcsin term in equation 7a becomes complex. This point is still under consideration. It is hoped that it will apply to the edge discontinuity that arises in most geodesic optical components.

A transformation of a very similar kind, closer in fact to an original proposition of Hamilton's (based on a previous paper by Gauss) is given by Budden (13).

5. THE MECHANICAL ANALOGUE

In classical mechanics, which besides optics and algebra, is the other major field in the works of both Newton and Hamilton, a characteristic function exists which plays a part analogous to the phase front in optics. It is defined in a phase-space, with canonical co-ordinates, q, the generalized position co-ordinate and p the generalized momenta. The function W(p,q) obeys an eikonal-like equation

$$(\nabla W)^2 = 2m(E - V) \qquad \ldots(12$$

where E is the total energy and V the potential.

Particle trajectories are given by

$$\delta \int ds \sqrt{2m(E - V)} = 0$$

If we compare these with Hamilton's optical equations

$$(\nabla S)^2 = \eta^2; \qquad \delta \int \eta \, ds = 0 \qquad \ldots(13$$

then we can see that $(2m(E - V)^{\frac{1}{2}}$ plays the part of a refractive index. The particle trajectories and the ray paths both derive from the Euler-Lagrange relations and the calculus of variations applied to the integrals. Buchdahl (14) has recently compared the basic problems in the two disciplines, the Maxwell fish-eye with the Kepler orbit. The generalized Snell's law for a spherical medium is Bouguer's theorem which states $r\eta(r)\sin\alpha$ = const. along any one ray (α being the angle between the ray tangent and the radius vector $\underset{\sim}{r}$). This is shown to be equivalent to the conservation of angular momentum, in this case for motion in a 1/r potential, the Kepler orbit. A second constant the analogue of the Runge-Lenz vector, is also found for the optical ray. The classical deflection function (Rutherford scattering) derives from an integral identical to the optical ray integral (15), and there is an inversion of the phase-space which gives new particle trajectories in an inverse potential distribution.

However, as with the theorem of Damien and the electromagnetic field, the connection is far from perfect. Kepler orbits are elliptical in real space but circular in phase-space. The fish-eye has circular rays in real space but elliptical (Kepler in fact) under the Legendre transformation. Legendre transformations among other effects, converts Lagrangians to Hamiltonians (16) and the components of E into εE (17). Thus the refractive space appears to be the correct phase-space for optical systems. The nearest approach in electromagnetic theory are a dual pair of "refractively loaded" Debye potentials, but ray analysis has not yet been applied in that area (18).

When the correct phase-space can be found for the electromagnetic field and the necessary transformation operators, the connection with ray optics and classical mechanics will lead to a large area of unification within the fields of physics. The more complex areas of field theory such as diffraction, evanescent fields and complex rays, may then be subject to a universal transformation theory.

REFERENCES

1. Cornbleet S. 1979. "A new geometrical method for the design of optical systems", Jour. I.E.E. MOA vol 3 p.78.
2. Cornbleet S and Jones MC. 1982. "The transformation of spherical non-uniform lenses", Jour. I.E.E. Pt. H vol 6 p.321.
3. Eaton JE. 1952. "The zero phase front in microwave optics", Trans. I.R.E. AP1 p.38.
4. Damien R. 1955. "Theoreme sur les surfaces d'ondes en optique geometrique", Gauthier Villars, Paris.
5. Maxwell JC. 1871-73. "On the condition that, in the transformation of any figure ... in three dimensions, every angle in the new figure shall be equal to the corresponding angle in the original figure", Proc. Lond. Math. Soc. vol 4 p.117-9.
6. Forsyth AR. 1903. "Differential Geometry", Cambridge University Press.
7. Bateman H. 1955. "Electrical and optical wave motion", Dover publications.
8. Cunningham E. 1910. "The principle of relativity in electrodynamics and an extension thereof", Proc. Lond. Math. Soc. vol 8 p.70.

9. Stavroudis ON. 1972. "The optics of rays, wavefronts and caustics", Academic Press pp.194-198.
10. Luneburg RK. 1964. "The mathematical theory of optics", University of California Press.
11. Kunz KS. 1954. "Propagation of microwaves between a pair of doubly curved conducting surfaces", Jour. App. Phys. vol 25 p.642.
12. Cornbleet S and Rinous PJ. 1981. "Generalized formulas for equivalent geodesic and non-uniform lenses", Jour. I.E.E. Pt. H vol 5 p.95
13. Budden KG. 1961. "Radio waves in the ionosphere", Cambridge University Press p.252.
14. Buchdahl HA 1975. "Conformal transformations and conformal invariance of optical systems", Optik vol 43 p.259.
Kepler problem and the Maxwell fisheye", Am. Jour. Phys. vol 46 p.840.
15. Klein A. 1978. "Alternative formula for the classical deflection function in a central field", Am. Jour. Phys. vol 46 p.1019.
16. Goldstein H. 1964. Classical Mechanics", Addison-Wesley.
17. Sommerfeld A. 1964. "Lectures on theoretical physics" vol IV p.150 Ed. P. ENZ.
18. Tai CT. 1971. "Dyadic Green's functions in electromagnetic theory", Intext Publishes Chap. 12.

HOLLOW WAVEGUIDES FOR LONG-WAVELENGTH I.R. RADIATION

M. BRENCI, R. FALCIAI, G.C. RIGHINI, V. RUSSO, A.M. SCHEGGI, S. SOTTINI
Istituto di Ricerca sulle Onde Elettromagnetiche, C.N.R., Florence, Italy

B. LOCARDI
Stazione Sperimentale del Vetro, Murano, Venezia, Italy

1. INTRODUCTION

The CO_2 laser has been extensively used both in industrial and medical applications; however the provision of a flexible delivery system of the I.R. laser light to a target has been the most difficult problem. Up to now the guidance has been carried out mainly by means of an articulated arm consisting of a number of metallic pipes with a mirror at each rotating joint, whose optical axis must be accurately aligned. However, due to the presence of some mechanical errors, the laser beam may shift from the optical axis with a resulting instability and degradation as the arm is manipulated. Further such arms are characteristically large, cumbersome, hence of not easy handling and not suitable for operating in cavities of difficult access. For these operations flexible waveguides are necessary, playing the same role as silica fibers for Ar or NdYag laser radiation transmission. Various waveguides have been proposed and tested which can be grouped into two main categories: long wavelength infrared fibers and hollow waveguides. Example of the first category are fibers made of single or polycristalline alkali or metallic halides such as KCl, AgCl, AgBr, CsI, CsBr, TlBr or TlBrI (KRS-5) and chalcogenide glasses (1-7). Metallic flexible structures and dielectric hollow waveguides belong to the second category.

Subject of the present paper is the description of some experimental investigations performed at IROE on two particular types of hollow waveguides.

2. HOLLOW METALLIC WAVEGUIDES

The possibility of using metallic waveguides for delivering high power CO_2 radiation has been and is under investigation in several laboratories.

A number of flexible structures have been taken into consideration, e.g. parallel-plate, single curved surface, cylindrical or helical tube waveguides (8-13). However, from a practical point of view, hollow rectangular metal waveguides appear today to represent the most interesting solution for some applications.

A theoretical analysis of losses in a perfect hollow planar waveguide can be carried out by using ray optics and assuming that walls have specular reflection and obey Fresnel's law of reflectivity (12). If one dimension of the cross-section (e.g. the width b in Fig. 1) is much larger than

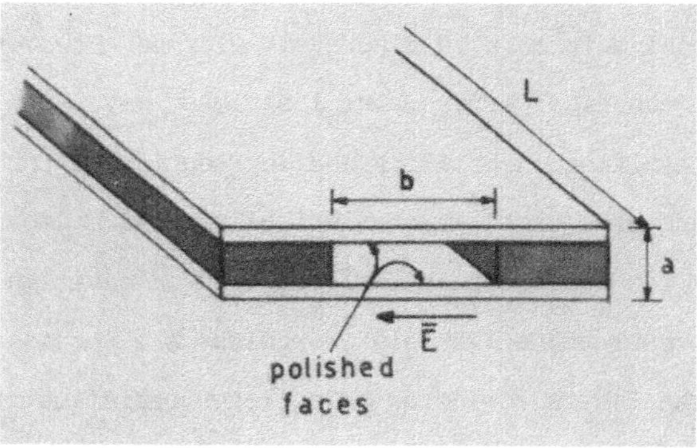

FIGURE 1. Rectangular metallic waveguide for I.R. radiation. Typical values are a = 0,5÷1 mm and b = 10÷20 mm, while L ≥ 1 m.

the other one, the total loss in a straight waveguide can be considered as the sum of the loss from the top and bottom walls and the loss from the sidewalls. If, moreover, the input laser beam is polarized parallel to the top and bottom walls, the two loss terms look like the TE and TM mode loss, respectively, in a planar waveguide. The loss of a single mode, either TE or TM, in a rectangular guide has an inverse cube dependence on the separation of the considered walls; however, the attenuation constant for TE modes is much smaller than for TM modes. As a consequence, to fabricate a low-loss waveguide one can reduce down to 0.5 mm the separation a of the two walls at which the electric field is parallel, while the distance b between the sidewalls cannot be made less than typically 10 mm.

To verify the dependence of guide losses on the width b or height a, we made measurements on several waveguides, fabricated with metal strips 10 to 20 mm wide and spaced 0.5 to 1 mm. In the measurements the beam from a cw CO_2 laser, with linear polarization aligned parallel to the wider guide walls, was focused into the input of the guide through a spherical Ge lens of 19 cm focal length, thereby exciting low loss TE modes. The results turned out to be in good agreement with theoretical predictions. This was true also for the results we obtained when bending or twisting the guides.

Once the design of the waveguide is optimized, it turns out that reflectivity and roughness of the two (wider) guiding plates are the most important parameters for a low attenuation. We tested both brass and aluminium metal surfaces, either as available or polished in the workshop. The best result so far has been obtained by using a brass waveguide of $1\times20\ mm^2$ cross-section, whose walls were accurately polished: the transmission was measured to be over 76% per meter, corresponding to 1.2 dB loss.

Table 1 reports the loss measurements relative to some of the straight and curved waveguides of different cross-sections we have tested. These values include coupling losses, that were typically of the order of 0.3 dB and that could also be further decreased.

Recently, the coupling problem has been considered with greater attention, in order to use the metal guides together with a commercially available domestic CO_2 laser for medical application (Valfivre LSS-100). The layout of the assembled laser and guide system is shown in Fig. 2. It includes a specially designed coupler, with a mirror mounted on a rotating joint, a waveguide tapered input, and an handpiece with focusing output lens. The mechanical joint with the mirror allows the guide to rotate in the vertical plane, thus reducing the necessity of twisting the guide itself and avoiding the consequent losses. Such a rotation, of course, makes the laser polarization to be no more aligned parallel to the wider walls at the waveguide input; however, no effective additional loss has been measured for a rotation angle of the joint up to 30^0. The focused output can provide a very narrow line of light as compared with the spot from an articulated arm or a cylindrical fiber. This line focus may be useful for distributing the laser energy over a large area or for cleaner and faster cuts over long sections.

Table 1. Measured loss of metal waveguides

WAVEGUIDE		TOTAL LOSS (dB)			
Metal	Cross-section	Straight guide	50^0-curved guide	90^0-curved guide	90^0-twisted guide
brass	1x20 mm^2	1.47	1.65	2.1	2.5
brass	0.5x20 mm^2	1.6	1.78	2.16	2.29
aluminium	0.5x20 mm^2	1.9	2.15	2.55	3.12
aluminium	0.5x10 mm^2	3.02	3.3	3.8	4.7

Note: The measurements were performed on 1 m-long metal guides. The 50^0 and 90^0 bends correspond respectively to a curvature radius of approximately 114 and 63 cm.

The efficiency of the device (the ratio of the CO_2 power at the output of the 1 m long guide to the laser output power) proved to be larger than 68%, corresponding to a total loss of 1.7 dB. This performance, however, should be further improved in a short time. In fact work is in progress to increase the coupling efficiency through the use of a cylindrical focusing lens and of a longer tapering of the input end of the guide. Tests are under way also to make simpler and more efficient the mechanical assembly of the guide itself.

An advantage of these metal waveguides is their power handling capacity, much larger than that of current IR fibers or even of other metal guides

(13): we have measured up to 50 watts of CO_2 laser power coming out from one of our guides without any damage of the inner walls. Moreover, the guide showed sufficient flexibility and easy handling, so that the laser guide system has been already used in experimental maxillo-facial surgery (14). These facts, together with the rugged structure and the very low cost, make rectangular metal waveguides attractive for different applications.

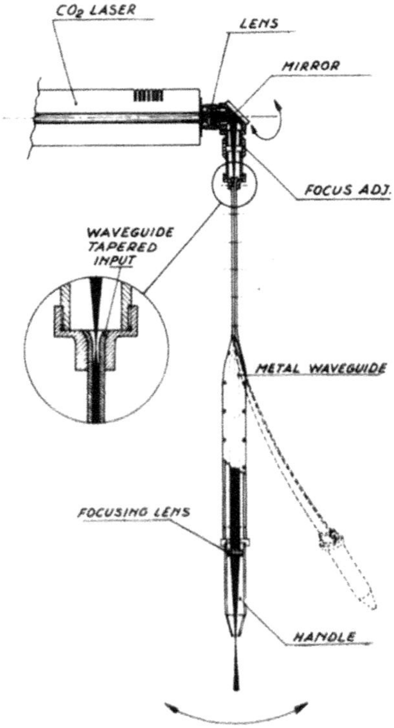

FIGURE 2. Sketch of the Laser-metal guide system for medical applications.

3. DIELECTRIC HOLLOW FIBERS

More recently a novel scheme was proposed of middle I.R. optical fiber using oxide glass as the cladding material to define the hollow core (15, 16). The working principle is based on the fact that most of all the glasses exhibit a complex refractive index with real part $n_r << 1$ in the wave number range 1000-1300 cm^{-1}. The laser light obliquely incident is guided by total reflection at the interface air-glass as in the usual optical fibers. The absorption losses can be made relatively low if an operative wavelength is choosen for which $n_r < 1$ but at the same time the imaginary part of the refractive index k is sufficiently small. In particular it has been shown that the presence in SiO_2 glass of heavy ions, such as Pb, shifts such operative wavelength towards $\lambda \simeq 10$ μm (1000 cm^{-1}). We have experimented on hollow fibers drawn from oxide glasses with different Pb percentages which present a minimum value of n_r around λ = 9.3 μm and used three emission lines (λ = 10.6, 9.5, 9.24 μm) of a c.w. CO_2 laser in order to achieve a "tuning" of the laser emission on the minimum transmission losses of the fiber. The maximum power obtainable from the laser at each wavelength was 10 W due to the introduction of a grating placed at one end of the cavity for tuning the laser on the three emission lines.The laser beam was focused onto the fiber input through a Ge lens of 12.7 cm focal length. Preliminary measurements showed that glasses with higher Pb percentage give rise to lower transmission losses; accordingly our tests were concentrated on a SiO_2-PbO-K_2O-Na_2O glass with 28% of Pb, and short lengths of hollow core fibers were fabricated with Kynar protection coating (∿20 μm thickness) applied in line during the drawing process, having different diameters (0.5÷0.7 i.d., 0.8÷1 mm e.d.) whose minimum curvature radius turned out to be ∿ 20 cm. Attenuation measurements were performed on 1 m long hollow fibers with 0.6 mm i.d. and 0.9 mm i.d. keeping the fiber straight or curved (∿ 1 m curvature radius). The loss values measured at the three different wavelengths are shown in Table 2.

These results are in good agreement with the values given in ref. (15) and suggest that an efficient CO_2 laser-fiber system could be obtained by maximizing the laser power on the lowest emission line. Note that the measured attenuation values include coupling losses and hence could be lowered by a better laser-to-fiber coupling. By a more suitable choice of the material the absorption losses could be reduced, while a protection coating will make such fibers stronger and more flexible.

Table 2. Measured loss of hollow fibers

λ (μm)	loss (dB) straight fiber	loss (dB) curved fiber
9.24	5	6
9.5	6.5	8
10.6	10	15

In conclusion these preliminary tests already indicate that it is possible to obtain dielectric waveguides for medium I.R. region with commercial glasses, cheaper than any I.R. transparent material. Such hollow waveguides present the same advantages of flexibility and reduced dimensions as the ordinary optical fibers and hence can find useful application also in endoscopic CO_2 laser surgery (at least for low power) for which the guidance along a fiber is essential.

REFERENCES

1. Pinnow D.A., Gentile A.L., Standles A.G., Timper A.J., 1978. Applied Physics Letters. Vol. 33, p.28.
2. Chen D., Garnfunkel J.H., Skogman R.A., Vora H., 1979. Fifth European Conference on Optical Fiber Communications (Amsterdam, Sept. 1979). Conf. Proc. p. 19-7-1.
3. Bridges T.J., Hasiak J.S., Strnad A.R., 1980. Optics Letters. Vol. 5, p. 85.
4. Okamura Y., Mimura Y., Komazawa Y., Ota C., 1981. Japanese Journal Applied Physics. Vol. 19, p. L649.
5. Mimura Y., Okamura Y., Komazawa Y., Ota C., 1981. Japanese Journal Applied Physics. Vol. 20, p. L17.
6. Boniort J.Y., Brehm C., Dpont P.H., Guignot D., Le Sergent C., 1980. 6-th European Conference on Opt. Communication (York, Sept. 1980). Conference Publication No.190, p. 61.
7. Brehm C., Cornbois M., Le Sergent C., Parant G.P., 1982. J. Non Crystalline Solids Nos. 1,2,47 (Part II), p. 251.
8. Nishihara H., Inone T., Koyama J., 1974. Applied Physics Letters. Vol. 25, p. 391.
9. Garmire E., McMahon T., Bass M., 1976. Applied Physics Letters. Vol. 29, p. 254.
10. Krammer H., 1978. Applied Optics. Vol. 17, p. 316.

11. Marhic M.E., Kwan L.I., Epstein M., 1978. Applied Physics Letters. Vol. 33, p. 609.
12. Garmire E., McMahon T., Bass M., 1980. IEEE J. Quantum Electron. Vol. QE-16, p. 23.
13. Kubo U., Hashishin Y., Natsakuta M., 1981. Proc. of the 4th Congress of Int. Society for Laser Surgery (Tokyo, 23-27 November 1981), p. 19-16.
14. Panzoni E., Clauser C., Sottini S., 1982. Proc. of the 8th Congress of Int. Association for Maxillo-Facial Surgery (Sorrento, 25-28 May 1982), paper 205.
15. Hidaka T., Morikawa T., Shimada J., 1981. Journal Applied Physics. Vol. 52, p. 4467.
16. Hidaka T., Kumada K., Shimada J., Morikawa T., 1982. Journal Applied Physics. Vol. 53, p. 5484.

ON THE EIGENSTATES OF POLARIZED GEOMETRICAL RAYS IN GRADED-INDEX MULTIMODE FIBERS

MASAHIRO HASHIMOTO

Department of Applied Electronic Engineering, Osaka Electro-Communication University, Neyagawa, Osaka 572, Japan

INTRODUCTION

Although an application of geometrical optics to the analysis of multimode fibers may lead to an unsatisfactory result, it has a great advantage of understanding fiber optics. In particular, the ray picture (see Fig.1), the projection of a ray path in the fiber cross-section or other sections, is useful for understanding the physical behaviors of propagation of modal fields. Use of ray optics for the fiber analysis will therefore be comprehensible. Arnaud [1] described the global characteristics of multimode fibers in terms of Hamiltonian optics, and he tried also to introduce the first-order effect of polarization into his fiber theory as the geometrical tortion of the ray [1, p.305].

The purpose of this paper is to generalize the polarized ray optics for use in a wide range of polarization phenomena which include the "anomaly" as reported recently [2,3]. This phenomenon contains the effect of changing the eigenstates of polarization from one to another (from the state of the circularly rotating polarization to the state of the opposite rotation) as the frequency varies. The effect is exhibited by the frequency-variant property of polarization which occurs in the fibers with certain special classes of refractive index [3], and its leading characteristic is normally involved in the higher-order terms of the asymptotic series solution whose first term is the geometric-optical solution. Hence, the problem of obtaining detailed information on this behavior is reduced to the mathematical problem of solving Maxwell's equations [4]. In the major part of this paper, however, we are devoted to presenting a simple model of propagation of polarized rays and explaining the physical aspects of polarization.

POLARIZATION OF RAYS

Fig.2 illustrates the paths of the TE and TM rays whose designations correspond to the definitions for TE and TM waves, respectively.

In the high frequency regime, the electric vector is directed toward the azimuthal axis for the TE ray and the radial axis for the TM ray (see Fig. 2), and the fields behave like a plane wave. In general, however, the modal rays are hybrid; the polarization of the hybrid ray is defined as the description of the temporal angular-variation of the combined electric vector at a point which the ray passes.

Apart from the fiber under consideration, we study the fundamental aspects of polarization which can be observed when a polarized ray is incident upon a dielectric surface. The result will later be applied to the fiber. A geometry for the scattering of this polarized ray is given in Fig.3. Top views of the incident rays with TE and TM polarizations and the reflected rays are given in Figs.4(a) and (b), respectively. From these pictures, we see that the TE-or TM-polarized incidence results in the TE-or TM-scattering, respectively. If, on the other hand, the ray direction is not parallel to the z-axis, the incidence of the TE-or TM-polarized ray results in the reflection and transmission of hybrid rays, as shown in Fig.5.

Let us return to the fiber. We replace the inhomogeneous core-medium by the multilayered medium with each layer uniform; the layer thickness is Δr and the difference between the values of the wavenumber k in ajacent layers is Δk. In such a model, the incident ray is refracted and partly reflected. The transmitted ray is thus multi-refracted, while the remainder is multi-reflected. By repeatedly applying the result obtained in Fig.5 at each interface, we obtain the desired hybrid-ray solution. The solution can indeed be evaluated as follows.

We assume the resulting ray to have the longitudinal electric (E_z) and magnetic (H_z) fields related to each other through the hybrid parameter P by

$$H_z = jP\sqrt{\frac{\varepsilon}{\mu}}\, E_z \qquad \text{or} \qquad E_\theta = -jP\sqrt{\frac{\mu}{\varepsilon}}\, H_\theta \tag{1}$$

where μ is the homogeneous permeability of the medium, $\varepsilon(\equiv\varepsilon(r))$ is the inhomogeneous permittivity of the medium, and E_θ and H_θ are the azimuthal electric and magnetic fields, respectively. $P=0$, ∞, -1, and $+1$ represent a TM ray, a TE ray, a left-handed circularly polarized ray, and a right-handed circularly polarized ray, respectively.

We now consider that the TE ray whose longitudinal magnetic field is

$$H_z = e^{-jk_r(r-r_i) + jm\theta - j\beta z}$$

is incident upon the surface $(r=r_i)$ of *ith* layer where m is an integer (= 0,1,2,...), β is the propagation constant, and k_r is the radial wavenumber $\sqrt{k^2-\beta^2}$ $(k\equiv k(r)=\omega\sqrt{\varepsilon\mu}$;ω the angular frequency). From the analysis of the scattering (the transmission in Fig.5(a)), we find that the induced TM-component of the transmitted ray, ΔE_z, has the form

$$\begin{aligned} \Delta E_z &= \Delta T_e \, e^{-jk_r(r-r_i) + jm\theta - j\beta z} \\ &\simeq \Delta T_e \, e^{jm\theta - j\beta z}, \quad (r_i \le r \le r_i + \Delta r) \end{aligned} \tag{2}$$

If the difference Δk between the values of k in *i-1th* and *ith* layers is small, the coefficient ΔT_e is given by

$$\begin{aligned} \Delta T_e &= -\sqrt{\frac{\mu}{\varepsilon}} \, \frac{m\beta}{rk_r} \, \frac{1}{k^2-\beta^2} \, \Delta k \\ &= -\sqrt{\frac{\mu}{\varepsilon}} \, \frac{m\beta / r}{(k^2-\beta^2)^{3/2}} \, \Delta k \\ &= -m\sqrt{\frac{\mu}{\varepsilon}} \, \frac{\partial}{\partial\beta} \left(\frac{k'}{r\sqrt{k^2-\beta^2}}\right) \Delta r \end{aligned} \tag{3}$$

where $k' = dk/dr$. From (1) and (2), we obtain

$$\begin{aligned} \Delta H_z &= jP\sqrt{\frac{\varepsilon}{\mu}} \, \Delta E_z \\ &\simeq jP\sqrt{\frac{\varepsilon}{\mu}} \, \Delta T_e \, e^{jm\theta - j\beta z} \end{aligned} \tag{4}$$

The total transmitted field becomes, at $r=r_{i+1} (\equiv r_i + \Delta r)$,

$$\begin{aligned} H_z + \Delta H_z &= e^{-jk_r\Delta r + jm\theta - j\beta z} + jP\sqrt{\frac{\varepsilon}{\mu}} \, \Delta T_e \, e^{jm\theta - j\beta z} \\ &\simeq e^{-j\left(\sqrt{k^2-\beta^2}\,\Delta r + mP \frac{\partial}{\partial\beta}\left(\frac{k'}{r\sqrt{k^2-\beta^2}}\right)\Delta r\right) + jm\theta - j\beta z} \end{aligned} \tag{5}$$

Similarly, the total transmitted electric field for the TM-ray incidence is

$$E_z + \Delta E_z = e^{-jk_r \Delta r + jm\theta - j\beta z} - jP^{-1}\sqrt{\frac{\mu}{\varepsilon}}\,\Delta T_h\, e^{jm\theta - j\beta z}$$

$$\approx e^{-j(\sqrt{k^2-\beta^2}\,\Delta r + mP^{-1}\frac{\partial}{\partial\beta}(\frac{k'}{r\sqrt{k^2-\beta^2}})\Delta r) + jm\theta - j\beta z} \qquad (6)$$

Here, ΔT_h is the coefficient of the induced magnetic field ΔH_z for the incident electric field E_z with unit magnitude,

$$\Delta T_h = \sqrt{\frac{\varepsilon}{\mu}}\,\frac{m\beta}{rk_r}\,\frac{1}{k^2-\beta^2}\,\Delta k$$

$$= m\sqrt{\frac{\varepsilon}{\mu}}\,\frac{\partial}{\partial\beta}\,(\frac{k'}{r\sqrt{k^2-\beta^2}})\Delta r \qquad (7)$$

Equations (5) and (6) are applied again as new incident fields at $r=r_{i+1}$. Repeat the procedure. At a result, the integrated phases for the TE and TM components are different. The Bohr's quantum-condition [4] must thus be altered as

$$\int_0^{r_1}\sqrt{k^2-\beta^2}\,dr + \frac{m}{2}\left\{\begin{matrix}P\\P^{-1}\end{matrix}\right\}\hat{D} = \pi(n+\frac{\alpha-1}{2}) \qquad (8)$$

for H_z and E_z, respectively. Here, $\hat{D}$ is a function of β,

$$\hat{D} = 2\frac{\partial}{\partial\beta}\int_0^{r_1}\frac{k'}{r\sqrt{k^2-\beta^2}}\,dr \qquad (9)$$

and α is an integer (note that $\hat{D}$ is related to the quantity D in [3] by $\hat{D} = D/k(0)$). From two equations in (8), we obtain

$$\frac{m}{2}(P-P^{-1})\hat{D} = 0 \qquad (10)$$

and we see that two possible values of P are ∓ 1 unless $\hat{D} = 0$. These states are designated as HE and EH and numbered by different indices, n-1 for HE and n for EH [1,2,4]. Since the lowest-order quantum-condition describes HE and EH states as completely degenerate states, we have [4]

$$\alpha = |\pm m - 1| \qquad (11)$$

for HE and EH states, respectively. Strictly speaking, the term in the left-hand side of (10) is not identical with zero but zero of order $1/k^2(0)$. Therefore, $P \approx \mp 1$. For parabolic profile media, as example, $\hat{D}$ does not vanish, so that the hybrid rays are in the states of nearly circular polarizations. However, it is not difficult to show that, for

$$k(r) = \frac{k(0)}{\sqrt{1 + A r^2}} \qquad (12)$$

with A constant, the polarizations are approximately linear because $\hat{D}$ vanishes exactly [3]. We note that (12) is close to the parabolic profile mentioned above.

If k is given by

$$k(r) \sim \frac{k(0)}{\sqrt{1 + A r^2}} \qquad (13)$$

$\hat{D}$ can vanish at certain frequencies ($\omega = \omega_i$) and the states of polarization may change from +1 to -1 or from -1 to +1 (see Fig.6). This change (between $HE_{m\ n+1}$ and EH_{mn} modal rays) greatly depends on the profile of refractive index, and it may be rapid or otherwise.

CONCLUSION

The eigenstates of polarized modal rays in graded-index fibers have been discussed in terms of the generalized ray optics. The frequency-variant property of polarization which may be observed in the fibers with certain special classes of refractive index has been well described. We have <u>not</u> begun with the vectorial wave equations to interpret the polarization effects. There is the possibility of further improvement to the model in which the higher-order effects of polarization as well as of propagation can be described.

REFERENCES

[1] J.A.Arnaud, Beam and Fiber Optics (Academic, New York, 1976).
[2] M.Hashimoto,"Hybrid modes of graded-index fiber", Electron.Lett.,17, 659-661, 1981; Erratum, ibid.,17, 987, 1981.
[3] M.Hashimoto,"Circularly polarized modal skew rays in graded-index optical fibers", J.Opt.Soc.Am.,72, 1147-1151, 1982.
[4] M.Hashimoto,"Asymptotic vector modes of inhomogeneous circular waveguides", Radio Sci.,17, 3-9, 1982.

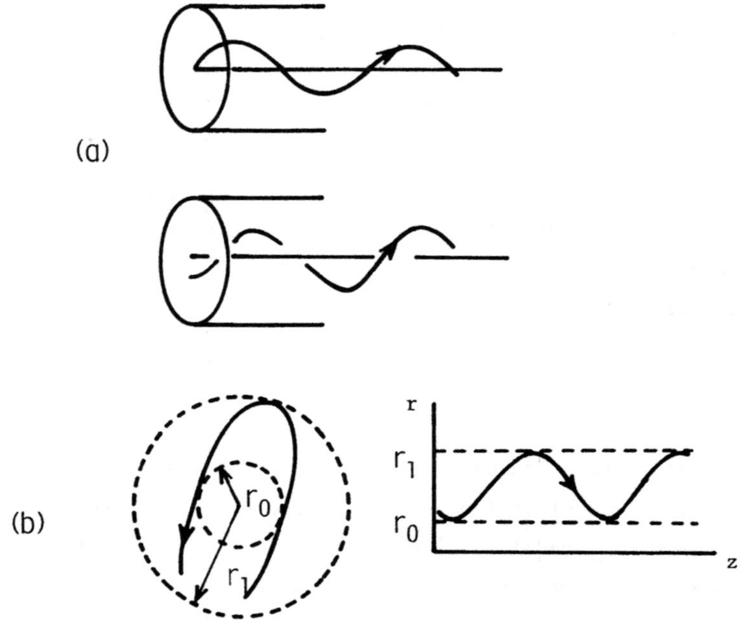

Fig.1. (a) Rays in fiber (b) Projection of ray paths.

$E_\theta \sim -\frac{\omega\mu}{\beta} H_r$ $E_r \sim \frac{\beta}{\omega\varepsilon} H_\theta$

Fig.2. TE and TM rays.

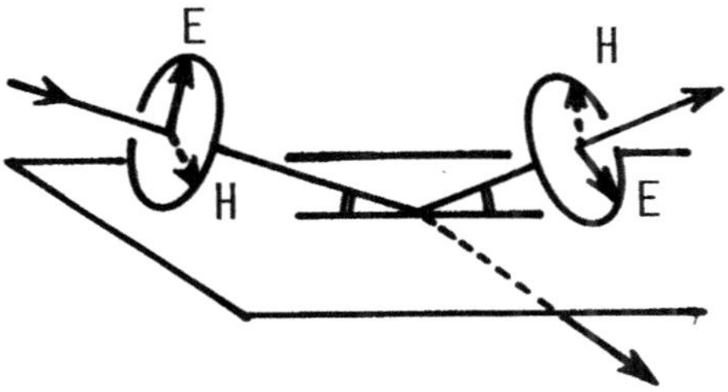

Fig.3. Reflection of linearly polarized plane-wave at dielectric surface.

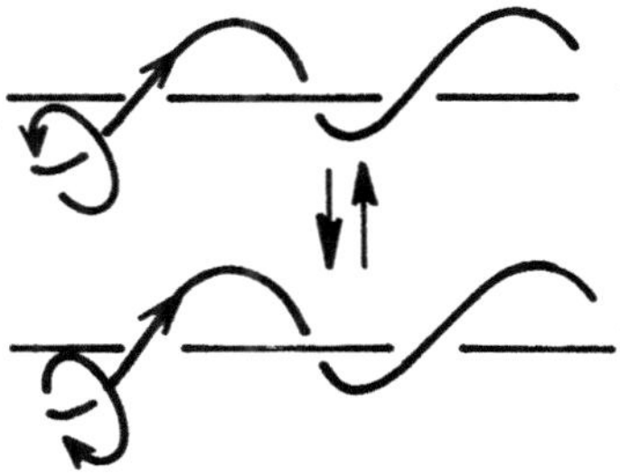

Fig.6. Change of polarization

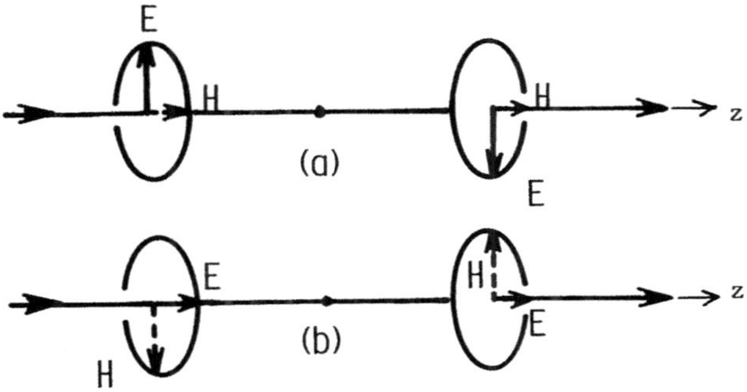

Fig.4. Top views of incident and reflected rays (the z-axis is parallel to the plane of incidence).

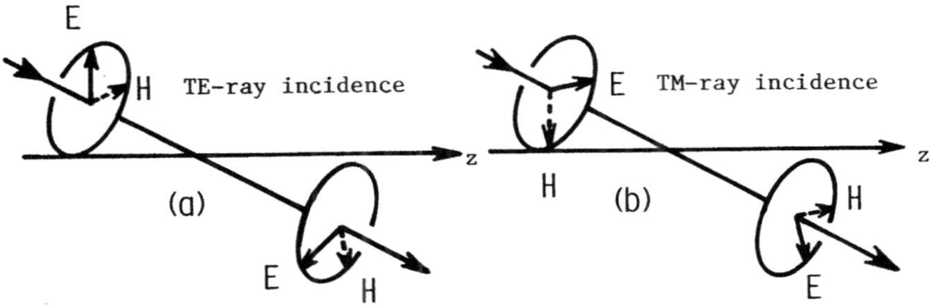

Fig.5. Top views of incident and reflected rays (the z-axis is not parallel to the plane of incidence).

MATRIX METHOD FOR DETERMINING PROPAGATION CHARACTERISTICS OF OPTICAL WAVEGUIDES

Enakshi K. Sharma, A.K. Ghatak and I.C. Goyal
Department of Physics, Indian Institute of Technology, New Delhi - 110016, India.

ABSTRACT: We have transformed the scalar wave equation into a matrix eigenvalue equation, the diagonalisation of which gives the guided modes and a discrete representation of the radiation modes. The method is simple and can be used for arbitrary refractive index profiles.

1. INTRODUCTION

In this paper we report a novel numerical procedure for solving the scalar wave equation to obtain the propagation characteristics of optical waveguides with an arbitrary refractive index profile. We have transformed the problem of solving the scalar differential equation (with appropriate boundary conditions) to a simple matrix eigenvalue problem. The diagonalisation of this matrix gives directly N eigenvalues, where N represents the order of the matrix. All these eigenvalues may correspond to the guided modes or a part may correspond to the guided modes and remaining would give a discrete representation of the radiation modes. In contrast, while obtaining the direct numerical solution of the wave equation, one usually carries out calculations for each mode separately[1-3]. Further, unlike the matrix procedure reported earlier[4] which requires a predetermined appropriate basis set of orthonormal functions and the matrix elements are given by integrals over these functions, the present procedure does not require a predetermined basis set of orthonormal functions to begin with and the matrix elements are given by simple algebraic expressions. We will present here the procedure as applied to planar waveguides; however, since the eigenvalue equation corresponding to round optical fibers can be transformed to a one dimensional equation, the present method can be extended to round optical waveguides.

2. Method

For a planar waveguide the following scalar wave equation need be solved (see, e.g., Ref. 5)

$$\frac{d^2\psi}{d\xi^2} + [\, k_o^2 a^2 n^2(\xi) - \beta^2 a^2 \,]\, \psi(\xi) = 0 \tag{1}$$

where $\xi = x/a$; a being a convenient linear dimension associated with the refractive index profile $n(\xi)$, k_o is the free space wave number, β the propagation constant and for TE modes, ψ represents the y component of the electric field. We define a new variable $y(\xi) = d^2\psi/d\xi^2$, and carry out two integrations to obtain

$$\psi = \int_o^{\xi} (\xi - \eta)\, y(\eta)\, d\eta + C_1\xi + C_2 \tag{2}$$

where C_1 and C_2 are constants to be determined from the boundary conditions which are given by

$$\psi(\xi = \xi_1) = 0 \quad \text{and} \quad \psi(\xi = -\xi_2) = 0 \tag{3}$$

where ξ_1 and ξ_2 are sufficiently large numbers where the model fields can be assumed to vanish.

We first consider a symmetric profile in which $n^2(-\xi) = n^2(\xi)$; for such a case, the modes are either symmetric or antisymmetric in ξ and instead of using the second boundary condition of Eq. (3) we may use $\psi'(0) = 0$ and $\psi(0) = 0$ for symmetric and antisymmetric modes respectively. Using these boundary conditions we obtain

$$\psi(\xi) = \int_0^{\xi} (\xi - \xi_1)\, y(\eta)\, d\eta + \int_{\xi}^{\xi_1} (\xi_1 - \eta)\, y(\eta)\, d\eta \quad \text{for symmetric modes} \tag{4}$$

$$= \left(\frac{\xi}{\xi_1} - 1\right) \int_0^{\xi} \eta y(\eta)\, d\eta + \xi \int_{\xi}^{\xi_1} \left(\frac{\eta}{\xi_1} - 1\right) y(\eta)\, d\eta \quad \text{for antisymmetric modes} \tag{5}$$

We next replace the integrals in the above equations by a sum at discrete points chosen so as to divide the domain $\xi = 0$ to $\xi = \xi_1$ into (N+1) equal intervals. Doing so, we can write $[\psi] = [T]\,[y]$ where $[\psi]$ and $[y]$ are column vectors of dimension N with elements $\psi_i = \psi(\xi_i)$ and $y_i = y(\xi_i)$ where $\xi_i = \frac{i\xi_1}{N+1}$ $(i = 1, 2, ---, N)$ and [T] is a symmetric $(N \times N)$ matrix defined as

$$T_{ij} = T_{ji} = \frac{\xi_1^{\,2}}{N+1} \left(\frac{i}{N+1} - 1\right) ; \quad j \le i \qquad \text{for symmetric modes} \tag{6}$$

$$= \frac{\xi_1^{\,2}}{(N+1)^2} \left(\frac{i}{N+1} - 1\right) j ; \quad j \le i \qquad \text{for antisymmetric modes} \tag{7}$$

Hence, Eq. (1) can be written at the N points $\xi_i (i = 1, 2, ---, N)$ in the form

$$[L][\psi] \equiv \left(\frac{[T]^{-1}}{k_o^{\,2} a^2} + [F]\right)[\psi] = \frac{\beta^2}{k_o^{\,2}}\,[\psi] \tag{8}$$

where [F] is a $N \times N$ diagonal matrix with elements $F_{ij} = \delta_{ij}\, n^2(\xi_i)$. Hence, the diagonalisation of the symmetric matrix [L] gives the eigenvalues. Further, our calculations show that the matrix [L] is a symmetric tridiagonal matrix.

We should mention here that the above analysis can be used even for an arbitrary asymmetric profile by shifting the origin ($\xi = 0$) to $\xi = \xi_2$. Thus, we define a new coordinate $\rho = \xi_1 + \xi_2$ and the boundary conditions [Eq. 3] can be written as $\psi(\rho = 0) = 0$ and $\psi(\rho = \xi_1 + \xi_2) = 0$. Because of these boundary conditions, the procedure for solving Eq. (1) for any asymmetric profile would be the same as that for solving for the antisymmetric modes of a symmetric profile.

3. Results and Discussion

We first apply the method to a symmetric profile with $n^2(x) = n_s^2 + \Delta n\, \mathrm{sech}^2\left(\frac{x}{a}\right)$ for which exact analytical solutions of Eq. (1) can be obtained and the propagation constants are given by simple algebraic expressions[7]. The values of the normalised parameter $b = (\frac{\beta^2}{k_0^2} - n_s^2)/\Delta n$ for bound modes as obtained by our calculations are tabulated in Table I for $k_0 a(\Delta n)^{1/2} = 4.0$ and 7.8 along with the exact results; m = 0, 2, 4, .. and m = 1, 3, 5 --- correspond to the symmetric and antisymmetric modes respectively. We may note that for modes far from cut off the propagation constants for bound modes are accurately determined by assuming $\xi_1 = 3.0$. However, in order to get an accurate description of modes near cut-off, it is necessary to take a large value of ξ_1 ; this is also that one expects physically because in such a case, the fields are appreciable even at large distances. It should be mentioned that to obtain good accuracy for large values of ξ_1, the order of the matrix (N) has to be large. It should be pointed out that for $k_0 a(\Delta n)^{1/2} = 4.0$, the exact calculations tell us that there are only four bound modes, two symmetric and two antisymmetric. However, the diagonalisation of a 40 × 40 matrix (corresponding to symmetric modes) gives 40 eigenvalues of which two would correspond to bound modes and the remaining 38 would give a discrete representation of the radiation modes; these are, however, not the true radiation modes because the use of boundary conditions of Eq. (3) implies that we are actually solving for a profile with metallic boundaries at $\xi = \xi_1$ and $\xi = \xi_2$. Nevertheless, these modes can be used to study the propagation of energy in the radiation modes by choosing sufficiently large values of ξ_1 and ξ_2.

We next consider an asymmetric profile given by $n^2(x) = n_s^2 + \Delta n \,\mathrm{sech}^2 \frac{x-d}{a}$ for $x > 0$ and $n^2(x) = 1.0$ for $x < 0$. Refractive index profiles for planar waveguides fabricated by the silver ion diffusion method can be written in the above form [see, e.g. Ref. 8]. The results of calculation of β/k_0 values for bound modes are tabulated in Table II for the following parameters:

$$n_s = 1.512, \quad \Delta n = 0.25884, \quad d = 2.6\ \mu\mathrm{m},$$

$$d/a = 1.03 \quad \text{and} \quad \lambda = 0.633\ \mu\mathrm{m} \tag{9}$$

We had earlier reported the exact results for the above profile and parameters[8]; these are also given for comparison in Table II.

In summary, we have presented a novel numerical procedure for solving the scalar wave equation to obtain the propagation characteristics of optical waveguides with an arbitrary refractive index profile. The problem of solving the scalar differential equation has been transformed into a simple matrix eigenvalue problem and we have shown that the values obtained by diagonalising the matrix are extremely accurate.

This work was partially supported by Department of Science and Technology (India).

Table I : Normalised Propagation constant **b** for the profile
$n^2(x) = n_s^2 + \Delta n \, \mathrm{sech}^2 \left(\frac{x}{a}\right)$

$k_o a\,(\Delta n)^{1/2}$	M	EXACT	$\xi_1 = 3.0$ N = 20	$\xi_1 = 3.0$ N = 40	N = 60	$\xi_1 = 5.0$ N = 40	$\xi_1 = 5.0$ N = 60
4.0	0	0.7793	0.7802	0.7795	0.7794	0.7799	0.7796
	1	0.4004	0.4030	0.4011	0.4007	0.4023	0.4013
	2	0.1465	0.1495	0.1454	0.1456	0.1492	0.1477
	3	0.0176		0.0135	0.0148	0.0168	0.0160
7.0	0	0.8670	0.8681	0.8672	0.8671	0.8677	0.8573
	1	0.6213	0.6254	0.6224	0.6218	0.6243	0.6226
	2	0.4165	0.4255	0.4187	0.4175	0.4226	0.4192
	3	0.2528	0.2634	0.2553	0.2537	0.2604	0.2560
	4	0.1294	0.1418	0.1321	0.1305	0.1379	0.1330
	5	0.0470	0.0532	0.0457	0.0443	0.0532	0.0496
	6	0.0055				0.0072	0.0063

Table II : β/k_o values for a planar waveguide with an index profile $n^2(x) = n_s^2 + \Delta n \; \mathrm{sech}^2(x-d)/a$ for $x > 0$ and $n^2(x) = 1.0$ for $x < 0$ the parameters used are given in Eq. (9).

EXACT	$\xi_1 = 3.0$ $\xi_2 = 0.11$ N = 20	N = 30	N = 40	N = 60	$\xi_1 = 4.0$ $\xi_2 = 0.21$ N = 30	N = 40	N = 60
1.5892	1.5893	1.5892	1.5892	1.5892	1.5892	1.5892	1.5892
1.5773	1.5777	1.5775	1.5774	1.5774	1.5777	1.5775	1.5774
1.5664	1.5674	1.5668	1.5666	1.5664	1.5672	1.5668	1.5665
1.5561	1.5580	1.5569	1.5566	1.5563	1.5577	1.5569	1.5564
1.5464	1.5494	1.5475	1.5471	1.5465	1.5488	1.5475	1.5467
1.5369	1.5414	1.5385	1.5330	1.5371	1.5405	1.5385	1.5374
1.5282	1.5340	1.5300	1.5295	1.5283	1.5328	1.5301	1.5288
1.5207	1.5271	1.5221	1.5213	1.5198	1.5257	1.5228	1.5213

REFERENCES

1. Sammut, R.A. and Pask, C. : Simplified numerical analysis of optical fibers and planar waveguides, **Electron. Lett.**, 1981, **17**, pp. 105-106.

2. Sharma, E.K., Sharma, A. and Goyal, I.C. : Propagation Characteristics of single mode optical fibers with arbitrary index profiles : A simple numerical approach, **IEEE J. Quant. Electron.**, 1982, (In Press).

3. Mammel, W.L. and Cohen, L.G. : Numerical Prediction of fiber transmission characteristics from arbitrary refractive index profiles, **Appl. Opt.**, 1982, **21**, pp. 699-703.

4. Meunier, J.P., Pigeon, J. and Massct, J.N. : A general approach to the numerical determination of model propagation constants and field distributions of optical fibers, **Opt. Quant. Electron.**, 1981, **13**, pp. 71-83.

5. Sodha, M.S. and Ghatak, A.K. : **Inhomogeneous Optical Waveguides**, Plenum Publishing Corporation, N.Y., 1977.

6. Sharma, E.K., Goyal, I.C. and Ghatak, A.K. : Calculation of Cut-off frequencies in Optical Fibers for arbitrary profiles using the matrix method, **IEEE J. Quant. Electron.**, 1981, **QE-17**, pp. 2317-2320.

7. Kumar, A. and Sharma, E.K. : A perturbation analysis for modes in diffused waveguides with a gaussian profile, **Opt. Commun.**, 1978, **27**, pp. 349-352.

8. Khular, E., Kumar, A., Sharma, A., Goyal, I.C. and Ghatak, A.K.: Modes in buried optical waveguides with graded index profiles, **Opt. Quant, Electron.**, 1981, **13**, pp. 109-116.

VARIABLE SECTION OPTICAL WAVEGUIDES

A.M. SCHEGGI, R. FALCIAI, M. BRENCI
Istituto di Ricerca sulle Onde Elettromagnetiche, C.N.R., Florence, Italy

1. INTRODUCTION

The study of propagation in longitudinally dependent dielectric waveguides and in particular in tapered waveguides is of interest in a variety of applications that include natural as well as man-made guiding environments. Examples of the former are tropospheric and ionospheric ducts, underwater acoustic ducts in the ocean, seismic channels in the earth's crust, etc. Examples of the latter are tapered dielectric antennas and optical fiber or slab waveguides. In particular these may be of interest for more efficient fiber to source coupling or for optical components as well as for high power radiation delivery systems (mechanical or medical applications) (1,2,3,4,5,6).

The present paper is to be concerned with a study of the radiation characteristics in the above-described waveguides.

In order to evaluate the near and far-field intensity patterns, one can solve the ray equation

$$\frac{d}{ds}\left(n\frac{d\bar{r}}{ds}\right) = \nabla n \qquad (1)$$

trace the rays, and then evaluate the flux change associated with geometrical spreading or concentration of the bundles.

We considered a multimode, tapered optical waveguide with homogeneous core and unlimited cladding. Under the hypothesis of incoherent illumination and keeping into account the cylindrical simmetry of the waveguide we denoted by $I_{in}(\rho,\theta) = I_{in}(\rho)I_{in}(\theta)$ the intensity at the input face. A Gaussian angular ray density and a uniform spatial density distribution were assumed, proportional to $I_{in}(\rho)$ and $I_{in}(\theta)$ respectively.

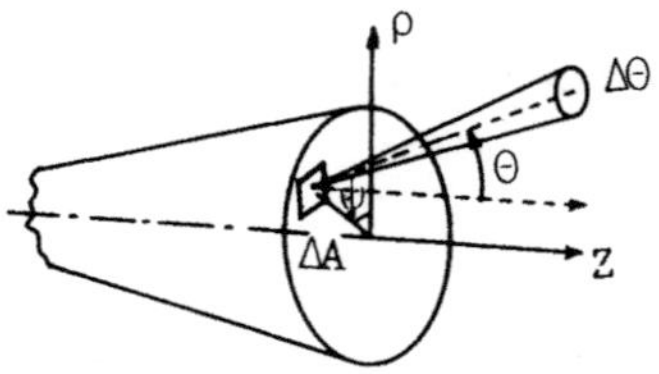

FIGURE 1. Coordinate system for the near and far field distributions.

The near and far field intensity patterns were obtained by evaluating at the exit face the ray density as a function of ρ or θ (Fig. 1)

$$I_{near}(\rho) = N(\rho)/\Delta A \qquad \Delta A = 2\pi\rho\Delta\rho$$
$$I_{far}(\theta) = M(\theta)/\Delta\Theta \qquad \Delta\Theta = 2\pi\theta\Delta\theta \qquad (2)$$

2. TAPERED SLAB WAVEGUIDE

As a first application we considered a slab tapered waveguide (Fig. 2) with a nearly step index distribution.

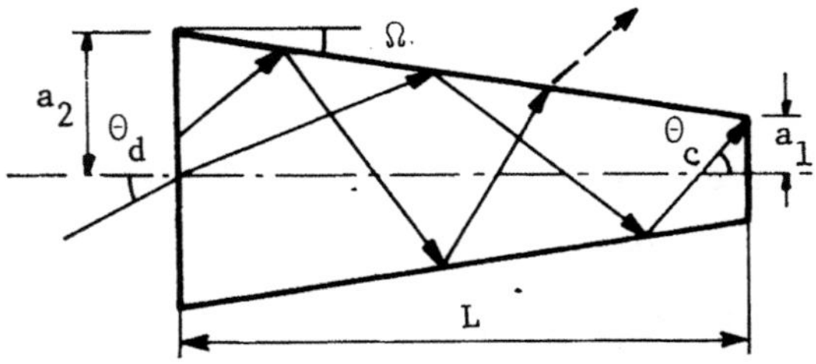

FIGURE 2. Ray paths in a tapered slab waveguide.

Figure 3 shows the near and far-field intensity patterns in the cases $a_1/a_2 = 0.75$, $L = 1$ mm ($\Omega = 17'$), and $a_1/a_2 = 0.5$, $L = 5$ mm ($\Omega = 28'$) compared with those from a 5 mm long uniform waveguide. The near-field patterns remain practically uniform, whereas the far-field patterns present half-intensity widths that increase inversely with the taper ratio.

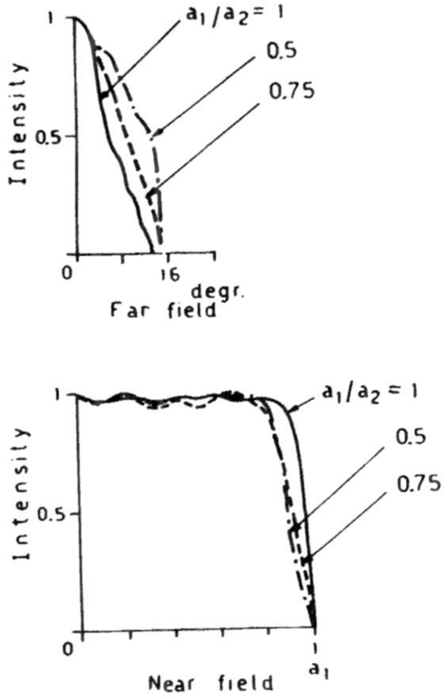

FIGURE 3. Near and far field intensity patterns from two slab tapered waveguides ($a_1/a_2 = 0.5, 0.75$) compared with those from a uniform waveguide ($a_1/a_2 = 1$).

When the taper angle increases, the number of rays that undergo multiple reflections and are accepted at the taper end decreases, whereas the number of rays that do not impinge on the taper walls increases. The far-field patterns are modified by the presence of side lobes, which are due to the contribution of those rays that undergo only one reflection and can emerge at very wide angles (larger than the critical angle θ_c). This effect is displayed in Fig. 4, in which some far-field intensity patterns are shown corresponding to increasing values of Ω compared with the input angular density distribution. The pattern corresponding to $\Omega = 1^{\circ}26'$ ($L = 1$ mm) is also shown for comparison with that corresponding to the same taper ratio (a_1/a_2= 0.5, L = 5 mm) but smaller taper angle Ω = 17'.

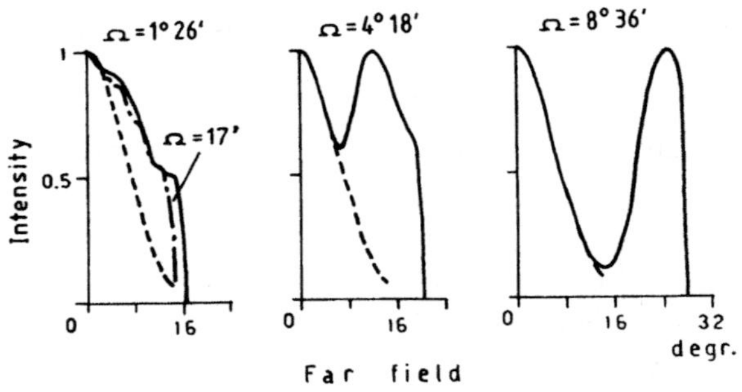

FIGURE 4. Far field intensity patterns from three tapered waveguides having increasing taper angles (Ω) compared with the pattern from the uniform waveguide (dashed line). The pattern for Ω = $1^{0}26'$ is also compared with that for Ω = 17' (dotted-dashed line) corresponding to a longer taper with same taper ratio.

The examples examined indicate that the near-field configurations remain practically unchanged with respect to the input spatial intensity distribution for each combination of the parameters a_1/a_2 and Ω. Until the taper angle becomes very small, the only parameter playing a role in the modification of the far-field intensity patterns is the taper ratio a_1/a_2. When Ω increases, the patterns become more and more modified and depend on both parameters a_1/a_2 and Ω.

3. ENLARGED TAPER ENDED FIBERS

Tapered conical ends in optical fibers apart from finding use as optical components can be of interest for either mechanical or surgical applications. In fact they can provide a more efficient coupling with large-spot high-power lasers and when used at the exit of the fiber present good collimation properties which are particularly useful for endoscopic applications.

We fabricated quasi-linear conical ends in silica-plastic fibers by introducing into the fiber drawing apparatus (5) an electronic control for varying the drawing speed with a prefixed law. In comparison with a previous study (7) these tapered fibers are much less fragile because of their plastic cladding, applied on line, during the drawing process, by a flexible plastic applicator with a suitable central hole accomodating fiber thickness variations.

Fibers were drawn with core diameters ranging between 50 and 500 μm and enlarged ends having taper length L varying from 2 to 100 cm, while the taper ratio a_1/a_2 (a_1 and a_2 are the radii of the smaller and larger end faces, respectively) can be decreased to 0.2. These fibers were characterized by using as a source a HeNe laser and measuring radiation patterns and losses. For completeness we evaluated the radiation characteristics of such fibers by applying the above described method to the threedimensional case and assuming also here a nearly step index profile distributions.

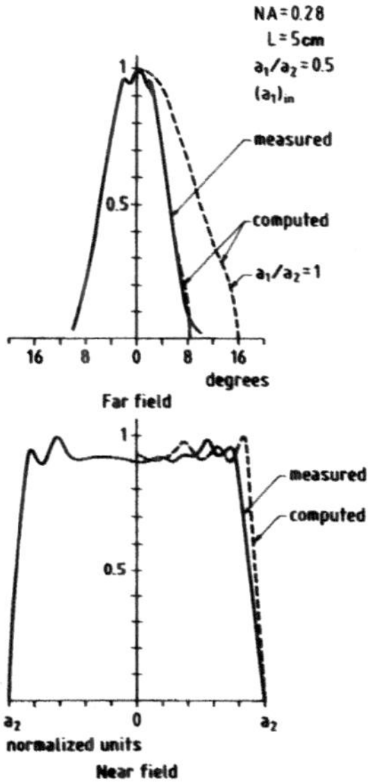

FIGURE 5. Comparison between measured and computed near and far field intensity patterns for an enlarged taper ended fiber. The computed far field intensity pattern for the uniform fiber is also shown.

Fig. 5 shows a comparison between the measured and computed radiation patterns from an enlarged taper fiber with $a_1/a_2 = 0.5$ L = 5 cm under maximum illumination NA. The beam turns out to be collimated by a factor a_1/a_2 with respect to that from the uniform fiber while the near field distributions remain substantially unchanged. When the fiber is illuminated at the enlarged end (again under maximum NA) the radiation pattern (Fig. 6) results broadened; in fact as the rays proceed on their zig-zag paths their propagation angles increase from reflection to reflection and eventually surpass the cut off angle, thus giving rise to leaky-tunneling rays. This is confirmed by the typical increasing behavior of the near field distribution at the fiber edge.

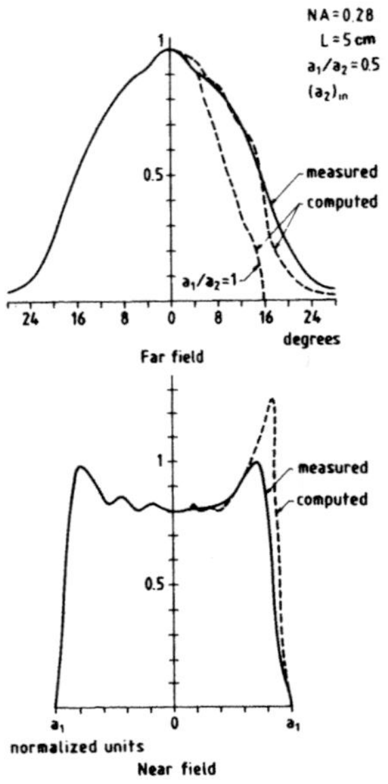

FIGURE 6. Comparison between measured and computed intensity patterns for a decreasing taper ended fiber. The computed far field intensity pattern for the uniform fiber is also shown.

The agreement between experimental and theoretical curves is quite satisfactory even if the contribution of the leaky rays in the measured patterns appears lower because the measurement was performed after a fiber length (although short) along which the leaky rays attenuate. As for the losses introduced by the taper the performed measurements indicated that by illuminating the fiber at the smaller end such losses result negligible. When the taper is illuminated at the larger end the losses increase for decreasing values of a_1/a_2 but decrease when the input NA decreases or when the taper is sufficiently long and smooth.

REFERENCES

1. Ozeki T., Ito T., Tamura T., 1975. Tapered Section of Multimode Cladded Fibers as Mode Filters and Mode Analyzers. Applied Physics Letters. Vol. 26, p.386.
2. Ozeki T., Kawasaki B.S., 1976. Optical Directional Coupler Using Tapered Sections in Multimode Fibers. Applied Physics Letters. Vol. 28, p.528.
3. Chang L.T., Auth D.C., 1978. Radiation Characteristics of a Tapered Cylindrical Optical Fiber. J. Opt. Soc. Am. Vol. 68, p.1191.
4. Kuwahara H., Furuta H., 1979. Efficient Light Coupling from Semi-Conductor Lasers into Tapered Hemispherical-End Fibers. Proc. IEEE. Vol. 67, p.1456.
5. Brenci M., Falciai R., Scheggi A.M., 1982. Tapered Enlarged Ends in Multimode Optical Fibers. Applied Optics. Vol. 21, p.317.
6. Scheggi A.M., Falciai R., Brenci M., in press. Radiation Characteristics of Tapered Slab Waveguides. Journ. Opt. Soc. Am.
7. Nath G., 1974. Device for Material Working by a Laser Beam and Method for its Production. U.S. Patent 3, 843, 865.

GUIDED MODES OF THE GRADED-INDEX OPTICAL FIBER WITH A PARABOLIC-INDEX CORE

H. IKUNO

1. INTRODUCTION

Most aspects of weakly-guiding optical fibers can be analyzed by the scalar wave approximation [1]. The difference between the scalar wave analysis and the vectorial wave analysis are significant for the mode dispersions at the near cutoff frequencies [2], [3],[4].

The purpose of this paper is to evaluate the guided modes of the square-law optical fiber with the homogeneous cladding in the sense of the vectorial wave analysis. The degeneracy of the LP-modes in the scalar wave approximation can be removed by the first-order vector corrections. The numerical results show that the vector solutions for the cutoff frequencies for the LP_{11}-mode are in good agreement with those gotten by the precise numerical method [4] and the effect of the vector correction on the mode dispersions is significant for the lower-order modes.

2. FORMULATION OF PROBLEM

We consider the optical fiber with the refractive-index distribution described by

$$n(r) = \begin{cases} n_1(\ 1 - h(gr))^{1/2}, & r < a \\ n_1(\ 1 - 2\Delta\ )^{1/2}, & r \geq a \end{cases} \tag{1}$$

where n_1, g, a, and Δ are positive constants and h is a monotonic function of r which satisfies h(0)=0. Here we use the cylindrical coordinate system (r,θ,z). The guided modes propagate along z-axis according to $\exp[j(\omega t+n\theta-\beta z)](n=0,1,2,...)$. First, we describe the hybrid modes ($n\neq0$) in the core region. By referring to the reference [5], the electromagnetic fields for the hybrid modes can be represented by solving the coupled differential equation (the vector wave equation)

$$\Phi'' + (1/r)\Phi' + (k^2(b-h)-(n-1)^2/r^2)\Phi + (f_a+f_b+f_c)\Phi + (f_a+f_b-f_c)\hat{\Phi} = 0$$
$$\hat{\Phi}'' + (1/r)\hat{\Phi}' + (k^2(b-h)-(n+1)^2/r^2)\hat{\Phi} + (f_a-f_b-f_c)\hat{\Phi} + (f_a-f_b+f_c)\Phi = 0 \qquad (2)$$

with

$$f_a = [(1/r)h'/(1-h)-h''/(1-h)-(3/2)h'^2/(1-h)^2]/4$$
$$f_b = (n/r^2)[(1-h)^{-1/2}-1] - (n/2r)h'(1-h)^{-3/2} \qquad (3)$$
$$f_c = (n/r^2)[(1-h)^{1/2}-1] \quad .$$

In Eq.(2), the primes denote the derivative with respect to r and the propagation constant β is expressed as

$$\beta = k(1-b)^{1/2}, \; k = k_0 n_1, \; k_0 = \omega(\varepsilon_0\mu_0)^{1/2} \qquad (4)$$

Introducing the transverse functions ϕ_1 and ϕ_2 defined by

$$\phi_1 = \Phi + \hat{\Phi} \; , \; \phi_2 = \hat{\Phi} - \Phi \quad , \qquad (5)$$

then we obtain the electromagnetic fields for the hybrid modes in the form

$$E_r = j(1-h)^{-1/2}C_1\phi_1$$
$$E_\theta = C_2\phi_2$$
$$H_r = -(1/\omega\mu_0\beta)[(n/r)(1-h)^{-1/2}U_1C_1\phi_1+(\beta^2+(n/r)^2)C_2\phi_2]$$
$$H_\theta = j(1/\omega\mu_0\beta)[(k^2(1-h)-(n/r)^2)(1-h)^{-1/2}C_1\phi_1-(n/r)U_2C_2\phi_2] \qquad (6)$$
$$E_z = (1/\beta)[(1-h)^{-1/2}U_1C_1\phi_1 + (n/r)C_2\phi_2]$$
$$H_z = j(1/\omega\mu_0)[(n/r)(1-h)^{-1/2}C_1\phi_1 + U_2C_2\phi_2]$$

with

$$U_1 = \phi_1'/\phi_1 + 1/r - (1/2)h'/(1-h)$$
$$U_2 = \phi_2'/\phi_2 + 1/r \qquad (7)$$

where C_1 and C_2 are unknown constants which are determined throug the boundary condition at the core-cladding interface.

In the case of n=0, we have electromagnetic fields for the TE- and TM-modes. For the TE-modes, we get

$$E_\theta = C\phi \; , \; H_r = -(\beta/\omega\mu_0)C\phi \; , \; H_z = j(1/\omega\varepsilon_0)C[\phi'+(1/r)\phi] \qquad (8)$$

where C is a normalization constant and ϕ satisfies

$$\phi'' + (1/r)\phi' + [k^2(b-h)-1/r^2]\phi = 0. \qquad (9)$$

For the TM-modes, we have

$$H_\theta = -j(\beta/\omega\mu_0)(1-h)^{1/2}D\psi/(1-b), \; E_r = -j((1-h)^{-1/2}D\psi$$
$$E_z = -(1/\beta)(1-h)^{-1/2}D[\psi' + (1/r - (1/2)h'/(1-h))\psi] \qquad (10)$$

where D is a normalization constant and ψ fulfills

$$\psi'' + (1/r)\psi' + (k^2(b-h)-1/r^2)\psi + 2f_a\psi = 0 \quad (11)$$

In the cladding, the electromagnetic fields for the guided modes can be represented by the modified Bessel function $K_n(\gamma r)$ and its derivative where γ is defined as $\gamma = k(2\Delta-b)^{1/2}$[1]. Matching the boundary condition at the core-cladding interface r=a, we have the characteristic equations for determining the guided modes of the graded-index optical fibers. The results are as follows: For the hybrid modes, we have

$$(1-2\Delta)n^2[(X_n{}^2-1)/(\gamma a)^2](U_1U_2-(n/a)^2) - $$
$$- (n/a)[(1-2\Delta)(X_nU_1-n/a) + (1-h)(X_nU_2-n/a)] + (1-h)\gamma^2 = 0 \quad (12)$$

with

$$X_n = -1 - (\gamma a/n)K_{n-1}(\gamma a)/K_n(\gamma a) \quad .$$

It follows from the boundary condition at r=a together with (12) that we have

$$C_2/C_1 = (n/a)(U_1 - (n/a)X_n)(1-h)^{-1/2}\phi_1$$
$$/[(n/a)X_nU_2 - (n/a)^2 - \gamma^2]\phi_2.$$

This relation specifies field factors as the excitation coefficients of a particular mode. For the TE- and TM-modes, we get

$$\phi'/\phi + 1/a + \gamma K_0(\gamma a)/K_1(\gamma a) = 0 \quad (13)$$

and

$$\psi'/\psi + 1/a - (1/2)h'/(1-h) + \gamma K_0(\gamma a)/K_1(\gamma a) = 0, \quad (14)$$

respectively.

It should be noted that Eqs.(9) and (13) indicate that the scalar wave analysis gives the exact solution for the TE-modes.

3. FIRST-ORDER SOLUTIONS

Now the problem is reduced to solve Eqs.(2), (9), and (11) in the core region. These equations can be solved by the perturbation method, provided that the corresponding scalar wave equations

$$U''_{n,m} + (1/r)U'_{n,m} + (1/\varepsilon^2)[E_{n,m}-h-\varepsilon^2(n-1)^2/r^2]U_{n,m} = 0 \quad (15)$$

and

$$\hat{U}''_{n,m} + (1/r)\hat{U}'_{n,m} + (1/\varepsilon^2)[\hat{E}_{n,m}-h-\varepsilon^2(n+1)^2/r^2]\hat{U}_{n,m} = 0 \quad (16)$$

to Eqs.(2),(9), and (11) can be solved exactly where we make the change of variable u=gr and ε=g/k. Such case can be found in the square-law profile as $h(u)= u^2$[1],[6].

Here we treat the square-law optical fiber and give the first-order corrections to the scalar wave analysis. The eigenfunctions and the eigenvalues of (15) and (16) are given as follows:

$$U_{n,m} = (2\Gamma(m+1)/\Gamma(m+n))^{1/2}(u^2/\varepsilon)^{(n-1)/2}\exp[-u^2/2\varepsilon]L_m^{(n-1)}(u^2/\varepsilon)$$
$$E_{n,m} = (4m+2n)\varepsilon \ , \ m=0,1,2,\ldots; \ n=1,2,3,\ldots \qquad (17)$$

and

$$\hat{U}_{n,m}=(2\Gamma(m+1)/\Gamma(m+n+2))^{1/2}(u^2/\varepsilon)^{(n+1)/2}\exp[-u^2/2\varepsilon]L_m^{(n+1)}(u^2/\varepsilon)$$
$$\hat{E}_{n,m} = (4m+2n+4)\varepsilon \ , \ m=0,1,2,\ldots; \ n=1,2,3,\ldots \qquad (18)$$

and

$$U_{0,m} = (2/(m+1))^{1/2}(u^2/\varepsilon)^{1/2}\exp[-u^2/2\varepsilon]L_m^{(1)}(u^2/\varepsilon)$$
$$E_{0,m} = (4m+4)\varepsilon \ , \ m=0,1,2,\ldots \qquad (19)$$

where $\Gamma(\cdot)$ and $L_m^{(n)}(\cdot)$ denote the Gamma function and the Laguerre polynomials[7].

Now, we can calculate the first-order perturbational solutions of the vector wave equation [2] with the aid of the conventional perturbation method [8]. In the calculation process, it is noted that the relations $E_{n,m}=\hat{E}_{n,m-1}$ hold except for m=0. The results are as follows:

(I) $HE_{n,m+1}$-modes

$$\Phi = U_{n,m} + ((5n+6)/16)((m+1)(m+n))^{1/2}\varepsilon^2 U_{n,m+1}$$
$$\hat{\Phi} = ((2n-3)/8)((m+n)(m+n+1))^{1/2}\varepsilon^2\hat{U}_{n,m} \qquad , \ m=0 \qquad (20)$$
$$b = (4m+2n)\varepsilon + n\varepsilon^2 + ((5n+6)/4)(2m+n)\varepsilon^3$$

and

$$\Phi = U_{n,m}$$
$$\hat{\Phi} = ((2n-3)/2n)(m(m+n))^{1/2}\varepsilon\hat{U}_{n,m-1} \quad , \ m=1,2,3,\ldots \qquad (21)$$
$$b = (4m+2n)\varepsilon + n\varepsilon^2$$

(II) $EH_{n,m+1}$-modes

$$\hat{\Phi} = \hat{U}_{n,m}$$
$$\Phi = ((2n+3)/2n)((m+1)(m+n+1))^{1/2}\varepsilon U_{n,m+1}, \ m=0,1,2,\ldots \qquad (22)$$
$$b = (4m+2n+4)\varepsilon + n\varepsilon^2$$

(III) $TE_{0,m+1}$-modes

$$\phi = U_{0,m}$$
$$b = (4m+4)\varepsilon \ , \quad m=0,1,2,\ldots \qquad (23)$$

(IV) $TM_{0,m+1}$-modes

$$\psi = U_{0,m} + (3/4)((m+1)(m+2))^{1/2}\varepsilon^2 U_{0,m+1}, \quad m=0$$
$$b = (4m+4)\varepsilon + 3(2m+2)\varepsilon^3 \tag{24}$$

and

$$\psi = U_{0,m} + (3/4)(m+1)^{1/2}\varepsilon^2[(m+2)^{1/2}U_{0,m+1} - m^{1/2}U_{0,m-1}]$$
$$b = (4m+4)\varepsilon + 3(2m+2)\varepsilon^3, \quad m=1,2,3,\ldots \tag{25}$$

Here, we do not calculate the higher-order perturbation, since the purpose of this paper is to evaluate the vector corrections to the scalar wave analysis.

For the cladded optical fiber, we must replace the integer mode-index m in Eqs.(20)-(25) by the unknown non-integer mode-index μ which includes the cladding effects [9]. Substituting these into Eqs.(12)-(14) and solving those numerically, we can determine the guided modes of the cladded square-law optical fibers. Now, we introduce two well-known parameters to evaluate the dispersion characteristics of the guided modes of the square-law optical fiber with the homogeneous cladding:

$$V = ka(2\Delta)^{1/2} \; ; \; B = 1 - b/2\Delta \quad .$$

First, we calculate the cutoff frequencies. Under the cutoff condition $b=2\Delta$, the characteristic equations (12)-(14) can be rewritten into more simplified forms. In Table I the cutoff frequencies for the LP_{11}-mode are shown. The cutoff frequencies obtained by the present method are in good agreement with those given by the precise numerical method [4]. Table II indicates that the vector correction to the cutoff frequencies is significant for the lower-order modes, as expected. Last, we calculate the mode dispersion for the HE_{11}-mode. Table III shows that the scalar wave approximation gives considerable errors in the low frequency regions.

4. CONCLUSION

We calculate the guided modes of the square-law optical fiber with the homogeneous cladding and remove the temporal degeneracy of the LP-modes. As a numerical example, we estimate the cutoff frequencies for several lower-order modes and the mode dispersion for the HE_{11}-mode, and discuss the the accuracy of the scalar wave analysis. The higher-order vector corrections will be done in the near future.

REFERENCES

1. H.-G. Unger, Planar optical waveguides and fibres, Clarendon Press, Oxford, 1977.
2. D.L.A. Tjaden, "First-order correction to 'weak-guidance' approximation in the fibre optics theory," Philips J. Res., vol.33, pp.103-112, 1978.
3. R.A. Summut et al., "Modal analysis of polarization effects in weakly-guiding fibres," Proc. IEE, vol.128, Pt.H, pp.173-187, 1981.
4. P. Danielsen, "Investigation of the difference between cut-off frequencies of scalar and vectorial modes in a single-mode fiber," J. Opt. Commun., vol.2, pp.101-104, 1981.
5. H. Ikuno, "Vectorial wave analysis of graded-index fibers," Radio Sci., vol.17, pp.37-42, 1982.
6. R. Yamada and Y. Inabe, "Guided waves in an optical square-law medium," J. Opt. Soc. Am., vol.64, pp.964-968, 1974.
7. A. Erdelyi et al., Higher transcendental function, vol.I, McGraw-Hill, New York, 1953.
8. L.I. Schiff, Quantum mechanics, McGraw-Hill, New York, 1968.
9. M. Hashimoto et al., " Analysis of guided waves along the cladded optical fiber: Parabolic-index core and homogeneous cladding," IEEE Trans. vol.MTT-25, pp.11-17, 1977.
10. T.I. Lukowski and F.P. Kapron, "Parabolic fiber cutoffs: A comparison of theories," J. Opt. Soc. Am., vol.67, pp.1185-1186, 1977.
11. Y. Kokubun and K. Iga, "Mode analysis of graded-index optical fibers using a scalar wave equation including gradient-index terma and direct numerical integration," ibid., vol.70, pp.388-394, 1980.

Table I: Comparison between various calculations of the cutoff frequencies V_c. $(ga)^2=2\Delta$, $2\Delta=0.01$.

MODE	present method	[4]	[11]
TE_{01}	3.518	3.518	3.518
TM_{01}	3.514	3.514	3.513
HE_{21}	3.522	3.522	3.521

Table II: Cutoff frequencies V_c. $2\Delta=0.01$.

MODE	$(ga)^2=2\Delta$ present method	$(ga)^2=2\Delta$ [10]	$(ga)^2=3\Delta$ present method	$(ga)^2=3\Delta$ [10]
TM_{01}	3.514		4.558	
TE_{01}	3.518	3.518	4.561	4.561
HE_{21}	3.522		4.567	
HE_{12}	5.067	5.068	6.931	6.929
EH_{11}	5.745	5.744	7.235	7.233
HE_{31}	5.748		7.239	
TM_{02}	7.449		9.6356	
TE_{02}	7.451	7.451	9.6364	9.636
HE_{22}	7.453		9.639	

Table III: Normalized frequencies B. $2\Delta=0.01$

V	$(ga)^2=2\Delta$ present method	$(ga)^2=2\Delta$ [10]	$(ga)^2=3\Delta$ present method	$(ga)^2=3\Delta$ [10]
2.0	0.1505	0.2687	0.0474	0.1400
3.0	0.3623	0.4025	0.2345	0.2680
4.0	0.5066	0.5183	0.3983	0.4067
5.0	0.6015	0.6051	0.5123	0.5146
6.0	0.6669	0.6682	0.5921	0.5928

APPROXIMATE DISPERSION CURVES OF SOME NONCIRCULAR DIELECTRIC OPTICAL WAVEGUIDES

Weigan Lin
Chengdu Institute of Radio Engineering

ABSTRACT

The dispersion curves of some noncircular dielectric optical waveguides are given approximately by averaging the upper and the lower limits given by the normalized propagation constants of the circumscribed and the inscribed fibers of appropriate cross-sectional shape.

Various noncircular dielectric waveguides such as elliptical fiber waveguides, rectangular fiber waveguides and triangular fiber waveguides are employed nowadays in fiber optics and integrated optics. It is generally known that no simple and satisfactory means exists in analyzing the problem of the guiding properties of such waveguides. The modified finite element approach gives very good results [1], but still there are discrepancies between the exact curve and the finite element curves. The multiple rod triangular-core fiber has the advantage of accurate control of the dimension and centering of the core, logical result has been obtained for the cut-off point for the higher-order modes [2]. We have solved the low frequency scattering problem [3] by finding the upper and the lower bounds on the normalized dipole moment and taking the mean value of these bounds as the final answer to our scattering problem of a dielectric rectangular cylinder. We now try to apply this same technique to solve the problem of the normalized propagation constants of these three noncircular dielectric waveguides.

Take the rectangular fiber with refractive index n_1, coated with a clad material of refractive index n_2, and of aspect ratio a:b, then take the inscribed elliptical fiber with major and minor axes of a and b and the circumscribed elliptical fiber with major and minor axes of $\sqrt{2}$ a and $\sqrt{2}$ b. Read out the values of the normalized propagation constant of these two elliptical fiber at $(2\pi/\lambda)b(n_1^2-n_2^2)^{\frac{1}{2}}$ and $\sqrt{2}(2\pi/\lambda)b(n_1^2-n_2^2)^{\frac{1}{2}}$, and take the geometric mean values of these two values of the normalized propagation constant $(\overline{\beta}^2-n_2^2)/(n_1^2-n_2^2)$ as the value of the normalized propagation constant $(\overline{\beta}^2-n_2^2)/(n_1^2-n_2^2)$ of this rectangular fiber at $(2\pi b/\lambda)(n_1^2-n_2^2)^{\frac{1}{2}}$. Fig.(1) gives the normalized propagation constant of aspect ratio 2:1 constructed in this manner from the existing curve of dispersion for an elliptical fiber. The values marked with cross are those from the normalized propagation constants of this rectangular fiber directly from finite element results. The agreement between the values from the mean values and those from the finite element results is good.

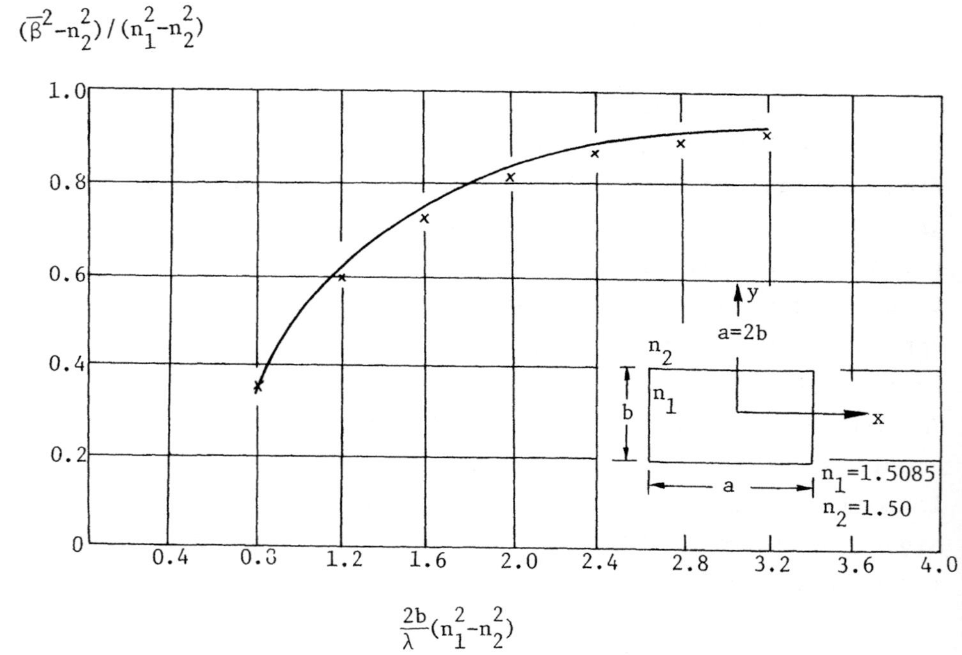

FIGURE 1. Dispersion curve for the dominant eHE_{11} mode in a rectangular fiber constructed by the averaging process, χ, values from numerical solutions of Yeh et al.

In the course of constructing the curve in Fig.(1) we discover that as far as the normalized propagation constant $(\overline{\beta}^2-n_2^2)/(n_1^2-n_2^2)$ is concerned, a rectangular fiber is equivalent approximately to an elliptical fiber of same aspect ratio a:b but with major and minor axes of $(1+\sqrt{2})/2$ a and $(1+\sqrt{2})/2$ b.

Since there exist already many dispersion curves for rectangular fibers [4], we may make use of them to construct those for elliptical fibers by taking the arithmetic mean of the normalized propagation constants of their own circumscribed rectangular fiber and inscribed rectangular fiber with sides reduced by a factor of $1/\sqrt{2}$. It is found that the arithmetic mean gives better results than the geometric mean. Fig.(2) shows such a construction of a dispersion curve for an elliptical fiber by means of those for the rectangular fiber [1]. Also, as far as the normalized propagation constant $(\bar{\beta}^2-n_2^2)/(n_1^2-n_2^2)$ is concerned, an elliptical fiber of aspect ratio a:b is equivalent to a rectangular fiber with sides of $(1+\sqrt{2})/(2\sqrt{2})a=0.854a$ and 0.854b.

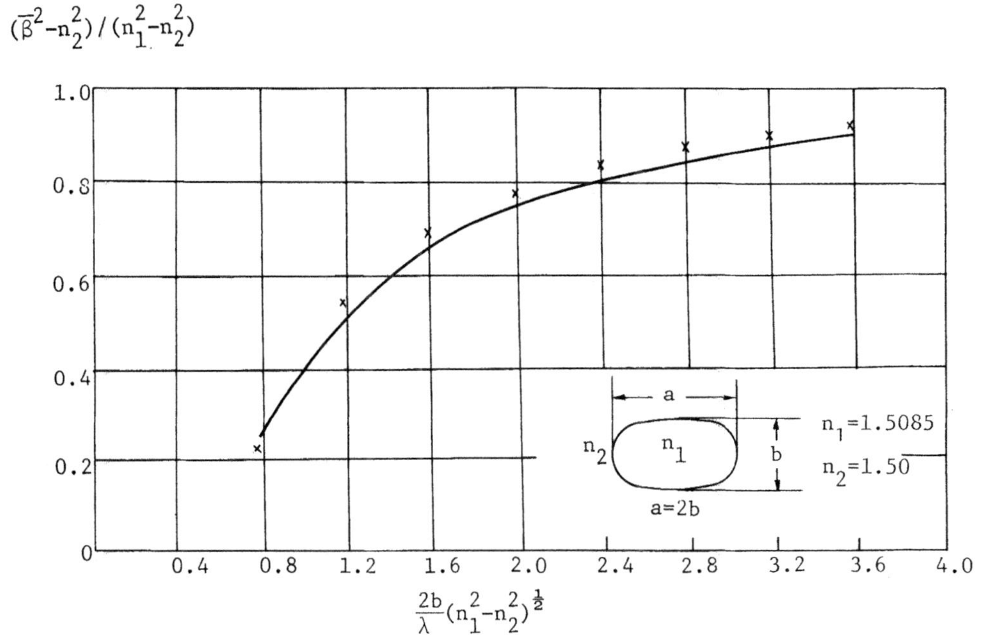

FIGURE 2. Dispersion curve for the dominant eH_{11} mode in an elliptical fiber constructed by the averaging process, χ, values from numerical solution of Yeh.

A triangular fiber of side a is then, as far as the normalized constant is concerned, equivalent to a circular fiber of radius $(1+2)/2\cdot(a/2\sqrt{2})=0.530a$. We here give the cut-off wavelength of the first high order mode of this triangular fiber from the equation

$$(n_1^2 - n_2^2)^{\frac{1}{2}} \frac{2\pi}{\lambda_o} q \left(\frac{\sqrt{3}}{4}a\right) = 2.405$$

where q=0.840 is a proportional constant [2]. giving

$$\frac{a}{\lambda_o} (n_1^2 - n_2^2)^{\frac{1}{2}} = 1.039$$

being higher than that given previously and so providing a better fit to the numerical result.

REFERENCES

1. Yeh, C., Ha K., Dong, S. B., and Brown, W. P., 'Single-mode Optical Waveguides', Applied Optics, Vol.18, No.10, 15 May, 1979, pp.1490-1540.
2. Dyott, R. B., Day, C. R., Brain, M. C., 'Glass-Fibre Waveguide With A Triangular Core', Electronics Letters, 26th June 1973, Vol.9, No.13, pp.288-290.
3. Lin, W. G., 'Low-Frequency Scattering of Dielectric Cylinders', IEEE Transactions on Microwave Theory and Techniques, Vol. MTT-28, No.11, November 1980, pp.1190-1204.
4. Barnoski, M. K. (Ed.); Introduction to Integrated Optics, (Plenum Press, New York and London, 1974).

PROCESSING OF NEAR-FIELD INTENSITY MEASUREMENTS IN OPTICAL FIBRES

G. COPPA, P. DI VITA and U. ROSSI

CSELT, Via G. Reiss Romoli, 274 - 10148 Torino (Italy)

ABSTRACT

The processing of near-field intensity measurements is discussed. In multimode fibres mode distributions can be derived; in monomode fibres the index profile and all the related quantities can be calculated.

1. INTRODUCTION

The simple and common near-field intensity (NFI) measurements are presently underestimated. They are generally used to obtain the refractive index profile in multimode fibres [1], and for the geometrical parameter measurements or the spot-size determination in monomode fibres. But a lot of information can be derived from an easy processing of such measurements (using very popular desk-top computers).

After a brief review of such processing techniques in the case of multimode fibres, in this work we present a new processing technique of the NFI for monomode fibres. This permits an easy and immediate determination of the refractive index profile, accurate enough to allow a precise deduction of the most important parameters.

2. MULTIMODE FIBRES

It has been recently shown [2] that the mode power distribution (versus the collective mode parameter x), R(x), is simply related to the NFI, I(r) (r - radial coordinate), through:

$$R(x) = \frac{1}{\pi} \frac{d\,I(r)}{dr} \Big/ \frac{d\,n^2}{dr} \quad , \quad x = f(r) \tag{1}$$

where $n^2(r) = n_0^2 - \Delta^2 f(r)$ is the squared refractive-index profile and f(r) is the profile function (n_0 = Max {n(r)} and Δ = maximum fibre numerical aperture).

Eq. 1 is valid under a number of hypotheses, that are verified in practice [2], the most important of which is the degeneracy versus the azimuthal mode number. Such a degeneracy can be caused by ellipticity (not only geometrical but also induced by stresses) [3] and it has been verified to be common in practice [4]. This degeneracy causes also a rapid extinction of leaky rays [3] that are ignored in Eq. 1. The problems that can arise from the numerical derivations (in Eq. 1) can be easily overcome by a number of known computing techniques and the results are quite reliable [2].

There is a large number of interesting applications of this elaboration of NFI measurements to obtain the mode power distribution:

- *Checking of launching conditions* by measuring R(x) after a short length of fibre in a given launching situation (including the effects of possible mode-filters, mode-scramblers, etc. [2]). This can result particularly useful when the equilibrium mode power distribution is required, e.g. for fibre attenuation measurements.
- *Differential mode attenuation measurements* by comparing the R(x)'s at two different fibre sections [5].
- *Characterization of joints and couplers* by measuring the R(x)'s before and after the device [5, 6].
- *Backscattering measurements* to check the mode power distribution in the forward and backward directions, in which the power decay measurements are performed [7]. This can also be useful to verify the mode filtering conditions required to separate the decay from the imperfection effects [8, 9].
- *Evaluation of mode coupling coefficients* by the measuring the R(x)'s in different launching situations after a short and after a long length of the same fibre and by processing such data according to the method suggested in Ref. 10.

Thus the NFI measurements can be exploited more deeply than merely to deduce the refractive index profile (for uniform excitations) for multimode fibres. In the following we are showing how such measurements are very meaningful also in the case of monomode fibres.

3. MONOMODE FIBRES

NFI measurements in monomode optical fibres are mainly used, at present, to obtain the spot-size. This is a fundamental quantity to characterize monomode optical fibres: e.g. microbending and joint losses depend strongly on it. It is also the basis for some sinthetic specifications such as ESI (Equivalent Step Index [11-13]) methods. But the ESI (and thus the spot-size) determination can only partially meet the desired precision on some important quantities such as the LP_{11} cut-off wavelength. A sufficiently accurate determination of the index profile of the fibre would permit to obtain all the relevant parameters with precision but this requires, at present, delicate and sophisticated measurements [14].

Now we demonstrate that a simple NFI measurement (at a single wavelength) permits such a sufficiently accurate determination of the index profile of monomode fibres. The evaluation of the profile in this case is not so immediate as for multimode fibres, since it requires an elaboration, which however is straightforward and very rapid with the desk-top calculators by now broadly used in the measurement equipments. But there are also some definite advantages for monomode fibres, due to the fact that NFI's are unambiguous: independent of the launching conditions and negligibly affected by the propagation effects and leaky mode contamination as far as higher order modes are not present at the output of the fibre.

Although the precision of the measurement can be very high, the spatial resolution of the refractive index profile may be smaller than in other traditional techniques, such as the refracted near-field method [14], due to the higher wavelengths used. However it must be stressed that the index profile evaluated from NFI measurements represents the index distribution actually "seen" by the propagating radiation: as a confirmation it is shown that, from the profile so determinated, all the relevant quantities (cut-off wavelength of LP_{11} mode, maximum numerical aperture, V-value, etc.) can be obtained with great accuracy.

The processing procedure [15] is based on the fact that in monomode fibres the NFI, I(r), is proportional to the square of a transverse field component, E(r), which fulfills the following scalar wave equation:

$$\frac{1}{r}\frac{d}{dr}\left(r\frac{dE(r)}{dr}\right) + \left[\left(\frac{2\pi}{\lambda}\right)^2 n^2(r) - \beta^2\right]E(r) = 0 \qquad (2)$$

(β is the longitudinal propagation constant and λ the operating wavelength)

The method consists in calculating the quantity (proportional to the index profile): $[(2\pi/\lambda)^2\ n^2(r) - \beta^2]$ from I(r) (considering that $E(r) \propto \sqrt{I(r)}$) directly from Eq. 2. Also in this case suitable techniques can be used to reduce the errors that can arise from numerical

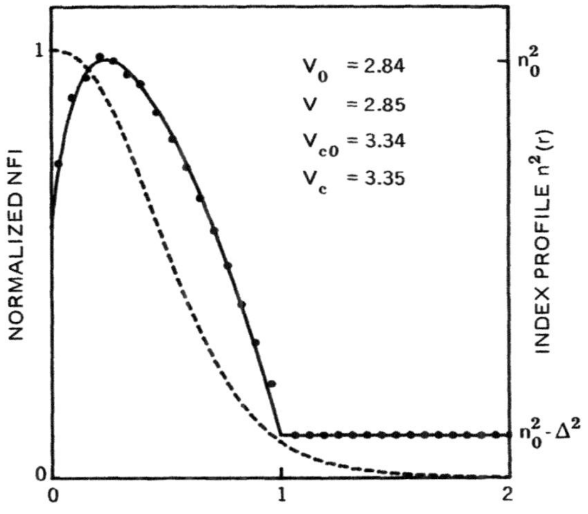

Fig. 1 - Near-field intensity (dashed line), original (full line) and recalculated (full points) refractive index profile for a monomode parabolic-index fibre with a dip. The original V-value (V_0) and LP_{11} cut-off V-value (V_{c0}) are reported together with the re-calculated values (V and V_c)

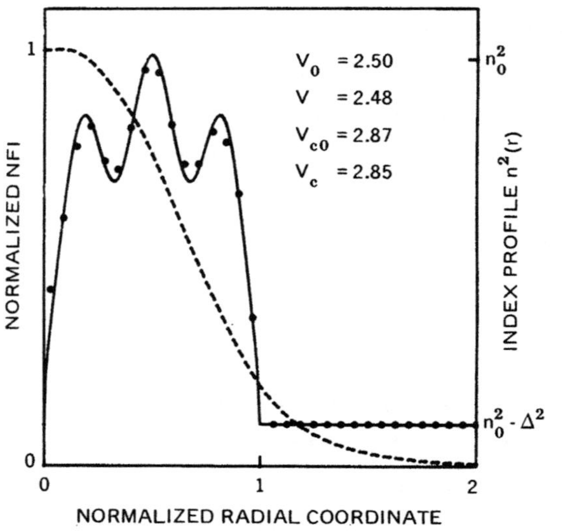

Fig. 2 - The same as Fig. 1, but for a different and more complicated profile with perturbations

derivations. The convergence of the algorithm is very quick and the results are quite reliable. Computer simulations have been performed by calculating the NFI corresponding to some given profiles. Figs. 1 and 2 show respectively a parabolic profile with a dip and a more complicated profile with perturbations (full line). The relative calculated NFI's are shown by dashed lines. In order to simulate the measurement conditions, a noise similar to the one of the measurement has been added to the calculated NFI's. Thus the index profiles have been recalculated with the present method, and the results (full points) are in excellent agreement with the original profile; also original and recalculated values of cut-off wavelength and V-value agree within 1%.

In Fig. 3 experimental results obtained on a single-mode fibre are reported. From the measured NFI pattern (dashed line, λ = 1300 nm) the refractive index profile (full points) has been computed. The refractive-index profile of the fibre has been measured independently by the known refracted near-field technique [14] and the results are shown in Fig. 3 by the full line. The excellent agreement between the two measured refractive-index profiles shows the reliability and the effectiveness of the present method.

Once the index profile has been derived all the relevant parameters of the fibre can be deduced. Fig. 4 shows, as an example, the spot-size of the fundamental mode calculated versus the wavelength. Note that such a parameter was deduced, at any wavelength, starting from only one simple measurement of the NFI at a specific λ. After a good agreement with direct experimental data is demonstrated, this method can therefore be considered a useful alternative to performing a set of scans at different wavelengths, as usually done [13].

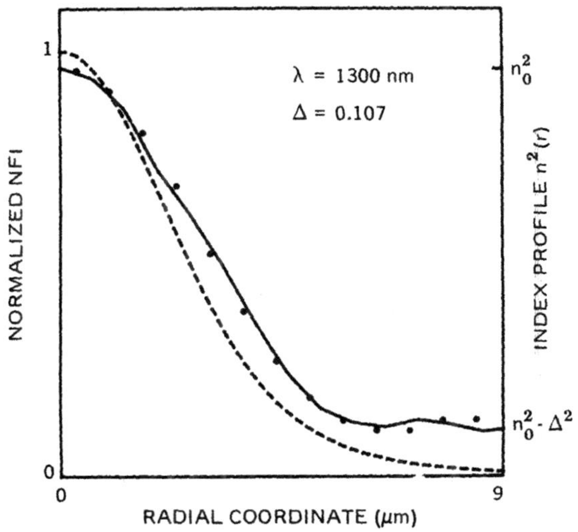

Fig. 3 - Measured near-field intensity (dashed line), calculated (from NFI) (full points) and measured (by refracted near-field) (full line) refractive-index profiles of a monomode fibre

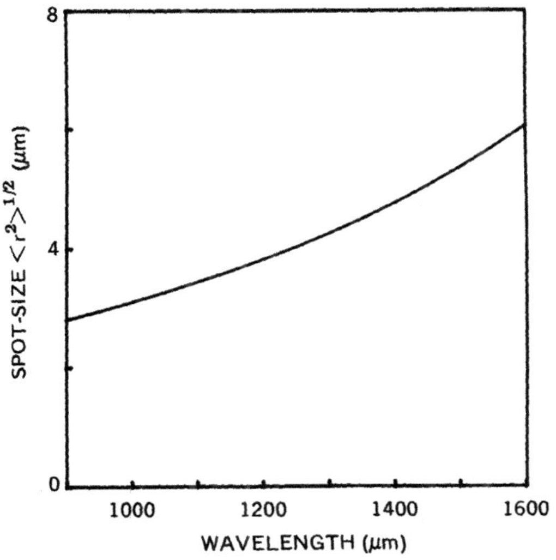

Fig. 4 - Fundamental mode spot-size versus wavelength for the fibre of Fig. 3

4. CONCLUSION

The simple NFI measurement has proved to be a very powerful technique. Relatively easy processings have permitted to derive a lot of information both for multimode and monomode fibres. In particular for monomode fibres a new technique has been suggested to derive, with sufficient precision, the index profile and to determinate, from this, very accurately all the relevant parameters of the fibre.

REFERENCES

1. D.Gloge and E.A.J. Marcatili, Bell Syst. Tech. J., Vol. 52 (1973) - pp. 1563-1578.
2. M.Calzavara et al., Electron. Lett., Vol. 17 (1981) - pp. 543-545.
3. G.Coppa et al., Opt. & Quantum Electron., Vol. 14 (1982) - pp. 283-309.
4. S. Kawakami, 7th Workshop on Optical Waveguide Theory, La Londe (France) - September 17-20, 1982.
5. M.Calzavara et al., 7th ECOC, Copenhagen, September 8-11, 1981, paper P7.
6. C.Marchesi and U.Rossi, Symp. on Opt. Fiber Meas., Boulder (USA), October 13-14, 1982, pp. 127-130.
7. B.Costa et al., 7th ECOC, Copenhagen, September 8-11, 1981, postdeadline paper 13.5.
8. P. Di Vita, Electron Lett., Vol. 17 (1981) - pp. 874-876.
9. A.R. Mickelson and M.Eriksrud, Appl. Opt., Vol. 21 (1982) - pp. 1898-1909.
10. K.Nagano and S.Kawakami, Appl. Opt., Vol. 19 (1980) - pp. 2426-2434.
11. A.W.Snyder and R.A.Sammut, J. Opt. Soc. Am., Vol. 69 (1980) - pp. 1663-1671.
12. W.J.Stewart, Electron. Lett., Vol. 16 (1980) - pp. 380-382.
13. C.A.Millar, Electron. Lett., Vol. 17 (1981) - pp. 458-460.
14. W.J.Stewart, IEEE J. Quantum Electron, Vol. QE-18 (1982) - pp. 1451-1466.
15. K.Hotate and T.Hokoshi, Appl. Opt., Vol. 18 (1979) - pp. 3265-3271.

ERROR ANALYSIS OF REFRACTIVE INDEX PROFILE MEASUREMENTS OF OPTICAL FIBRE AND PREFORM

P.L. CHU and T. WHITBREAD

School of Electrical Engineering and Computer Science
University of New South Wales
P.O.Box 1, Kensington
N.S.W. 2033
AUSTRALIA

ABSTRACT

It is shown that many of the existing index profiling methods for optical fibre and preform are related to one another through the deflection function from which the index profile is reconstructed. These methods are classified into direct deflection function measurement, integral (or phase) measurement, and differential (or intensity) measurement. The deflection function from the integral method is obtained by measuring the phase distribution of the exit rays followed by a numerical differentiation. Thus the reconstructed index profile has high frequency errors. The deflection function from the differential method is obtained by measuring the light intensity distribution followed by a numerical integration. This process enhances low frequency errors but suppresses high frequency errors. The direct method has errors of all frequencies but of lesser magnitude.

1. Introduction

Several methods have been developed for the measurement of the refractive index profile of optical fibre and preform. For fibres, there are the near field scanning[1], refracted near field[2], interferometric[3] and focussing methods[4]. For preforms, there are the focussing[4], spatial filtering[5], holographic[6] and direct display[7] methods. Except for the two near field methods, all the above-mentioned methods are based on the knowledge of the deflection function $\psi(y)$ which is defined as the exit angle of the probing ray as a function of its incident position y as shown in Fig.1[8]. The deflection function is related to the refractive index profile n(r) by the integral:-

$$n(r) = n_0 + \frac{n_0}{\pi} \int_r^a \frac{\psi(y)\, dy}{\sqrt{y^2 - r^2}} \qquad \ldots .(1)$$

where a is the radius of the fibre or preform and n_0 is the refractive index of the liquid in which the specimen is immersed.

The various methods mentioned above differ from one another only in the manner of the measurement of $\psi(y)$. The interferometric method measures the phase function $\phi(y)$ that each ray suffers in traversing the fibre cross-section. The focussing method measures the light intensity distribution P(y) across an image plane in the fibre or preform. The spatial filtering, holographic and direct display methods measure the deflection function $\psi(y)$ directly. It can be shown that $\psi(y)$ is equal to the derivative of $\phi(y)$ and is also equal to the integral of P(y), i.e.

$$\psi(y) = K \frac{d\phi(y)}{dy} \qquad \ldots .(2)$$

$$\psi(y) = B \int_0^y P(y')dy' \qquad \ldots .(3)$$

where K and B are constants of proportionality. Thus a unified error analysis of all these measurement methods is possible. The purpose of this paper is to report such an analysis.

2. Errors due to Profile Reconstruction

The integral in eq.(1) has to be evaluated numerically for a given deflection function $\psi(y)$ which is usually in tabulated form. The simplest numerical integration is by means of trapezoidal rule in which two consecutive data points are represented by a straight line i.e.

$$\psi(y) = a_i y + b_i \qquad \ldots .(4)$$

where $a_i = \{(N-1)/a\}\{\psi_{i+1}(y_{i+1}) - \psi_i(y_i)\}$

$b_i = i\psi_i(y_i) - (i-1)\psi_{i+1}(y_{i+1})$

N = total number of data

$\psi_i(y_i)$ is the measured deflection function at point y_i.

Substituting eq.(4) into eq.(1), the index $n(r_i)$ at radius r_i may be calculated as:-

$$n(r_1) - n_o = \sum_{i=1}^{N} \frac{n_o}{\pi} \int_{r_i}^{r_{i+1}} \frac{a_i y + b_i}{\sqrt{y^2 - r_1^2}} dy$$

$$= \sum_{i=1}^{N} \frac{n_o}{\pi} \left\{ a_i \left(\sqrt{r_{i+1}^2 - r_1^2} - \sqrt{r_i^2 - r_1^2} \right) \right.$$

$$\left. + b_i \ln \frac{r_{i+1} + \sqrt{r_{i+1}^2 - r_1^2}}{r_i + \sqrt{r_i^2 - r_1^2}} \right\} \quad \ldots .(5)$$

The error involved in this process of reconstruction is mainly due to the linear interpolation of the deflection function in eq.(4). This equation may be re-written as the convolution of the sampled data $\psi_i(y_i)$ and a triangular function

$$\Lambda(y) = 1 - \frac{|y|}{\Delta y} \qquad |y| \leq \Delta y \quad \ldots .(6)$$

i.e. $$\psi(y) = \sum_{i=1}^{N} \psi_i(y_i) \Lambda(y - y_i) \quad \ldots .(7)$$

This is shown in Fig.2. To appreciate the error involved, we take the Fourier transform of eq.(7). The spectrum of the reconstructed deflection function $\psi(y)$ is equal to the product of the spectra of the sampled data $\psi_i(y_i)$ and of $\Lambda(y)$ as shown in Fig.3. Since

$$FT\{\Lambda(y)\} = \frac{4\sin^2 \omega\Delta y/2}{\Delta y \omega^2} \quad \ldots .(8)$$

the side lobes of $FT\{\Lambda(y)\}$ for frequencies larger than $2\pi/\Delta y$ are negligible compared to the main lobe. Thus the effective bandwidth of $\psi(y)$ is $2\pi/\Delta y$. Furthermore the function in eq.(8) decays rapidly as ω approaches $2\pi/\Delta y$. Therefore the high frequency components in the sampled data $\psi_i(y_i)$ will be attenuated. Consequently the index profile reconstructed from the deflection function thus obtained has negligible high frequency components near the limit of the bandwidth.

At the other extreme, the deflection function $\psi(y)$ may be reconstructed by the convolution between the sampled-data $\psi_i(y_i)$ and a sinc function as (Fig.4):-

$$\psi(y) = \sum_{i=1}^{N} \psi_i(y_i)\ \mathrm{sinc}(y-y_i) \qquad \ldots .(9)$$

In Fourier plane, the sinc function transforms into a rectangular pulse of width $4\pi/\Delta y$ as shown in Fig.5. Thus, although the bandwidth of $\psi(y)$ remains the same as in the previous case, its high frequency components are not attenuated. It is therefore expected that the resultant index profile is of higher resolution. A difficulty with this representation is that the resultant integral of eq.(1) cannot be readily evaluated.

A good compromise between these two extremes is perhaps a parabolic interpolation in which $\psi(y)$ is represented by a second order polynomial.

Now if the sampled-data $\psi_i(y_i)$ has an error of constant standard deviation σ and zero mean, a compact expression for the standard deviation σ_n of the index profile n(r) may be derived in terms of Δy and σ directly from eq.(1) in the spirit of Presby et.al.[3]:-

$$\sigma_n(r) = \frac{\sigma}{\pi}\sqrt{\frac{\Delta y}{r}\ \ln\frac{2r}{\Delta y}} \qquad \ldots .(10)$$

For example, if σ=0.0025 rad., Δy=30μm for a preform, we find σ_n=0.00032 at r=700μm. It is noted from eq.(10) that the error gets worse as r decreases. To appreciate the reason for this deterioration of accuracy as r reduces, let us return to eq.(1) and note that the reconstructed refractive index at r depends on that part of the deflection function $\psi(y)$ in which $y \geq r$. Thus a measurement error of $\psi(y)$ at y_1 will affect the accuracy of the reconstructed n(r) for all $r \leq y_1$ while in the region $r > y_1$, n(r) is not affected by the error of $\psi(y_1)$. This effect accummulates as the number of measurement errors increases.

3. Noise in Reconstructed Index Profiles

To study the noise in the reconstructed index profiles, we need firstly to understand the noise characteristics in the deflection function. As mentioned in the introduction, the deflection function may be obtained in three different ways. If it is measured directly, its noise spectrum would be that of Gaussian white noise as shown in Fig.6 the bandwidth of which is limited by the measuring instrument. The autocorrelation function of the noise is a sinc function as shown in Fig.7. The effective correlation length in this case is that of one sampling period Δy. Thus eq.(10) can be applied directly to estimate the standard deviation of the reconstructed index profile. Fig.8 shows a good agreement between this estimation and the standard deviation calculated directly from the results of ten reconstructed profiles of the same preform from the holographic method[6].

If the deflection function is obtained through the differentiation of the phase data $\phi(y)$ as given by eq.(2), the white noise in $\phi(y)$ will be transformed into a noise rich in high frequency components. Fig.9a shows such a noise spectrum obtained from the interferometric measurement of a single mode fibre. Its autocorrelation function is shown in Fig.9b and the corresponding error curve is shown in Fig.10 which also shows a good agreement between the estimation from eq.(10) and that calculated directly from experimental results.

If the deflection function is obtained through the integration of the intensity data $P(y)$ as given by eq.(3), the white noise in $P(y)$ is now transformed into a noise rich in low frequency components as shown in Fig.11. Its autocorrelation function, shown in Fig.12, is significantly different from the previous two cases. Consequently eq.(10) is not applicable without appropriate modifications. Fig.13 shows the error curve obtained directly from experimental results for the same single mode fibre as in the previous case except that it is now measured by the modified focussing method[10].

Comparison between Figs.10 and 13 shows that the integration process of obtaining the deflection function such as the focussing method incurs larger errors in the index profile than the differentiation or direct processes. The large low frequency noise tends to cause the reconstructed index profile to sway as shown in Fig.14 which is the profile of a single mode fibre obtained from the modified focussing method. The profile itself is smooth, indicative of the absence of high frequency noise. However, the absolute values of the indices have large errors because of the swaying.

On the other hand, the same fibre was measured again by the interferometric method. The reconstructed index profile is shown in Fig.15. The high frequency noise is clearly visible but the local average index agrees better with the expected value.

The direct methods of measuring the deflection function[5,6,7] are applicable to preforms only while the focussing method is valid for both fibres and preforms. Fig.16 shows the comparison of the profiles of the same preform measured by a direct method[7] and by the modified focussing method. As expected, both the fine structures and high frequency noise in Fig.16b are much smaller than those in Fig.16a.

When there is a sharp change in refractive index such as at the interface between the preform surface and matching liquid, the direct methods of measuring the deflection function may not be accurate because the deflection function at the sharp transition is usually a large spike with rapidly decreasing intensity. Consequently the reconstructed profile in the cladding region shows a gradual decrease in index as the radial distance decreases as shown in Fig.17.

4. Conclusions

The errors involved in several of the index profiling methods have been analysed in a unified manner. All these methods are related to one another through the deflection function. Some methods measure the deflection function directly, while others obtain its derivative or integral from which the function itself is derived. It is shown that if the derivative of the function is measured such as the focussing method, high frequency noise in the reconstructed index profile is attenuated but low frequency noise is accentuated causing the profile to sway thus introducing large standard deviation in the indices. On the other hand, if the integral of the deflection function is measured such as the interferometric method, high frequency noise are accentuated but low frequency noise are suppressed. The resulting profile is one that has rugged appearance but its standard deviation error is small. The direct method of obtaining the deflection function produces results that are the approximate average of the previous two cases. However, an additional error is introduced in this method. It results from the difficulty of determining the deflection angle at sharp boundaries.

5. References

(1) F.M.E. Sladen, D.N. Payne, M.J. Adams "Determination of optical fibre refractive index profiles by a near field scanning technique" Appl. Phys. Letts. Vol.28,pp.255-258, 1976

(2) W.J. Stewart "A New Technique for measuring the refractive index of graded

optical fibres" Technical Digest 1977 Int. Conf. Integrated Optics and Optical Fibre Comm. (IOOC) Tokyo, Japan, July 18-20, 1977

(3) H.M. Presby, et. al. "Rapid Automatic Index Profiling of whole fibre samples: Part II" B.S.T.J. Vol.58, No.4, pp.883-902, April 1979

(4) D. Marcuse "Refractive index determination by the Focussing Method" Appl. Opt. Vol.18, No.1, pp.9-13, January 1979

(5) I. Sasaki, D. Payne, M.J. Adams "Measurement of refractive index profiles in optical fibre preforms by spatial filtering techniques" Electronics Letters Vol.16, No.6, pp.219-221, 1980

(6) P.L. Chu, D. Peri "Holographic measurement of refractive index profile of optical fibre preform" Electronics Letters Vol.16, No.23, pp.876-877, November 1980

(7) D. Peri, P.L. Chu, T. Whitbread "Direct display of the deflection function of optical fibre preform" Appl. Opt. Vol.21, No.5, pp.809-814 March, 1982

(8) P.L. Chu "Non-destructive measurement of index profile of an optical fibre preform" Electronics Letters Vol.13, No.24, pp.736-738, November 1977

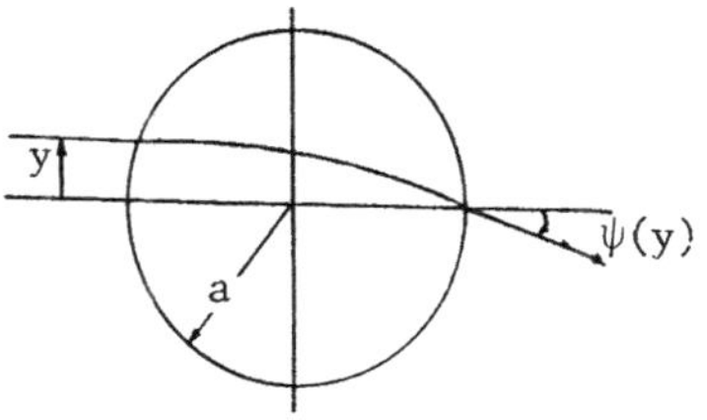

Fig.1

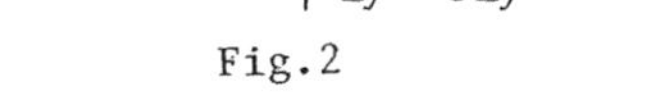

Fig.2

Fig.3

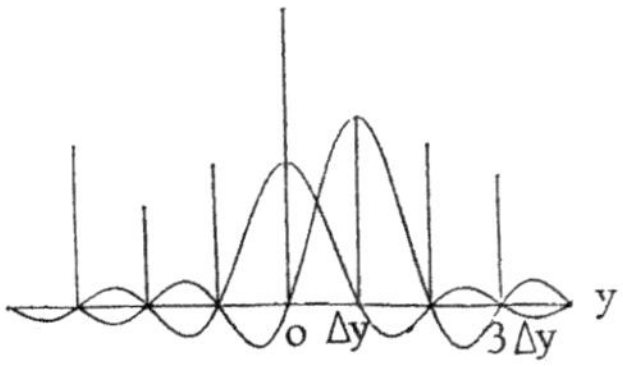

Fig.4

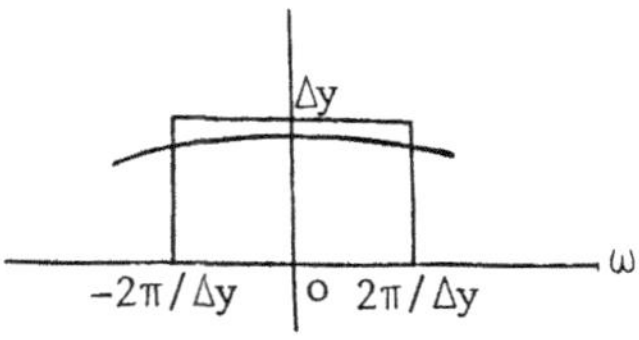

Fig.5

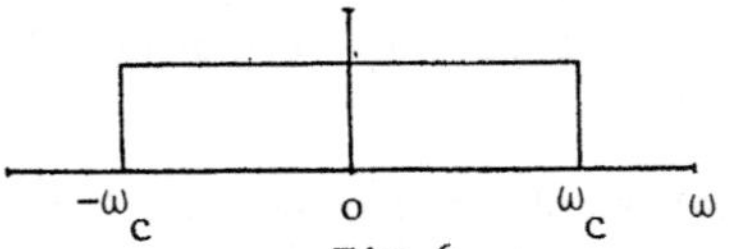

Fig.6

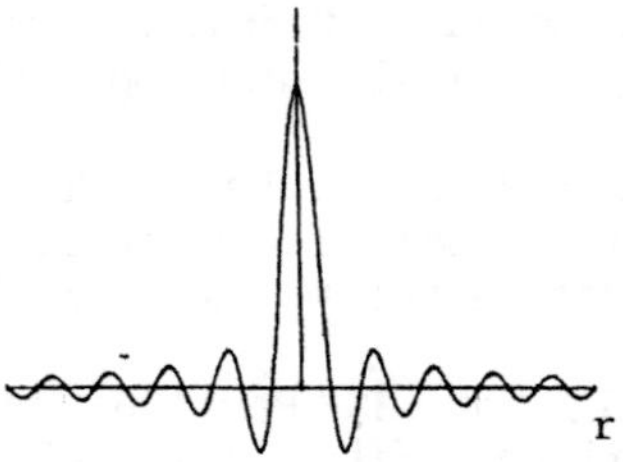

Fig.7

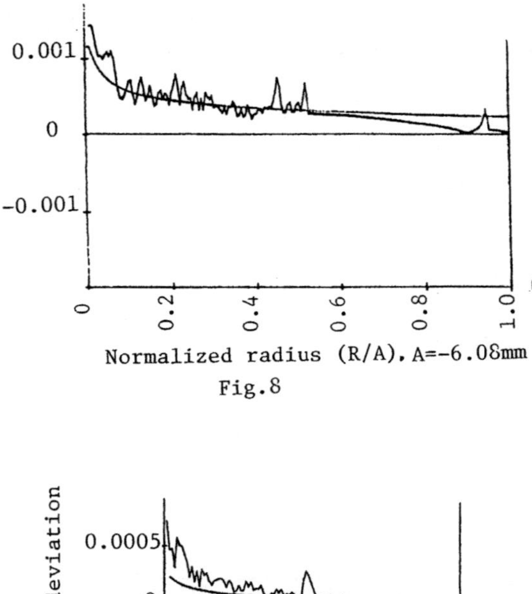

Fig.8

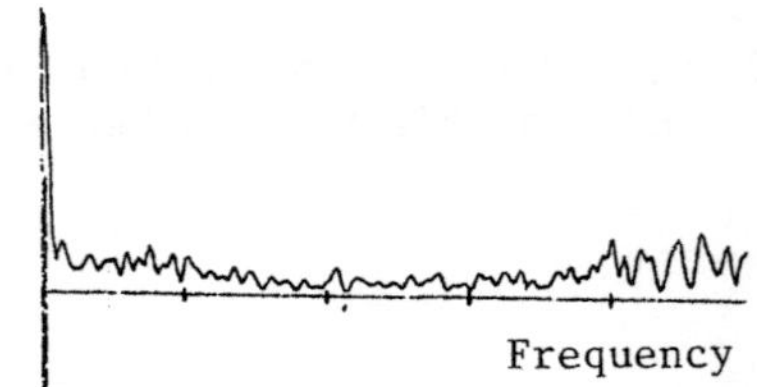

Fig.9a

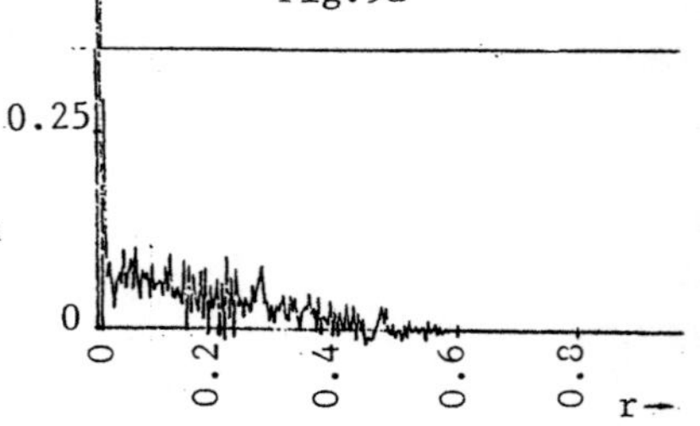

Fig.9b

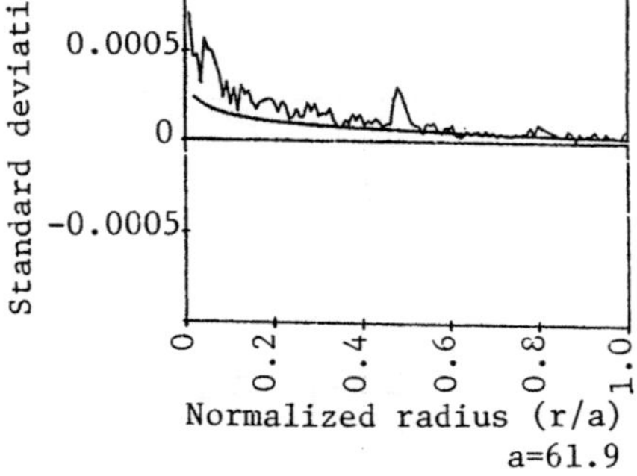

Fig.10

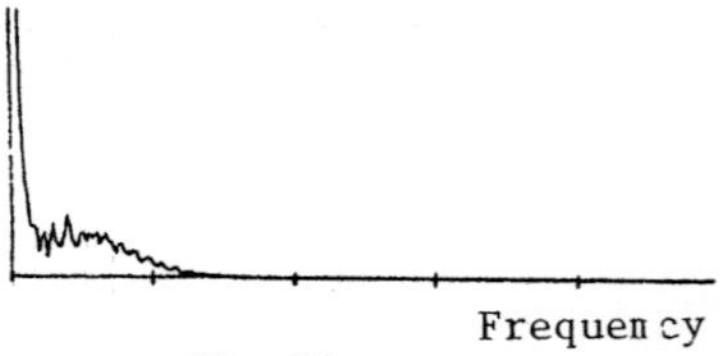

Fig.11

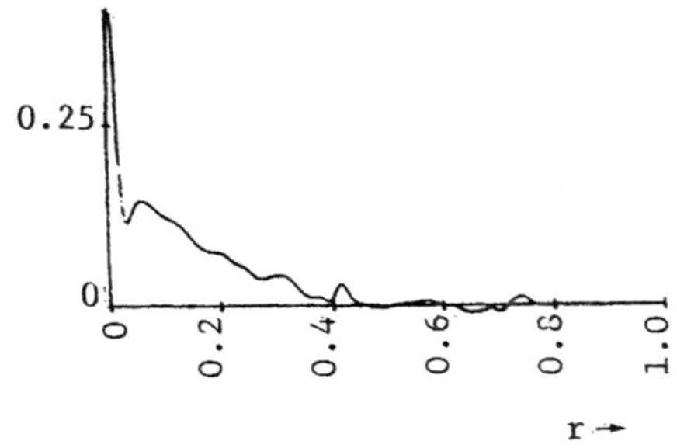

Fig.12

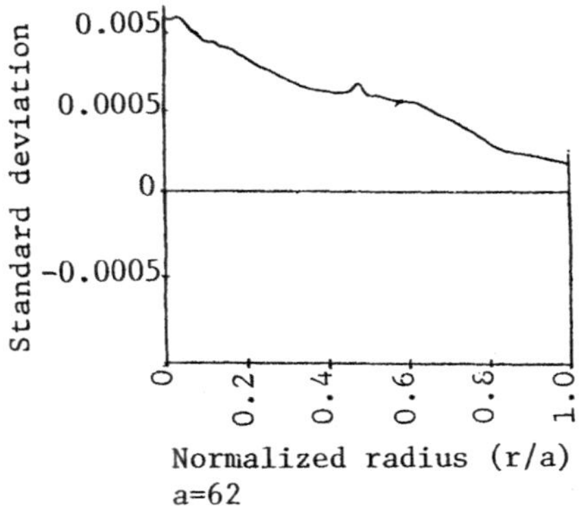

Fig.13

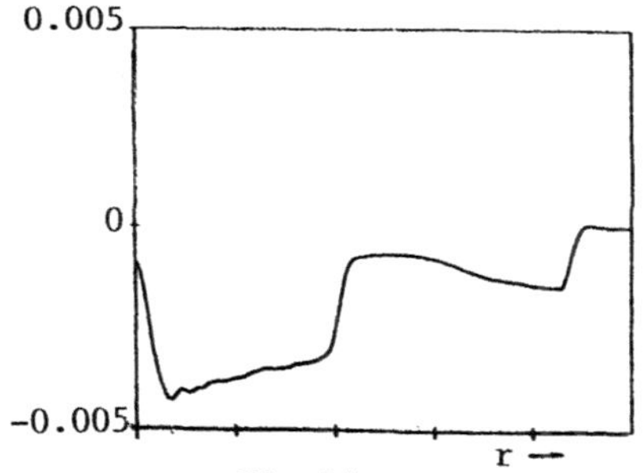

Fig.14

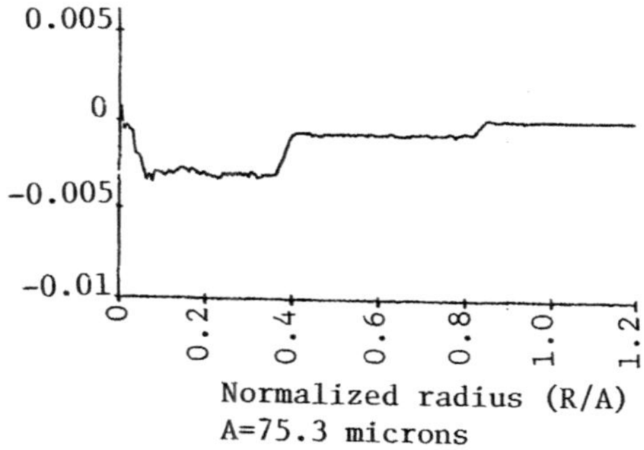

Fig.15

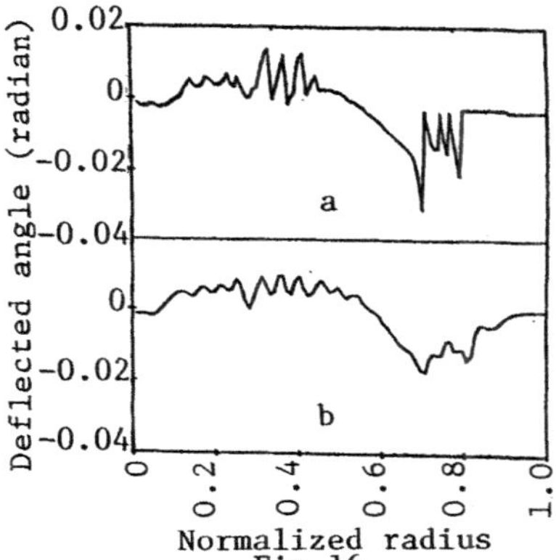

Fig.16

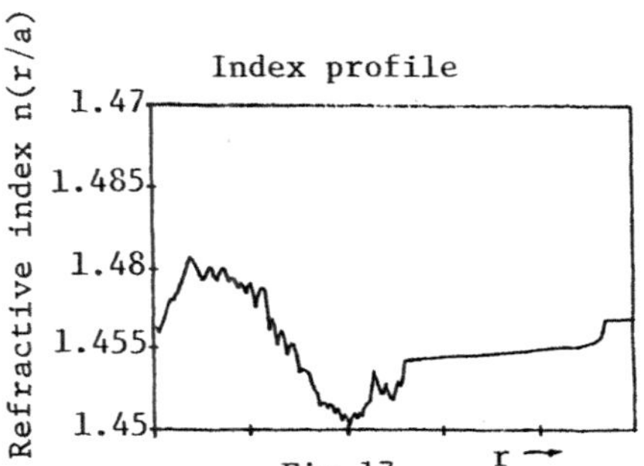

Fig.17

CONCATENATION PHENOMENA IN MULTIMODE FIBERS

CARLO G. SOMEDA+

1. INTRODUCTION

Many experimental results have shown that the behavior of a link, which consists of a sequence of multimode fibers, can not be calculated in a straightforward deterministic way from the knowledge of the so-called base-band power transfer function of the individual fibers.
These results are simple to explain qualitatively; but, on the contrary, a reliable prediction of the overall link bandwidth can not yet be considered as a completely solved problem [1, 2]. This viewpoint is applicable to two, strictly related problems. The first question is what tolerances on slow longitudinal profile fluctuations must be satisfied by a fiber manufacturing process, in order to yield a specified fiber bandwidth. The second one is how one must characterize, and then combine, individual fibers, in order to achieve a specified link bandwidth. The aim of this paper is to review the historical evolution of the theoretical models which have been developed as succcessive attempts to fulfil these purposes. The final goal is to assess the degree of confidence one can have in the present results.

2. BACKGROUND

The evolution of the theoretical models as a function of time has been, to a large extent, a consequence of changes in the relative weight of the parameters and phenomena, from whose interplay stem the bandwidth limits. As well known, the most important among these parameters are the following ones.
A) The index profiles of the individual fibers. They determine

(+) Istituto di Elettrotecnica ed Elettronica, Università di Padova, Italy.

the group delays of the guided modes.
At present, we can consider it as a well established point, that the WKBJ method is adequate for computing group delays for the present purposes. On the contrary, there can be an open question on which is the appropriate mathematical model for the profile. For most authors, the so-called α-profiles:

$$n^2(r) = n_o^2 \left[1 - 2\,\Delta(r/a)^{\alpha}\right] \qquad (1)$$

are suitable, but a recent analysis [3] indicates they might lead to an excess of optimism is estimating the bandwidth. At any rate, there is strong experimental evidence that suitable combinations of "undercompensated" ($\alpha > \alpha$ optimum) and "overcompensated" ($\alpha < \alpha$ optimum) profiles can be quite beneficial, in terms of link bandwidth.

B) Coupling among guided modes in each individual fiber. This type of coupling is distributed along the fiber length, and can be caused by various mechanisms: geometrical irregularities, lateral stress, microbending, etc..
It has been checked experimentally [4] that the so-called selection rules, i.e. the details of the individual mode-to-mode coupling mechanism, depend on the type of fiber distortion. However, these details should not have a large weight on our problem, at least as long as all the fibers have nearly optimum profiles, which minimize the spread of modal group delays.

C) Mode mixing at the joints. This type of localized coupling depends on fiber misalignments, and also on the difference between the index profiles of two consecutive fibers. In the extreme case of several closely spaced joints, and for fibers whose index profiles are far from optimum, the influence of this type of mode coupling on the bandwidth can be very large [5]. On the contrary, if one wants to emphasize the effects of small changes in the profile near the optimum one, then it is reasonable (and realistic) to consider a model of the link, where the joints are assumed to be perfect

3. THEORETICAL MODELS

When the first basic experimental results on concatenation were published [6], the state of the art in the cabling and splicing processes was such, that theoreticians were induced to attribute a large weight to the items labelled B and C in the previous Section. This led to the development of a theory [7] that was considerably influenced by previous work on step-index fibers, a case where, obviously, the problem labelled A in Section 2 did not exist. The key feature of this theory was to describe the performance of two jointed fibers by means of a statistical parameter, called the correlation coefficient, r_{12}, which shows up in the formula:

$$\sigma^2 = \sigma_1^2 + \sigma_2^2 + 2\sigma_1\sigma_2 r_{12}, \qquad (2)$$

where σ_1, σ_2 and σ are the variances of the travel time for a photon in fiber 1, in fiber 2 and in the whole link, respectively. Similar lines of thought have been pursued by several other authors, and eventually led to the introduction of the so-called "gamma factor", i.e. to the relationship [8]:

$$B = (\Sigma_i \; B_i^{-1/\gamma})^{-\gamma} \qquad (3)$$

where B_i= bandwidth of the i-th fiber, with $\gamma<1$.
A common feature of all these models in that they are, to a large extent, phenomenological. Consequently, confidence in their actual capability of predicting link bandwidth has always been rather scarce.
More recently, technological improvements in fibers, cables and splices led to a decrease in the weight of items B and C of the previous Section, compared to A. Theoreticians were thus induced to study in depth the effects of pure profile changes. The first models developed along this line of thought [9, 10, 3] were deterministic, i.e., the profile fluctuations were completely specified by single - valued functions of the longitudinal coordinate, z. A major consequence was the possibility of defining an "equivalent profile" [9], whose optimization depends on the amplitude and shape of the z-dependent perturbation, but

which allows, under optimum conditions, a bandwidth which is essentially the same as that of a uniform fiber.
Furthermore, this type of approach shows [10] that a "tandem connection", i.e. a sequence of an over-compensated fiber and an under-compensated one, made on purpose, can be beneficial under the very practical viewpoint of manufacturing tolerances. See, e.g., Figs. 1 and 2.
A major question, though, is whether these optimistic conclusions reached with deterministic models are confirmed or not, in the presence of random profile fluctuations. Previous work which included, to some extent, statistical variables for the profile [11] , was rather encouraging, but not decisive. Quite recently [12], the problem has been tackled again in statistical terms by means of a Monte-Carlo simulation, where the individual fiber lengths and profiles were generated as pseudorandom sequences. The results obtained in this way depend radically on some cru-

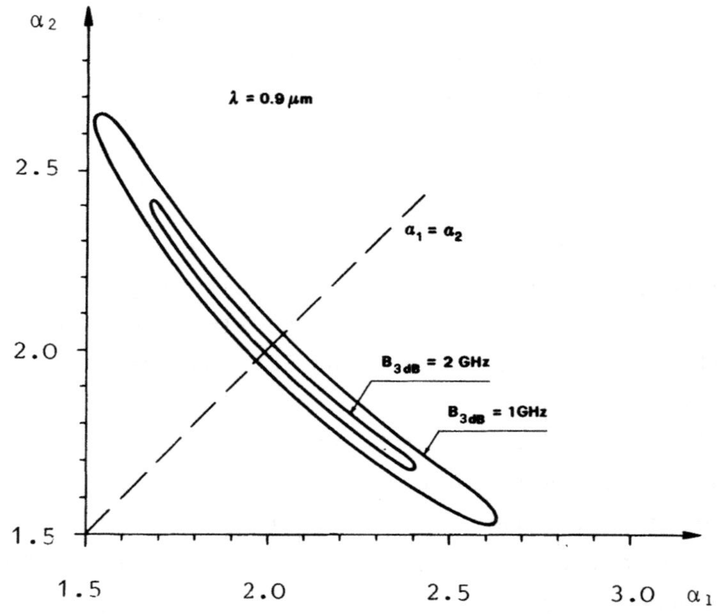

FIGURE 1. Example of bandwith vs. profile exponent for tandem fibers. The parameter α_1 is the profile exponent for the first half of the fiber; the abscissa α_2 is the exponent for the second half (from ref. 10).

cial assumptions one makes in the Monte-Carlo approach. If the random sequences are highly correlated among themselves, i.e. if one simulates a situation where sistematic errors are dominant compared with purely random ones, then the link bandwidths are very similar to those given by the deterministic models. On the other hand, if correlation among pseudorandom sequences is completely destroyed (i.e. if one considers the complementary case, where random errors are dominant), then the results are radically different. Knowing the average values of the profile and the r.m.s. value of its fluctuations is not enough, in order to infer the link bandwidth. Furthermore, the possible values of this bandwidth are spread over large intervals, with nonnegligible probabilities of ending up with very poor links. An example is provided in Fig. 3.

4. CONCLUSIONS

At the present time, the conclusion that one grasps from known results is that systematic errors in the profile can be compensated for. This means that very broad bandwidths may be obtained in fibers which exhibit profile fluctuations, but only under special circumstances. Also, this has an imortant impli-

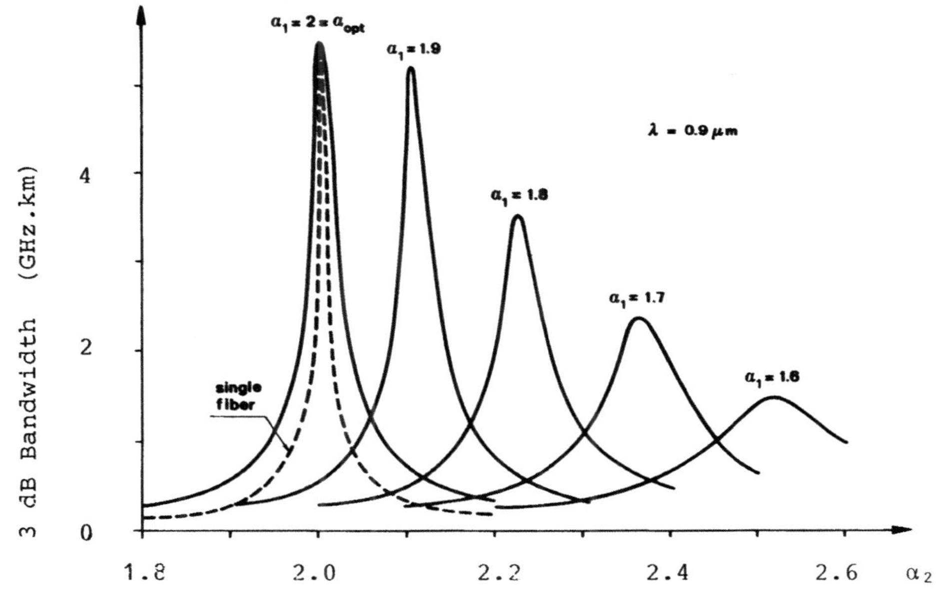

FIGURE 2. Constant bandwidth contours on the (α_1, α_2)-plane for tandem fibers (from ref. 10).

cation on how one selects the fibers to be connected together for a long link. This point can be very important in practice, since purely random errors in the profile must be looked at as a potential cause of very severe bandwidth deterioration.

REFERENCES

1. Reitz P R. Tech. Digest Symp. on Opt. Fiber Meas.1982, NBS Publication 641, pp. 1-8
2. Bouillie R., Bizeul J C. Ibid., pp. 17-20.
3. Tanifuji T. IEEE J. Quantum El., QE - 18, 1982, pp.844-849.
4. Kawakami S. Presented at 7th Workshop on Opt. Waveg. Theory, La Londe 1982.
5. Rizzoli V., Someda C G. Electron. Lett., 17, 1981, pp.906-907.
6. Eve M. et al. Proc. 3rd ECOC, Muenchen, 1977, pp.53-55.
7. Eve M. Opt. Quantum El., 10, 1978, pp.41-51.
8. Suzuki S. et al. Proc. Opt. Comm. Conf., Amsterdam, 1979, 12.6.
9. Pask C. Opt. Quantum El., 12, 1980, pp. 281-290.
10. Galeati P. et al. J. Opt. Comm., 2, 1981, pp. 54-58.
11. Miyamoto M. et al. Proc. 3rd IOOC, San Francisco, 1981, WG3.
12. Rizzoli V., Someda C G. To appear in Opt. Quantum Electr.

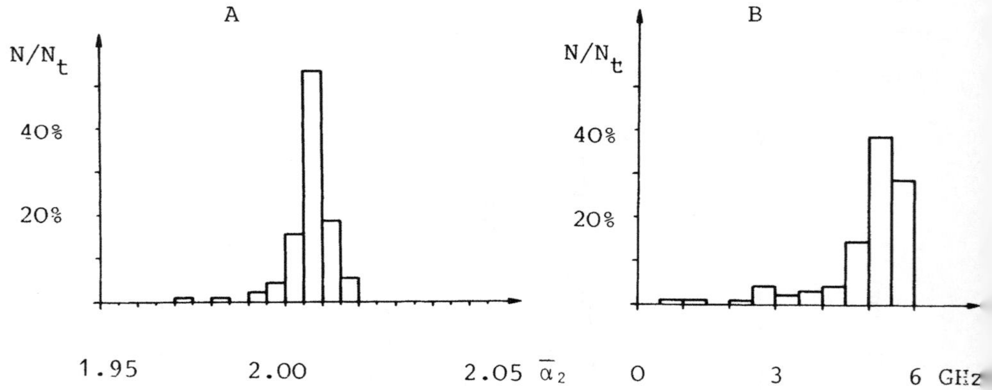

FIGURE 3. Examples of histograms given by a Monte-Carlo simulation of random profile (from ref. 12). A) Optimum average exponent. B) Maximum bandwidth. No. of fiber sections with different profiles: n = 20. R. m. s. deviation of profile exponents: σ_α = 0.15.

ANALYSIS AND DESIGN OF ANGULAR SCRAMBLING STAR COUPLER

Huang Shang-yuan
Wang Twen-shang

Jiao Tong University, Shanghai

ABSTRACT

Based on light propagation in a cylindrically homogeneous and square-law medium, loss and light-power assignment of an angular scrambling star coupler are studied. The concepts of inherent loss, optimal length of the lenses and optimal length of the scrambler (mixer) are presented. Our analysis and calculation show that the optical length of the lenses is not exactly a quarter of the period of a paraxial path, and the power assignment ratio is controlled by the length of the scrambler which can be calculated from the material and structural parameters of the coupler.

1. INTRODUCTION

A star coupler is an important device in optical fiber communication systems. Up to now many types of star coupler have been proposed. The angular scrambling star coupler is one of them(1). It is a passive micro-optical device with a configuration shown in Fig. 1. A cylindric mixing rod with a homogeneous index is in the middle, with a selfoc-lens (a square-law medium) joined to either end, while each of the outer ends of the lenses is linked with several optical fibers which are arranged in a circle. All parts of the coupler are coaxially connected. With the lens I, the light from any of the input fibers is collimated into a beam of parallel rays at an axial angle Θ, then it goes into the mixer. By means of the scrambling action of the mixer, the beam emerges at the other end as a howwlow cone of light with a semi-angle Θ. Then these

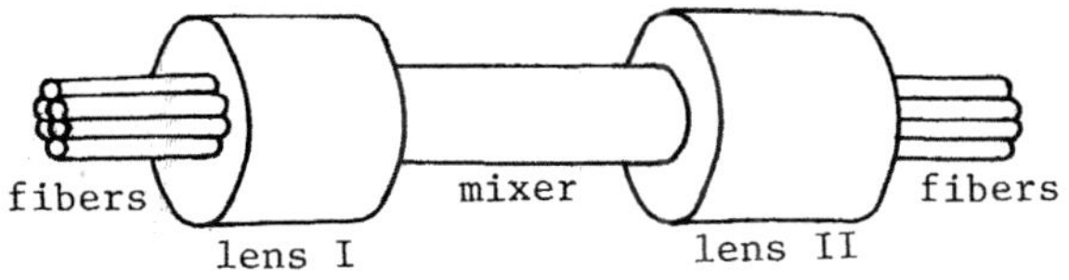

Fig.1. Configuration of the angular scrambling star coupler

rays are focused by the lens II, and form a circular light ring at its output end-face. The light power is then coupled to the output fibers which are aligned with the light ring.

Though the device has been developed, its detailed analysis and design methods have so far not been found. The aim of this paper is to study the loss and light distribution uniformity of the angular scrambling star coupler and find the factors controlling the loss and the uniformity so as to determine the lengths of the lenses and the mixer. It should be pointed out that the method of geometric optics is adopted here, because the size of the coupler is considerably large compared with light wavelengths.

II. LAW OF LIGHT PROPAGATION IN A CYLINDRIC MEDIUM

A. Homogeneous Medium Rod

Any ray in a cylindric rod of a homogeneous medium maintains its reflection angle Φ (i.e. incident angle) and axial angle Θ in its transmission process. And they satisfy[2]:

$$\cos\Phi=\sin\Theta\cos\gamma \tag{1}$$

Here, γ is the semi-angle of the projection trace on the cross-section (see Fig. 2).

Assuming that the coordinate axis Z coincides with the axis of the cylinder, and the axis X coincides with the initial direction of the projection of the incident ray (see Fig. 2), after passing a distance Z in Z-direction, the number of reflections is given by:

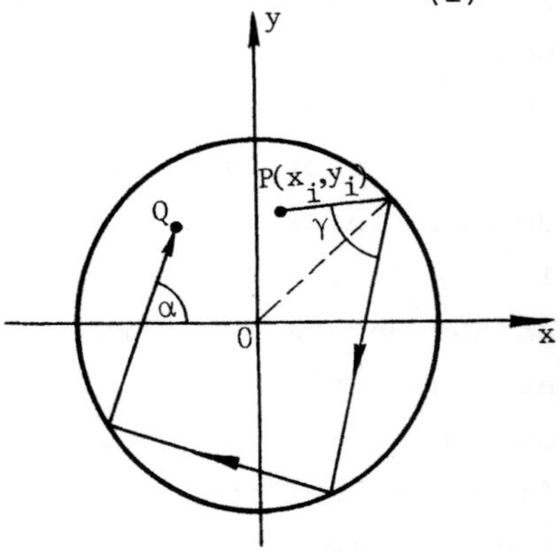

Fig.2. Projection of the path of a skew ray on the cross-section of a cylindric homogeneous medium

$$m=\left[\frac{Z\,tg\Theta+R\cos\gamma+x_i}{2R\cos\gamma}\right] \tag{2}$$

where the sign [ ] denotes the integral part, $\gamma=\sin^{-1}(y_i/R)$, x_i, y_i are the coordinates of the starting point P. R is the radius of the rod.

B. Square-law Medium Rod

The relative index of a square-law varies, as

$$n(\gamma)=n_o(1-Ar^2)^{\frac{1}{2}} \tag{3}$$

Starting from the basic formula of geometric optics, we can obtain the exact ray-trace in the square-law medium.

The trace is given by its projections on XOZ and YOZ planes:

$$\begin{cases} x(z)=x_o\cos\frac{\sqrt{A}z}{c}+x_o'\frac{c}{\sqrt{A}}\sin\frac{\sqrt{A}z}{c} \\ y(z)=y_o\cos\frac{\sqrt{A}z}{c}+y_o'\frac{c}{\sqrt{A}}\sin\frac{\sqrt{A}z}{c} \end{cases} \tag{4}$$

where $c=\frac{n(x_o,y_o)\cos\theta_o}{n_o}$. x_o y_o are the coordinates of the ray and θ_o is its axial angle at z=0, and $x_o'=\frac{dx}{dz}\Big|_{z=0}$, $y_o'=\frac{dy}{dz}\Big|_{z=0}$.

From (4) we see that the period of a ray is $\frac{2\Pi c}{\sqrt{A}}$ which is related to its starting position and axial angle. So the image of any object point in the medium has several kinds of aberration. Only for the meridional and paraxial rays, the period P has nothing to do with its initial condition.

$$P=\frac{2\Pi}{\sqrt{A}} \tag{5}$$

III. INHERENT LOSS OF THE COUPLER AND OPTIMAL LENGTH OF LENSES

The insertion loss of an angular scrambling star coupler results mainly from the packing loss ℓ_p. Assume that the power of the light ring is uniformly distributed, then

$$Lp=-10lg(\frac{n}{4}\ \frac{d}{D}\ \sin\frac{\Pi}{m}) \tag{6}$$

where n is the number of terminals, d,D are the inner and outer diameters of the fiber respectively.

From Section II it can be seen that there exists no such an ideal lens-length for collimation and focusing. There are more or less some rays out of the light-ring or beyond the aperture limits of the output fibers even if they are focused on the ring. These lead to an inherent loss L_{INH} which is related to the lengths of the lenses and the mixer, the diameter of the mixer and the fiber position etc. L_{INH} is another important loss besides Lp.

The relationship between L_{INH} and the length of the lenses has been calculated using ray-tracing procedures. We divide the end-face of an input fiber into 61 equal areas. The light power on each area is considered to be carried by two hundred rays from the centre. These rays are distributed over several cones with the central point as their vertex (These cones are within the limits of the fiber aperture).

If $(\bar{x},\bar{y})$ and $\bar{\Theta}$ are the calculated coordinates and axial angle of a ray at the output end-face of the lens II, then all rays which satisfy $\gamma_1^2 \leqslant x^2+y^2 \leqslant \gamma_2^2$ (γ_1 γ_2 are the inner and outer radii of an ideal light-ring respectively) and $n(\bar{x},\bar{y})\sin\bar{\Theta} \leqslant NA_f$ (NA_f is the numerical aperture of the fiber) are classified into useful range, while the rest into inherent loss.

The calculation is performed for three lengths of the mixer (z=16.5, 24.5, 32.5mm) with 11 different lens-lengths:

$$K=0.95,\ 0.96 \ldots\ldots 1.04,\ 1.05 \qquad (7)$$

The results are shown in Fig. 3. As Z=24.5 mm and 32.5 mm we obtain ℓ_{opt}=0.99 P/4, the correlative minL_{INH} is 0.22dB and 0.23dB respectively. For Z=16.5 mm, we get ℓ_{opt}=1.00 P/4

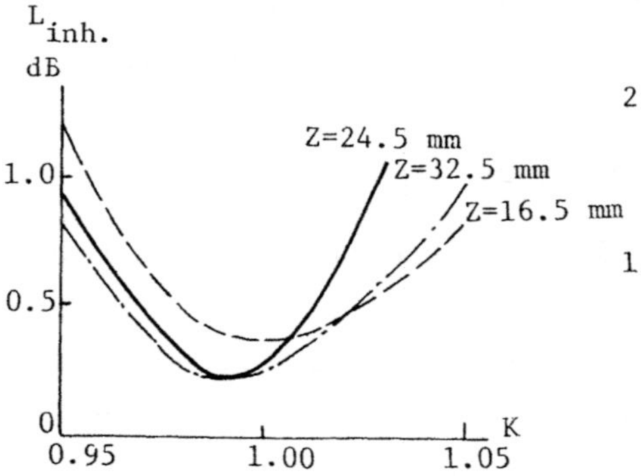

Fig.3 Calculated inherent loss versus lens-length factor K for 3 mixer lengths with $R=R_c$=0.35mm

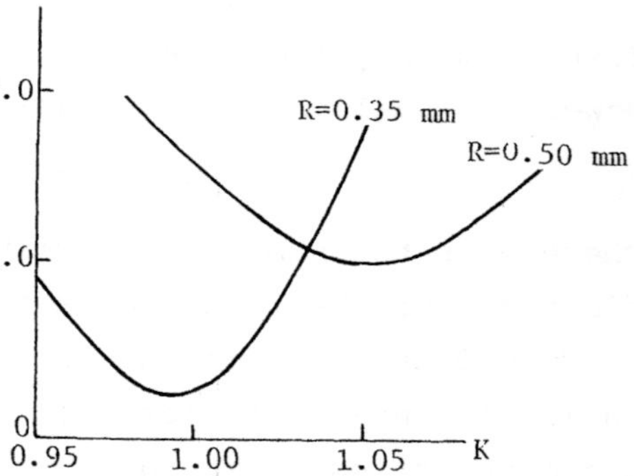

Fig.4 Calculated inherent loss vs lens-length factor K for $R=R_c$ and $R>R_c$ with mixer length Z=24.5mm, R_c=0.35mm

and $\min L_{INH}$ is 0.33dB. This shows that it is not certain for ℓopt to be P/4. Its value is determined by the material parameters, the length of the mixer and the location of the fibers.

The above calculation is performed on condition that the diameter of the mixer R equals that of the collimated beam R_C. If the former is larger than the latter, the inherent loss will increase greatly. Fig. 4 shows two $L_{INH} \sim K$ curves corresponding to R=0.35 mm and R=0.50 mm respectively with Z=24.5 mm, R_C= 0.35 mm. Obviously the inherent loss of the latter increases greatly and the value of K corresponding to $\min L_{INH}$ changes as well.

Though L_{INH} is much smaller than Lp, it is the key factor for a star coupler with a polygonal mixer, which has no packing loss[1].

IV. POWER DISTRIBUTION UNIFORMITY AND OPTIMAL LENGTH OF MIXER

It can be seen from Fig. 2 that a different γ leads to a different α , which is the ultimate azimuth of the ray. Furthermore a different α corresponds to a different position on the light-ring.

For every ray in the collimated beam with Z=∘ we have α=∘ (The assumption is permissible in discussing the uniformity). For rays at the upper half area of the collimated beam travelling along the mixer with a length Z, α is written as

$$\alpha=-m(\Pi-27)\times\frac{180°}{\Pi} -[\frac{-m(-27)}{2\Pi}]\times 360°+360° \qquad (8)$$

where α has been turned into its principal value within 0°∿360°, and m is determined by (2). Sign [ ] means taking the integral part.

For 21 mixer-lengths from Z=∘ to Z=33.6 mm, the uniformity has been calculated. In operation the collimated light spot is divided into many meshes, each of which corresponds to a ray. We find α of every ray by (24), then classify weighted α into the azimuth array. Thus the power distribution in the light-ring can be obtained. In order to find the power assignment ratio for n output fibers we need still make a power summation with respect to each fiber.

The results show that the power assignment ratio can be controlled by the mixer length.

For a 6×6 coupler, let the deviation of P_i (the accepted power by i-th fiber) from the average power $\overline{P}$ ($\overline{P} = \sum_1^6 P_i/6$) be Δ_i:

$$\Delta_i = 10\lg(P_i/P) \qquad i=1\sim6 \tag{9}$$

$$\Delta_{max} = |\max_{i=1\sim6} \Delta_i| \tag{10}$$

Fig. 5 shows the calculated results of Δ_{max} versus Z when

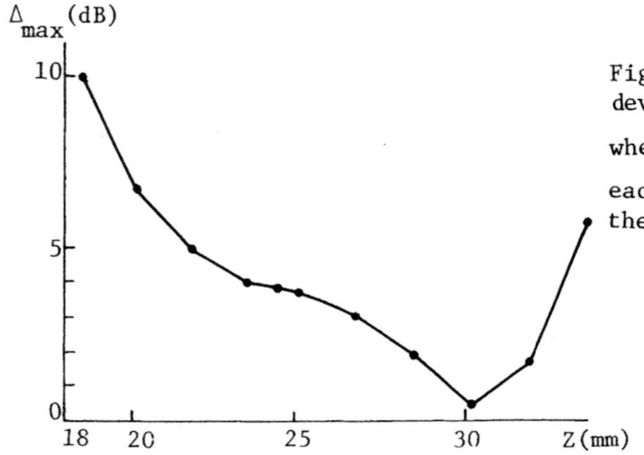

Fig.5 Calculated results of max. deviation Δ_{max} vs mixer-length Z, when tgθ=0.0417, $R=R_c$=0.35 mm and each fiber core occupies 40° on the light-ring.

tgθ=0.0417, $R=R_c$=0.35 mm and each fiber core occupies 40 degrees on the light-ring. It can be seen that the optimal mixer length Z_{opt}, which leads to the most uniform power assignment, is about 30.2 mm in this case. Note that if any of the above mentioned parameter changes, Z_{opt} will change correspondingly.

The other calculation shows that the scrambling effect will weaken as $R>R_c$. Therefore R should be equal to R_c for both a low inherent loss and an efficient scrambling.

V. CONCLUSION

We have shown that the inherent loss, being another important loss of the coupler besides packing loss, depends on the length of the lenses. The optimal length of the lenses is not exactly one quarter of the period of a paraxial path, its value is determined by the material parameters, the mixer length and the fibers' location. Secondly, the power assigneent for every output fiber can be controlled by the mixer length. If the optimal length Z_{opt} calculated from the material and structural parameters is used, we can obtain the best uniformity. Finally, the diameter of the mixer must be equal to that of the collimated beam in order to get a lower loss and a better scrambling effect.

A sample of a 6×6 star coupler has been made. Its material and structural parameters are the same as those used in calculation. The adopted mixer length is 24.5 mm, length of lenses is 0.99 P/4, i.e., 3.62 mm. The average insertion loss of the sample measured is 4.7dB. The light power assignment ratio of the six fibers is 1:0.45:0.25:0.31:0.31:0.43, which is very close to the result of the calculation 1:0.44:0.19::0.20:0.19:0.44.

ACKNOWLEDGEMENT

The authors wish to thank Cao Jie-yuan for his support in the sample fabrication and measurements.

REFERENCE

1. Yoshio Miyake etc. "Angular scrambling star coupler" 5th European Conference on optical fiber communication (1979) 6.5
2. W. B. Allen, Fiber optics theory and practice, p.25

AN ACCURATE ANALYSIS OF THE EMISSION FROM A SEMICONDUCTOR LASER AND ITS APPLICATION TO EXCITATION OF AN OPTICAL WAVEGUIDE

T. TAKENAKA, O. FUKUMITSU, AND M. TANAKA

1. INTRODUCTION

The launching problems of an optical waveguide have been studied by many investigators. However, almost all of their works[1,2] have treated the problem for an incident Gaussian beam. Since most semiconductor lasers typically have an active layer whose thickness is much smaller than its width, the divergence angle of the emission is very large in the plane perpendicular to the p-n junction. This large divergence angle of the emission is one of the main factors of the large coupling loss. As is well known,[3] a light beam with a large divergence angle cannot be represented by a paraxial light beam such as a Gaussian beam. For the excitation problems of an optical waveguide by a semiconductor laser, a more accurate expression of the laser emission must be investigated in the plane perpendicular to the p-n junction.[4,5]

In this paper, we present an accurate numerical analysis of the emission from a semiconductor laser, which incorporates the iterative procedure into Lewin's method. The analysis requires very short execution time on a computer because of the fast convergence of the solutions. The power coupling from a semiconductor laser into an optical waveguide is also discussed.

The time factor $\exp(j\omega t)$ is assumed.

2. LASER EMISSION

2.1. Mode matching in spectral form

In a typical semiconductor laser, the stripewidth is considerably larger than the thickness of the active layer and it is justified that the structure of the laser is taken as a dielectric slab waveguide as shown in Fig. 1.

Let the lowest order TE_0 mode be incident on the laser-air interface z=0. The incident mode is reflected and radiated.

Applying the Fourier transform to the expressions of continuity of the transverse electric and magnetic field components (E_y, H_x), we get

$$\begin{Bmatrix} e_0(\mu) \\ \beta_0 e_0(\mu) \end{Bmatrix} \pm \sum_{m=0}^{M} R_m \begin{Bmatrix} e_m(\mu) \\ \beta_m e_m(\mu) \end{Bmatrix} \pm \begin{Bmatrix} F(\mu) \\ G(\mu) \end{Bmatrix} = \begin{Bmatrix} F^a(\mu) \\ G^a(\mu) \end{Bmatrix} \quad (1)$$

where $e_m(\mu)$ and β_m are the spectrum and the propagation constant of the m-th guided mode of the laser. The spectrum of the backward radiated electric field and of the forward radiated electric field are denoted by $F(\mu)$ and $F^a(\mu)$, respectively. Those corresponding to the magnetic field are denoted by $G(\mu)/\omega\mu_0$ and $-G^a(\mu)/\omega\mu_0$, where ω and μ_0 are the angular frequency of the field and the permiability of free space. The coefficient R_m is the amplitude of the guided mode. The maximum order of the guided mode supported in the laser is indicated by M.

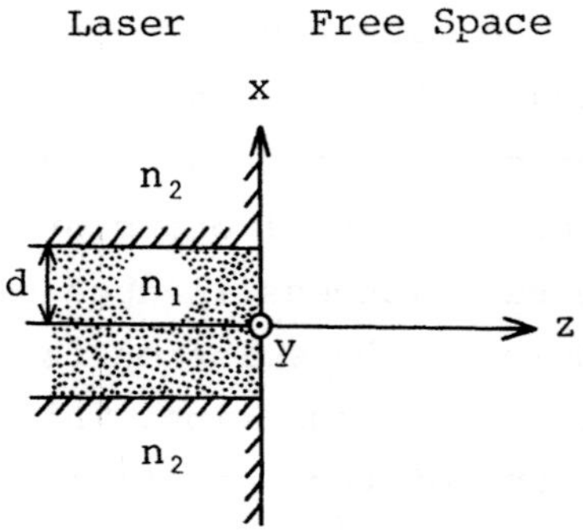

FIGURE 1. Geometry of a semiconductor laser.

As $F^a(\mu)$ and $G^a(\mu)$ are represented by a superposition of the spectra of the radiation modes in the air, the following relation is derived:

$$G^a(\mu) = q(\mu)F^a(\mu) \quad (2)$$

where

$$q(\mu) = \begin{cases} \sqrt{k^2-\mu^2}\,, & |\mu| \leq k \\ -j\sqrt{\mu^2-k^2}\,, & |\mu| \geq k \end{cases} \quad (3)$$

and where k is the wave number in free space.

Following Lewin's method[6], we use a quantity $p(\mu)$ which is related to the ratio of the spectrum $G(\mu)/\omega\mu_0$ of the radiated magnetic field to the spectrum $F(\mu)$ of the radiated electric field in the laser

$$p(\mu) = G(\mu) / F(\mu). \tag{4}$$

Use of (2) and (4) in (1) gives the spectra

$$F(\mu) = \frac{\beta_0 - q(\mu)}{p(\mu)+q(\mu)} e_0(\mu) - \sum_{m=0}^{M} R_m \frac{\beta_0 + q(\mu)}{p(\mu)+q(\mu)} e_m(\mu) \tag{5}$$

$$F^a(\mu) = \frac{\beta_0 + p(\mu)}{p(\mu)+q(\mu)} e_0(\mu) - \sum_{m=0}^{M} R_m \frac{\beta_0 - p(\mu)}{p(\mu)+q(\mu)} e_m(\mu). \tag{6}$$

By using the orthogonality between the guided modes and the radiated fields in the laser, we get two different simultaneous linear equations both of which determine the amplitude R_m

$$\Delta_{em} = \int_{-\infty}^{\infty} F(\mu) h_m^*(\mu) d\mu = 0, \qquad m = 0,1,\cdots,M \tag{7}$$

$$\Delta_{hm} = \int_{-\infty}^{\infty} [G(\mu) / \omega\mu_0] e_m^*(\mu) d\mu = 0, \quad m = 0,1,\cdots,M \tag{8}$$

where $h_m(\mu)$ is the spectrum of the magnetic field of the guided mode. The asterisk denotes the complex conjugate. Let R_m determined from (7) and that from (8) be R_{em} and R_{hm}. If the exact value of $p(\mu)$ was known, R_{em} would be identical with R_{hm}. Then, substitution of the value of $p(\mu)$ and of R_m into (5) and (6) gives the exact values of the spectra $F(\mu)$ and $F^a(\mu)$. Once, the spectrum $F^a(\mu)$ has been known, the radiated fields (E_y^a, H_x^a) in the air are obtained by the plane wave spectrum representation. In the limit $k\sqrt{x^2+z^2} \gg 1$, the integrals in the spectrum representation are evaluated approximately by the steepest-descent method. Therefore, the problem which we are left with is to give a method for determining the exact value of $p(\mu)$.

2.2. Iterative method for $p(\mu)$

As is well known, in the limit when refractive index n_1 of the core approaches index n_2 of the cladding, the radiation mode of the laser reduces to that of the uniform dielectric of refractive index n_2. Lewin[6] has regarded the laser as a uniform dielectric of refractive index n_2 when considering the radiated field in the laser and has given the approximate form for the ratio

$$p(\mu) \simeq \begin{cases} \sqrt{(n^2k)^2 - \mu^2}\ , & |\mu| \leq n_2k \\ -j\sqrt{\mu^2 - (n_2k)^2}\ , & |\mu| \geq n_2k. \end{cases} \quad (9)$$

We determine numerically the accurate value of $p(\mu)$ by the following iterative method using (9) as a initial approximate value of $p(\mu)$.

Let the i-th order renewed value in the iterative procedure be indicated by superscript (i). If $p^{(i)}$ is known, $F^{(i)}$ is a function of μ with a parameter $R_m^{(i)}$ as found from (5). The value of $R_m^{(i)}$ is chosen so that $F^{(i)}$ is orthogonal to the spectra $\{h_m(\mu)\}$ of the transverse magnetic field components of all guided modes, for $F^{(i)}$ should be expressed by a superposition of the radiation modes only. Namely, $R_{em}^{(i)}$ determined from (7) is taken for $R_m^{(i)}$. According to the definition of p, multiplication of $F^{(i)}$ by $p^{(i)}$ gives $G^{(i)}$. However, $G^{(i)}$ obtained in this way is not generally orthogonal to the spectra $\{e_m(\mu)\}$ of the electric field of all guided modes as a result of using the approximate value for p. Therefore, we take anew the quantity $\tilde{G}^{(i)}$, given by subtracting the component which is not orthogonal to the spectra of the guided modes from $G^{(i)}$, as the i-th order renewed value of G:

$$\tilde{G}^{(i)}(\mu) = G^{(i)}(\mu) - \sum_{m=0}^{M} (\pi / P)\beta_m \Delta_{hm}^{(i)} e_m(\mu) \quad (10)$$

where P represents the power carried by each mode in the z-direction per unit length along the y-direction. The (i+1)-th order renewed value $p^{(i+1)}$ is represented by the ratio of $\tilde{G}^{(i)}$ to $F^{(i)}$:

$$p^{(i+1)}(\mu) = p^{(i)}(\mu) - [\sum_{m=0}^{M} (\pi/P)\beta_m \Delta_{hm}^{(i)} e_m(\mu)] / F^{(i)}(\mu). \quad (11)$$

This procedure is repeated untill the modulus of the relative difference between $R_{em}^{(i)}$ and $R_{hm}^{(i)}$ becomes lower than some specified accuracy. At the final iteration, power conservation is confirmed.

2.3. Numerical results

Numerical calculations have been carried out for the laser

having n_1=3.50, n_2=3.23, and 2d=0.4μm. The wavelength λ is 1.3μm.

Fig. 2 shows the amplitude and phase distributions of the laser emission with the distance z away from the laser-air interface as a parameter. The amplitudes of the fields are normalized with respect to the values of those on the z axis. The phases of the fields are relative values to those on the z axis. In the excitation problem of a laser emission into an optical waveguide, the emission is often approximated by a Gaussian beam with a smallest spot size at the laser facet.

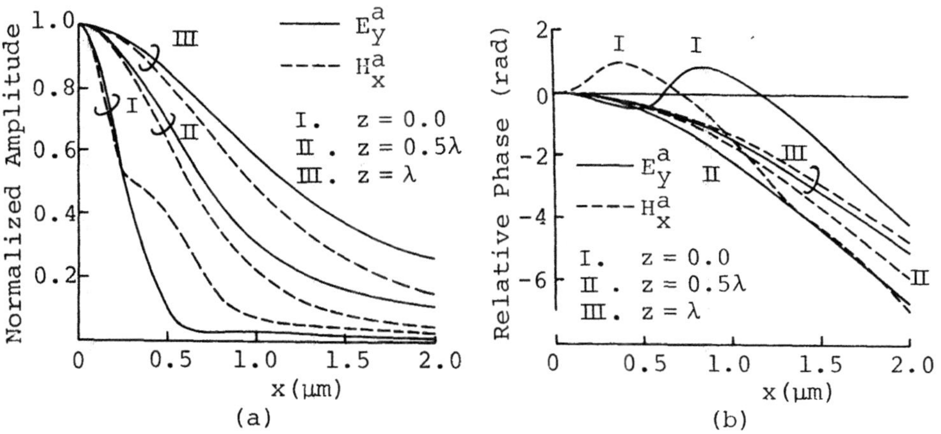

FIGURE 2. Field distributions. (a) Amplitude. (b) Phase.

Table 1. Comparison of the fields (superscript f) based upon the far field representation with those (superscript a) based upon the numerical calculation of the plane wave spectrum representation.

x (μm)	2.0	4.0	6.0	8.0	10.0
$\|E_y^f\|$	0.861	0.619	0.436	0.319	0.243
$\|E_y^a\|$	0.865	0.623	0.438	0.320	0.244
$\|H_x^f\|$	0.823	0.527	0.320	0.201	0.133
$\|H_x^a\|$	0.828	0.532	0.323	0.202	0.134
$\angle E_y^f, \angle H_x^f$ (rad)	-1.454	-5.472	-11.338	-18.405	-26.230
$\angle E_y^a$ (rad)	-1.433	-5.424	-11.280	-18.344	-26.172
$\angle H_x^a$ (rad)	-1.425	-5.400	-11.242	-18.295	-26.112

From Fig. 2, however, the approximation is found to be not so well, especially, for the magnetic field.

The normalized amplitudes and the relative phases of the emission at a distance $z=5\lambda$ are given in Table 1. It is found that the values of the amplitude and phase based upon the far field representation given by the steepest-descent method agree fairly well with those based upon the numerical calculation of the plane wave spectrum representation.

The error of power conservation is less thatn 0.1 percent for the present numerical results.

3. EXCITATION OF AN OPTICAL WAVEGUIDE

In this section, we discuss the efficiency of coupling the emission from a semiconductor laser into an optical waveguide (Fig. 3). A lens used to improve the efficiency is assumed to act as a wavefront converter but not to influence the amplitude of the field. For simplicity, we neglect the multiple reflections between the end facet of the laser and the waveguide.

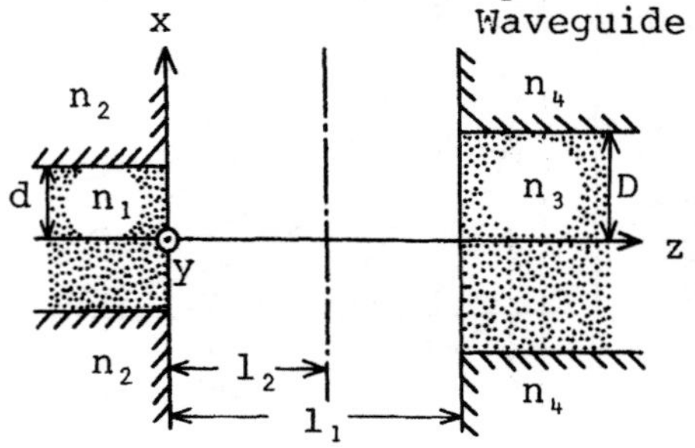

FIGURE 3. Geometry of an excitation problem.

Multiplication of the electric field of the emission immediately to the left of the lens with the phase term due to the lens focusing effect yields the field immediately to its right. Applying the Fourier transform to the field and multiplying the phase term $\exp[-jq(\mu)\cdot(l_1-l_2)]$, we get the spectrum $A(\mu)$ of the electric field incident on the waveguide end facet.

Continuity of the transverse electric and magnetic field components at the air-waveguide interface $z=l_1$ is expressed in spectral form as

$$\left\{\begin{matrix} A(\mu) \\ q(\mu)A(\mu) \end{matrix}\right\} \pm \left\{\begin{matrix} A^r(\mu) \\ q(\mu)A^r(\mu) \end{matrix}\right\} = \sum_{n=0}^{N} C_n \left\{\begin{matrix} e_n^w(\mu) \\ \beta_n^w e_n^w(\mu) \end{matrix}\right\} + \left\{\begin{matrix} A^w(\mu) \\ B^w(\mu) \end{matrix}\right\} \qquad (12)$$

where $A^r(\mu)$ is the spectrum of the reflected electric field. The spectrum and the propagation constant of the n-th guided mode of the waveguide are denoted by $e_n^w(\mu)$ and β_n^w. The spectra $A^w(\mu)$ and $-B^w(\mu)/\omega\mu_0$ represent the sepctrum of the radiated electric field and of the radiated magnetic field in the waveguide, respectively. The maximum order of the guided mode supported in the waveguide is indicated by N. Introducing the ratio $p^w(\mu)$ of $B^w(\mu)$ to $A^w(\mu)$ and following nearly the same procedure as used in the previous section, we can derive the iterative formula for $p^w(\mu)$ to determine the coupling efficiency C_n of the n-th guided mode numerically.

As is well known, the HE_{11} mode of a step index optical fiber is linearly polarized, so that the iterative method described above may be useful for analysis of the excitation efficiency of the HE_{11} mode of a single mode optical fiber.

4. CONCLUSIONS

The iterative method has been proposed to analyze rigorously the radiation pattern of the emission from a semiconductor laser beyond paraxial region. Nearly the same procedure in the method has been applied to the excitation problem of an optical waveguide by a laser emission. Numerical results show that the method converges very fast and that power conservation is satisfied sufficiently.

The method can be extended to analyze asymmetric structures.

REFERENCES

1. L.G. Cohen, Bell Syst. Tech. J., vol. 51, pp. 573-594, 1972.
2. J. Sakai and T. Kimura, IEEE J. Quantum Electron., vol. QE-16, pp. 1059-1067, 1980.
3. M. Lax, W.H. Louisell, and W.B. Mcknight, Phys. Rev. A, vol. 11, pp. 1365-1370, 1975.
4. T.E. Rozzi and G.H. in'tVeld, IEEE Trans. Microwave Theory Tech., vol. MTT-28, pp. 61-73, 1980.
5. Ph. Gelin, M. Petenzi, and J. Citerne, IEEE Trans. Microwave Theory Tech., vol. MTT-29, pp. 107-114, 1981.
6. L. Lewin, IEEE Trans. Microwave Theory Tech., vol. MTT-23, pp. 576-585, 1975.

BACKSCATTERING IN OPTICAL FIBERS

Alan Rolf Mickelson

Electronics Research Laboratory, Norwegian Institute of Technology

ABSTRACT

The effect of source coherence on backscattered signals is discussed. It is found that there are certain advantages associated with use of coherent sources and other advantages associated with the use of incoherent sources.

Backscattered signals contain a wealth of information concerning the variation of intrinsic parameters, Rayleigh scattering loss, absorption, etc. of the test fiber. The problem facing the experimentalist is to sift out the information which he desires from all that is present in a given signature. The present work addresses this signal interpretation problem within a general theoretical framework.

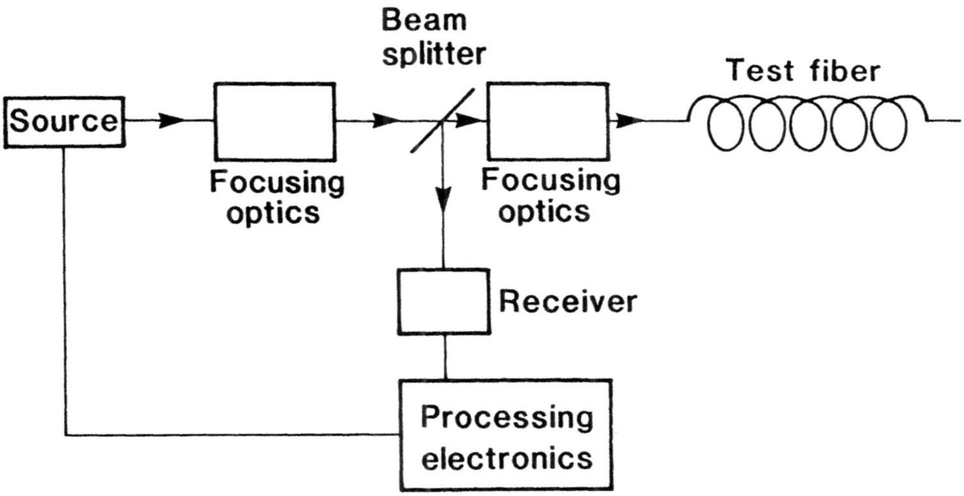

Fig.1 Schematic of an archetypal backscattering apparatus

Fig.1 illustrates a usual backscattering set-up. Such apparatus can vary greatly in signal processing technique, but the basic building blocks remain; pulsed source, focusing optics, beam splitted and receiver. The variable quantity in the backscattering apparatus which is most germane to the present discussion is the source coherence. As is known, if the sources' dimensionless bandwidth exceeds the fibers' dimensionless mode spacing the propagating mode structure within the fiber can be characterized as a continuum(1). As will presently be demonstrated, the fiber's backscattering characteristics are very different in the mode continuum limit than they are for excitation with a fully coherent source.

Clearly a system which consists of a continuum of modes is best described by a Boltzmann equation, that is, an equation of radiative transfer. One can take guidance into account in such a description by ignoring derivatives with respect to transverse coordinate. This approach, then, yields equations which possess unperturbed (no mode-coupling) solutions which decay exponentially according to the attenuation coefficient of the mode in which the power was launched. The system can be written in the form

$$\frac{\partial p_f(R,z,t)}{\partial z} + \frac{1}{c(R)} \frac{\partial p_f(R,z,t)}{\partial t} = \eta_f(R,z,t) - \chi_f(R,z,t)\, p_f(R,z,t) \qquad (a)$$

$$\frac{\partial p_b(R,z,t)}{\partial z} + \frac{1}{c(R)} \frac{\partial p_b(R,z,t)}{\partial t} = \eta_b(R,z,t) - \chi_b(R,z,t)\, p_b(R,z,t) \qquad (b) \qquad (1)$$

where p_f (p_b) is the forward (backward) travelling modal power distribution, R is the mode continuum parameter, z is the longitudinal coordinate of the assumed r,Θ,z coordinate ysytem, t is the time, c(r) is the group velocity of the mode labeled by R, η_f (η_b) is the function which describes the distribution of secondary sources in the forward (backward) direction and χ_f (χ_b) is the modal attenuation coefficient in the forward (backward) direction. Note that eq. (1) is implicitly coupled through the secondary source terms η_f and η_b, as η_f must be a function of the backward travelling wave, and η_b, of the forward travelling wave. One can indicate these dependencies explicitly by writing

$$\begin{aligned} \eta_f(R,z,t,) &= \int A(R,R';z,t)\ p_f(R')R'dR' \\ &+ \int S(R,R';z,t)\ p_b(R')R'dR' \qquad \text{(a)} \\ \eta_b(R,z,t) &= \int A(R,R';z,t)\ p_b(R')R'dR' \\ &+ \int S(R,R';z,t)\ p_f(R')R'dR' \qquad \text{(b)} \end{aligned} \qquad (2)$$

where A represents a mode-coupling redistribution operator and S a backscattering operator. If one defines an incoherent reflection coefficient $\sigma(R,R';z,t-t')$ by

$$p_b(R,z,t) = \iint R'dR'dt'\sigma(R,R';z,t-t')p_f\ (R,z,t') \qquad (3)$$

one finds immediately that

$$\begin{aligned} &\frac{\partial\sigma(R,R';z,t-t')}{\partial z} + \frac{2}{c(R,R')}\frac{\partial\sigma(R,R';z,t-t')}{\partial t} = S(R,R';z,t)S(t) \\ &- 2\ \chi(R,z,t)\ \sigma(R,R';z,t-t') \\ &+ \int R''dR''\ A(R,R'',z,t)\ \sigma(R'',R';z,t-t') \qquad (4) \\ &+ \int R''dR''\ \sigma(R,R'',t,t-t')\ A(R'',R';z,t) \\ &+ \int R''dR''R'''dR'''dt'\ \sigma(R,R'',z,t-t'')\ S(R'',R''')\ \sigma(R''',R';z,t''-t') \end{aligned}$$

where $c(R,R')$ can be defined by

$$c(R,R') = 2\frac{c(R)c(R')}{c(R)+c(R')} \qquad (5)$$

Although eq. (4) appears formidable, it will be shown that simple solutions can be obtained in most cases of practicle interest and that, indeed, these solutions agree with previous theory[2] in the limits where they should.

Marcuse's coupled mode theory[3] can well serve as a basis for a coherent theory of backscattering. The coupled mode equations serve as the analog of eq. (1). One can define a coherent reflection coefficient $\rho_{\mu\nu\ \mu'\nu'}(z,t)$

$$B_{\mu\nu}(z,t) = \sum_{\mu'\nu'} \rho_{\mu\nu\ \mu'\nu'}(z,t)\ F_{\mu'\nu'}(z,t) \qquad (6)$$

where F(B) represents a forward (backward) travelling wave and $\mu(\nu)$ is the radial (azimuthal) mode number. It should be noted here ρ is of a different nature than ρ. eq. (6) relates field amplitudes whereas eq. (3) relates power flows. To find the power flow associated with eq. (6) one must take squares of the amplitudes. If one substitutes eq. (6) into the coupled mode equations one indeed finds an equation which is the matrix analog of eq. (4). The resulting equation can be explicitly solved in the limit of a few propagating modes. Such solutions exhibit coherence effects which are absent from the solutions of eq. (4).

There are advantages associated with the use of low coherence sources and advantages associated with the employment of high coherence sources. For measurement of parameter fluctuations it is clearly preferable to operate in the mode continuum limit. However, high coherence sources offer the advantage that source modulation can enhance the precision of certain measurements, an advantage which is not available in the mode continuum limit.

REFERENCES

1. A. R. Mickelson and M. Eriksrud, Mode-continuum approximation in optical fibers, Opt. Lett. 7, 572-574 (Nov. 1982).
2. See, for example, the theory presented and verified in: A. R. Mickelson and M. Eriksrud, Theory of the backscattering process in multimode optical fibers.
3. D. Marcuse, Theory of Dielectric Optical Waveguides, (Academic Press, London, 1974). See Chap. 3.

HAMILTONIAN ANALYSIS OF THE PROPAGATION OF BEAM MODES IN MULTI-MODE GRADIENT INDEX FIBERS

WALTER K. KAHN and SHUWEN YANG*
George Washington University, Washington, D.C., U.S.A.

ABSTRACT

The formal analogy between classical mechanics/quantum mechanics and geometrical optics/physical optics is reviewed. Application to propagation of Gaussian beams in a primitive system of coupled gradient-index fibers yields some new results.

FOREWORD

This review of the application of Hamiltonian theory to propagation of beams in optical fibers is substantially based on papers the authors have submitted for publication [29,30].

PART I - GENERAL THEORY

Hamiltonian methods in geometrical optics and physical optics provide a consistent framework within which ray and wave results and interpretations articulate in an understandable way. Thanks to the labor and insights of many keen minds it has long been established that the principal features of propagation within a gradient index fiber (or a periodic sequence of focusing elements) are related more or less immediately to families of rays within the fiber [1-12]. The approach of this paper draws on formulations of classical and quantum physics noted for their power and elegance.

When the time is replaced by the axial coordinate of the fiber and the potential function by variation in the refractive index, the classical particle trajectories correspond to

*On leave from Chengdu Institute of Radio Engineering, Chengdu, Sichuan, People's Republic of China

rays and the stationary wave functions to modes of the fiber [13-14]. The principal developments center on elaboration of a further analogy between coherent states and beam-modes [15-20]. Off-axis beam-modes are generated from the (lowest) mode of the fiber by a unitary transformation, the parameters of which are the average position and average optical direction cosine (momentum). The latter are (classical) ray coordinates, Fig. 1. The off-axis mode is an eigenfunction of the operator analogous to the annihilation operator. The (complex) eigenvalue of the annihilation operator can be employed as usual to label the beam-mode and as the variable of the representative function. The real and imaginary parts of this complex variable again evolve like ray coordinates. A Gaussian beam different from the modes of the fiber may readily be represented in terms of beam-modes belonging to the fiber profile, and its evolution is then easily calculated.

TABLE I

coordinate	conjugate momentum
$\underline{x} \to x$	$\underline{p} \to \frac{\lambda\!\!\!^{-}}{i}\,\frac{\partial}{\partial x}$
$\underline{z} \to z$	$-\mathcal{H} \to \frac{\lambda\!\!\!^{-}}{i}\,\frac{\partial}{\partial z}$

The relationship between the geometric optical rays and the quasi-plane waves of physical optics has its counterpart in the relation between classical and wave mechanics. In quantum mechanics the classical coordinates x and moments p are replaced by operators $\underline{x}$ and $\underline{p}$ on an abstract state space. When the eigenvectors of $\underline{x}$ are selected as the basis, the operators have representations listed in TABLE I [13,18,23]. The result is the (Schroedinger) wave mechanics.

For the square-law gradient index

$$N(x;z) = \eta x^2 + N_o \tag{1.1}$$

The paraxial wave equation, $\exp\{iN_o z/\lambda\}$ omitted, becomes

$$-\frac{\lambda}{i}\frac{\partial}{\partial z}\psi = \left[N_o \eta x^2 - \frac{\lambda^2}{2N_o}\frac{\partial^2}{\partial x^2} \right]\psi\ , \tag{1.2}$$

which is identified with the harmonic oscillator [18].

Dirac has shown how all significant relations involving the harmonic oscillator modes can be obtained by algebraic operators techniques [17,23]. We employ his notation $|n>$ for the z independent Hermite-Gaussian modes. The term "beam mode" describes a field pattern similar to the Gaussian which constitutes the lowest true mode of the fiber, $|0>$, but is displaced from the axis of the fiber. Although such a beam mode is actually a complex superposition of the true modes of the fiber, it was discovered at an early stage [3] that these patterns meander through the fiber while retaining their shape much as true modes do. The center of the beam pattern traces the same path as a ray through the fiber. In fact the complete propagation of the beam mode was predicted by judicious combination of geometrical and physical optics which now follows naturally.

The unitary operators [17,23] $\varepsilon^{+i\bar{\xi}\underline{P}}$ and $\varepsilon^{-i\bar{\rho}\underline{Q}}$ produce respective normalized displacements $\bar{\xi}$ and $\bar{\rho}$ of any state in phase space $\sqrt{2}\ \bar{\alpha} = \bar{\xi} + i\bar{\rho}$.

The eigenvectors $|\bar{\alpha}>$ of the annihilation operator $\underline{a}$ with eigenvalue $\bar{\alpha}$ are displaced versions of the fundamental mode. The real and imaginary parts of the eigenvalue $\bar{\alpha}$ is proportional to the displacements. Since it will often be convenient to display these directly, we will use as equivalents $|\bar{\alpha}> \equiv |\bar{\xi},\bar{\rho}>$.

As pointed out at the start, the beam "modes" are actually superpositions of true modes. The precise constituents of $|\alpha>$ are [16,18]

$$|\alpha> = \exp\left(-\tfrac{1}{2}|\alpha|^2\right) \sum_{n=0}^{\infty} \frac{\alpha^n}{(n!)^{\frac{1}{2}}} |n> \tag{1.3}$$

The relative amplitudes of the modal constituents in $|\bar{\xi},0>$ are shown as a function of displacement, $\bar{\xi}$ in Fig. 2.

Finally, we apply wave theory to the propagation of a beam mode in the fiber. We assume any state at a plane $\zeta \equiv z/\lambda\!\!\!^{-}$ of the fiber may be related to an initial state at a plane ζ_0 by the transfer operator $\underline{T}(\zeta,\zeta_0)$. In terms of the Hamiltonian

$$|\alpha;\zeta> = \underline{T}(\zeta,\zeta_0)|\alpha;\zeta_0>$$

$$= \varepsilon^{-i\underline{H}(\zeta-\zeta_0)} \exp(-\tfrac{1}{2}|\alpha|^2) \sum_{n=0}^{\infty} \frac{\alpha^n}{(n!)^{\frac{1}{2}}} |n;\zeta_0>$$

$$= \varepsilon^{-\frac{i\hbar\omega_c}{2}(\zeta-\zeta_0)} |\alpha\varepsilon^{-i\hbar\omega_c(\zeta-\zeta_0)};\zeta_0> \qquad (1.4)$$

PART II - A PRIMITIVE SYSTEM OF COUPLED FIBERS

We consider the index profile shown in Fig. 3. The structures are invariant in the y and z directions. The equation satisfied by physical optics scalar wave field is again cast into the form of the Schroedinger equation, this time with potential shown in the lower half of the figure. This is solved for the normal modes by expansion in Hermite-Gaussian functions employing a coupled mode technique. Five modes in each guide were taken into account. A limiting case of this potential was treated more directly by Merzbacher [27]. Related calculations have been made by Feit and Fleck [28]. Numerical results for the propagation constants for the even and odd normal modes

$$\beta_{\substack{en\\o}} \simeq k_o N_o - \omega_{\substack{en\\o}} \qquad k_o = 2\pi/\lambda_o \qquad (2.1)$$

of the fibers shown in Fig. 3 were found.

A beam mode with ray normalized displacement parameter $\bar{\alpha}$ can be represented as a superposition of the Hermite-Gaussian modes of the individual fibers. The required relative mode amplitudes for the nth mode of fiber # 1 follow from the recognition of beam-modes as coherent states. They are given by eq (1.3),

$$A_n^{(1)} = \frac{(\bar{\alpha})^n}{(n!)^{\frac{1}{2}}} . \qquad (2.2)$$

Conversely we may inspect the relative modal amplitudes in fiber # 1 at any cross-sectional plane to discover if $|\bar{\alpha}|$, i.e. $\bar{\xi}$ and $\bar{\rho}$, have been reduced. In 1966 Marcuse [25] showed that, as a consequence of Liouville's theorem of statistical mechanics, it is impossible to achieve this desired effect in a single fiber [26]. It has been shown that through the introduction of new degrees of freedom, in the form of a second fiber loosely coupled to the main fiber, the desired result becomes consistent with the constraint of Liouville's theorem and can, in fact, be achieved in a substantial degree.

REFERENCES

1. Pierce, J.R., "Modes in Sequences of Lenses," Proc. Nat'l Acad. Sci., 47, November 1961, pp. 1808-1813.
2. Deschamps, G.A., and P.E. Mast, "Beam Tracing and Applications Proceedings of the Symposium on Quasi-Optics, Vol. XIV, Polytechnic Press, N.Y., 1964, pp. 379-395.
3. Tien, P.K., J.P. Gordon, and J.R. Whinnery, "Focusing of a Light Beam of Gaussian Field Distribution in Continuous and Periodic Lens-Like Media," Proc. IEEE 53, February 1965, pp. 129-136.
4. Marcatili, E.A.J., "Modes in a Sequence of Thick Astigmatic Lens-Like Focusers," B.S.T.J., 43, November 1964, pp. 2887-2904.
5. Kogelnik, H., "Imaging of Optical Modes-Resonators with Internal Lenses," B.S.T.J., 44, March 1965, pp. 455-494.
6. Kurauchi, N., and W.K. Kahn, "Rays and Ray Envelopes within Stable Optical Resonators Containing Focusing Media," Applied Optics, 5, June 1966, pp. 1023-1029.
7. Arnaud, J.A., "Nonorthogonal Optical Waveguides and Resonators," B.S.T.J., 49, November 1970, pp. 2311-2348.
8. Arnaud, J.A., "Beam and Fiber Optics," Academic Press, New York, 1976.
9. Marcuse, D., "Light Transmission Optics," Van Nostrand-Rheinhart Co., New York, 1972.
10. Shin, S.Y. and L.B. Felsen, "Gaussian Beam Modes by Multipoles with Complex Source Point," J. Opt. Soc. of Am., 67, May 1977, pp. 699-700.
11. Stoler, D., "Operator Methods in Physical Optics," J. Opt. Soc. Am. 71, March 1981, pp. 334-341.
12. Nazarathy, M., and J. Shamir, "First Order Optical Systems - A Generalized Canonical Operator Group Representation," Part I- IV J. Opt. Soc. Amer. 72, 1982.
13. Gloge, D. and D. Marcuse, "Formal Quantum Theory of Light Rays," J. Opt. Soc. Amer., 59, December 1969, pp. 1629-1631.
14. Eichmann, G., "Quasi-Geometric Optics of Media with Inhomogeneous Index of Refraction," J. Opt. Soc. Am., 61, February 1971, pp. 161-168.
15. Siegman, A.E., "Hermite-Gaussian Functions of Complex Argument as Optical-beam Eigenfunctions," J. Opt. Soc. Am. 63, September 1973, pp. 1093-1094.
16. Glauber, R.J., "Quantum Optics and Electronics, Les Houches Lectures 1964," C. DeWitt, A. Blandin and C. Cohen-Tannoudji eds., Gordon and Breach, New York, 1965.
17. Klauder, R.J. and E.C.G. Sudarshan, "Fundamentals of Quantum Optics," W.A. Benjamin, Inc., New York, 1968.
18. Cohen-Tannoudji, C.B. Diu, F. Laloë, "Quantum Mechanics," Vol. I & II, John Wiley & Sons, New York, 1977.

19. Shah, H.K., "An Investigation in Hamiltonian Optical Theory of the Propagation of Coherent State Modes in Gradient Index Fibers," D.Sc. Dissertation, The George Washington University, Washington, D.C., May 4, 1980.
20. Kahn, W.K., and H.K. Shah, "Gaussian Beam-Modes as Coherent States," Proceedings of the International U.R.S.I. Symposium, Munich, August 26-29, 1980, pp. 345B/1-345B/4.
21. Born, M. and E. Wolf, "Principles of Optics," Pergamon Press, 1959.
22. Goldstein, H., "Classical Mechanics," 2nd Ed., Addison-Wesley, Reading, MA, 1980.
23. Dirac, P.A.M., "The Principles of Quantum Mechanics," Fourth Ed., Oxford at the Clarendon Press, 1958.
24. Marcuvitz, N., "Waveguide Handbook," Vol. 10, Radiation Laboratory Series, McGraw-Hill, N.Y., 1951.
25. Marcuse, D., "Physical Limitations on Ray Oscillation Suppressors," Bell Syst. Tech. Journ. 45, pp. 743-751, May-June 1966.
26. Marcuse, D., "Compression of a Bundle of Light Rays," Applied Optics, 10 No. 3, pp. 494-497, March 1971.
27. Merzbacher, E., Quantum Mechanics, John Wiley & Sons, New York, 1961, Ch. 5, Sect. 5, pp. 64-76.
28. Feit, M.D. and J.A. Fleck, Jr., "Propagating Beam Theory of Optical Fiber Cross Coupling," Journal of the Opt. Soc. of America, Vol. 71, No. 11, pp. 1361-1372, November 1981.
29. Kahn, W.K. and S.W. Yang, "Hamiltonian Methods in Optics: Optical Slab-Guide," accepted for publication, J. Opt. Soc. Amer., May 1983.
30. Kahn, W.K. and S.W. Yang, "Beam Oscillation Suppressor for Multimode Gradient Index Fibers," accepted for publication, Optics Letters, April 1983.

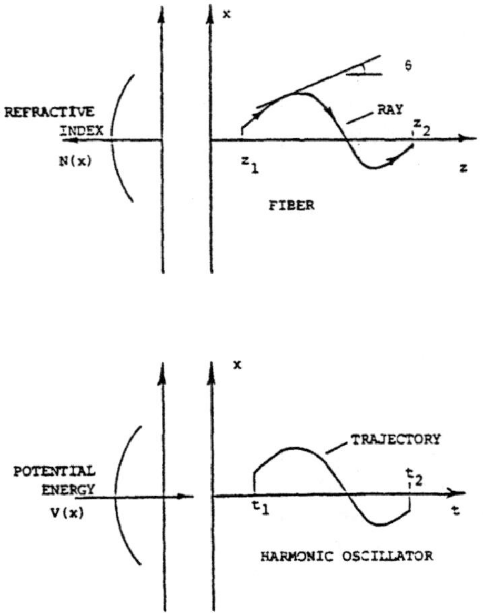

FIGURE 1. Analogous Optical & Mechanical Systems

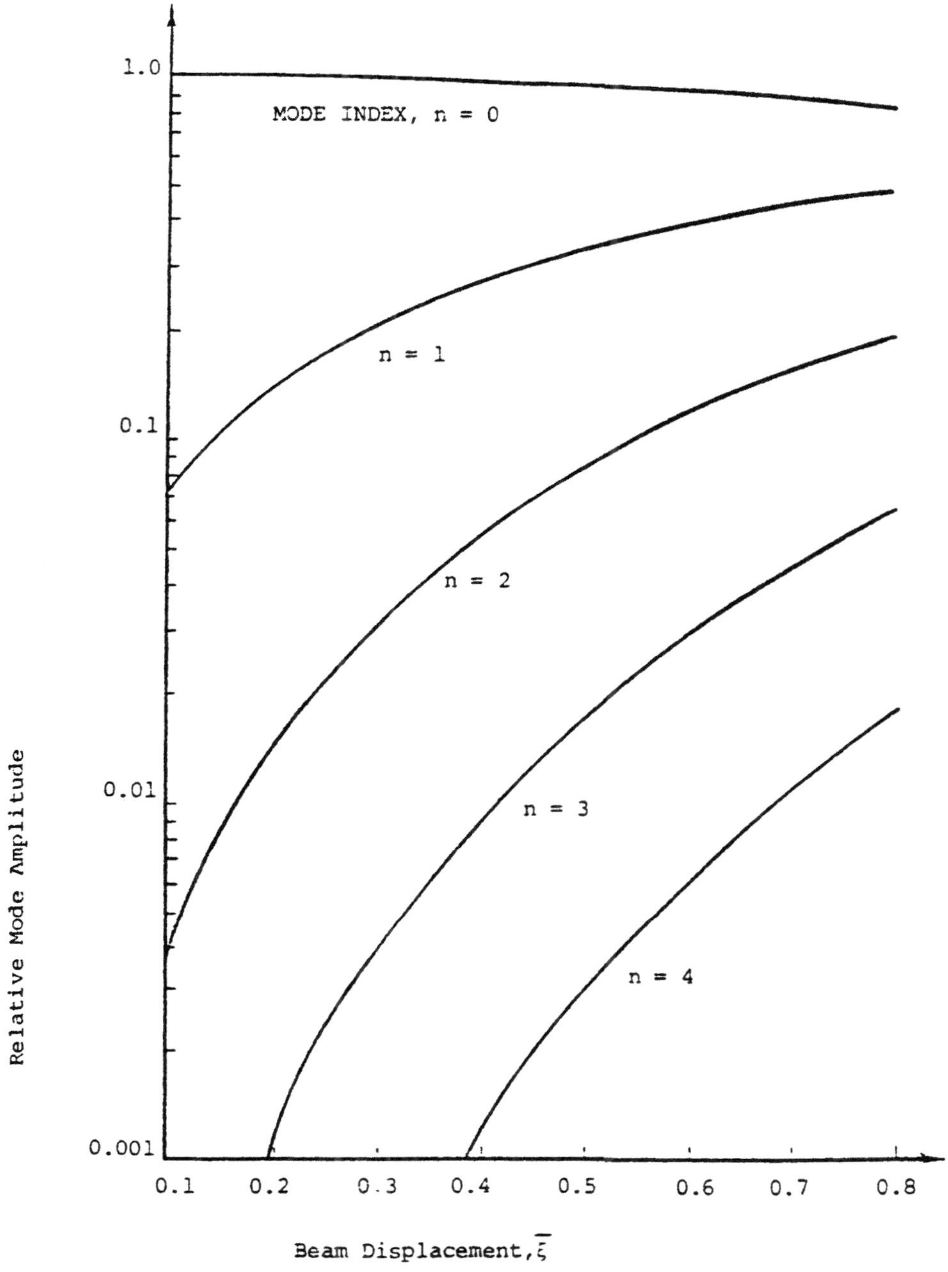

FIGURE 2. Relative amplitude of true modal constituents for off axis beam "mode".

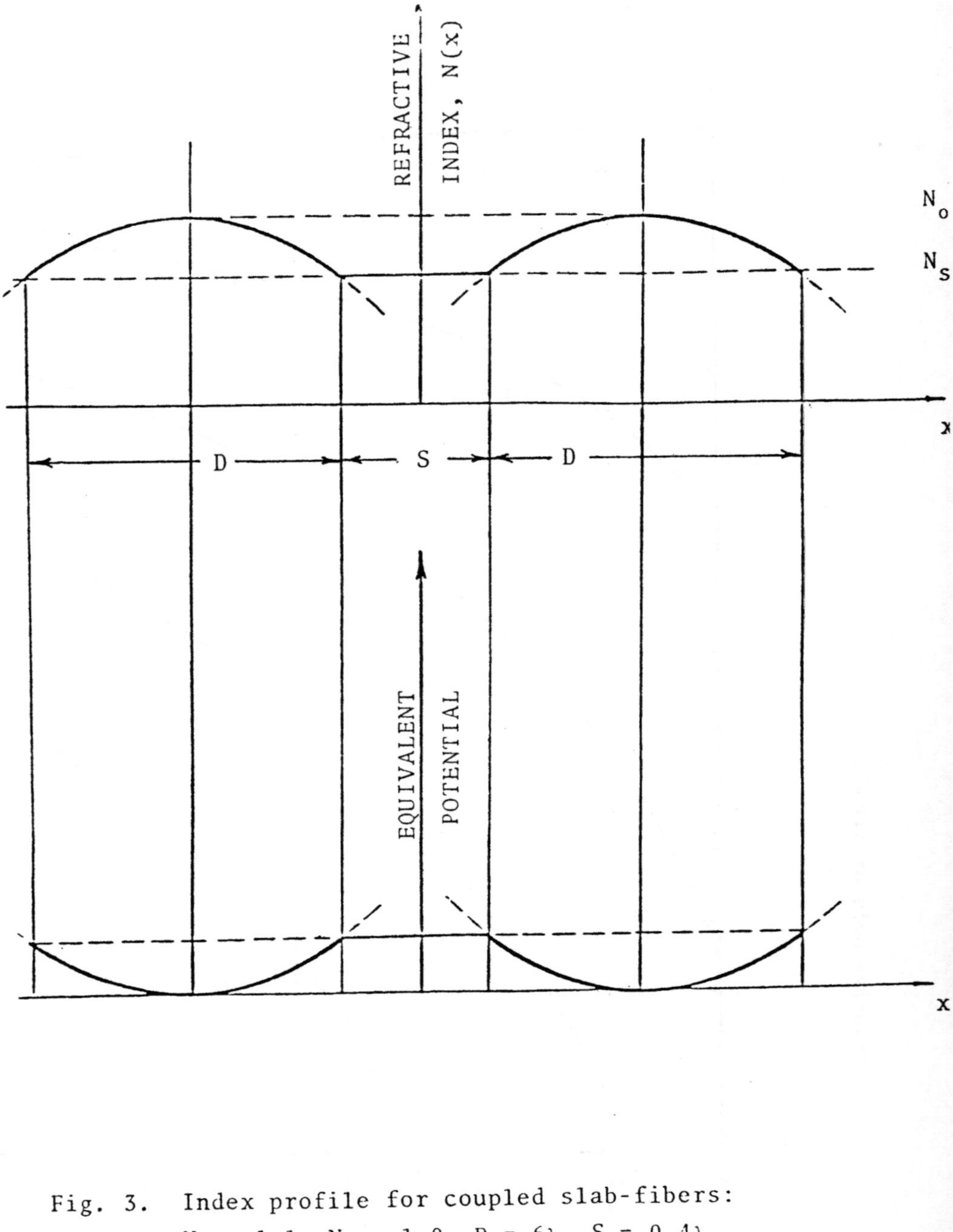

Fig. 3. Index profile for coupled slab-fibers: $N_o = 1.1$, $N_S = 1.0$, $D = 6\lambda$, $S = 0.4\lambda$, $\lambda = 0.55\text{x}10^{-6}$ m.

ASYMPTOTIC MODE EXPANSION IN DIFFUSED OPTICAL WAVEGUIDES WITH GAUSSIAN PROFILE

Fan Jun-qing

Changchun Institute of Physics, Academia Sinica

ABSTRACT

Gaussian profile of diffused optical waveguides is approximated by a suitable polynomial function. Asymptotic splutions of mode effective refractive indices in an analytical form for guided modes are derived. Also discussed is the use of asymptotic solution equations in determining the surface index and diffusion depth of waveguides from measured values of mode effective indices. We shall give some examples. And we believe that our approach yields more accurate results than those published in previous papers.

The propagation characteristics of diffused optical waveguides with a Gaussian index profile are often evaluated by the approximate assumption of a parabolic[(1)] or a secant hyperbolic profile[(2)]. In this approximation, the error is large, especially for modes near cut-offs.

In this paper, a suitable polynomial is used to approximate the Gaussian function. The asymptotic expansion method[(3)] is adopted and the asymptotic solutions of mode effective indices are derived. With the aid of the asymptotic solution equations, the surface index and the diffusion depth of waveguides are determinable from measured values of mode effective indices.

The refractive index profile of optical waveguides is assumed as the following:

$$n^2=n_b^2+2n_b\Delta n\exp(-x^2/d^2), \quad x\geq 0 \qquad x<0 \tag{1}$$

where n_b is the substrate index, $\Delta n=n_s-n_b$, n_s is the surface index ($n_s>N_b>1$), and d is the diffusion depth. For the index profile given by Eq. (1), closed form solution for mode effective indices do not exist. In our approach, we consider the following profile as an approximation of Eq. (1)

$$n^2=n_s^2-2n_b\Delta n f(x) \qquad (2)$$

where

$$f(x)=(x/d)^2-a(x/d)^4+b(x/d)^6 \qquad (3)$$

Suitable values of a and b are found to be 0.4820 and 0.1141, respectively. It can be shown that these values of a and b equalize the areas under the two curves given by Eqs. (1) and (2) up to the diffusion depth and exp(-1)=1-f(d).

We assume an abrupt decay of the evanescent field at the surface (x=0) of the diffused waveguides; thus, the field at x=0 is assumed zero. Hence, the propagation constants are the antisymmetric solutions of the following eigen-equation, for the profile given by Eq. (2):

$$\frac{d^2\Phi}{dx^2} - (K^2 f-K\delta)\Phi=0, \quad \Phi\to 0 \text{ when } |x|\to\infty \qquad (4)$$

$$K=k_o(2n_b\Delta n)^{\frac{1}{2}} \qquad (5)$$

$$\delta=(n_s^2k_o^2-\beta^2)/K \qquad (6)$$

where $K_o=2\Pi/\lambda$ is the free space wavenumber, β is the progation constant, and Φ is the electric field for TE modes.

Using the asymptotic expansion method[3], the asymptotic solutions for δ in Eq. (4) can be obtained

$$\delta=\delta_o+\frac{\delta_1}{K}+\frac{\delta_2}{K^2} \qquad (7)$$

where

$$\delta_o=\frac{2m+1}{d}, \quad m=0,1,2,... \qquad (8)$$

$$\delta_1= -\frac{3}{8}a\left(\delta_o^2+\frac{1}{d^2}\right) \qquad (9)$$

$$\delta_2= -\left(\frac{17}{64}a^2 - \frac{5}{16}b\right)\delta_o^3-\left(\frac{67}{64}a^2 - \frac{25}{16}b\right)\frac{\delta_o}{d^2} \qquad (10)$$

Substituting Eqs. (5), (6) and (8)-(10) into Eq. (7) and letting $m=2\mu+1$, the mode effective indices for the asymmetric waveguides are obtained

$$N_\mu^2=n_s^2-(2n_b\Delta n)\quad \frac{(4\mu+3)}{k_o d}+\frac{3}{8}\,a\left(\frac{4\mu+3}{k_o d}\right)^2+\frac{3}{8}\,a\frac{1}{(k_o d)^2}$$
$$+\left(\frac{17}{64}a^2-\frac{5}{16}b\right)\left(\frac{4\mu+3}{k_o d}\right)^3\frac{1}{(2n_b\Delta n)^{\frac{1}{2}}}$$
$$+\left(\frac{67}{64}a^2-\frac{25}{16}b\right)\frac{(4\mu+3)}{(k_o d)^3(2n_b\Delta n)^{\frac{1}{2}}} \qquad (11)$$

where $N_\mu=\beta_\mu/k_o$, $\mu=0,1,2,\ldots$, are the mode effective indices.

It can be easily shown that Eq. (11) coincides with the result obtained in Ref. 4.

We have calculated the mode effective indices by using Eq. (11) for a typical set of parameters: $n_b=2.20$, $\Delta n=0.05$ and $\lambda=0.6328\mu m$. Fig. 1 shows our results as compared with others' results.

If the measured values of mode effective indices in a diffused optical waveguide are given, the surface index and diffusion depth of the waveguide can be determined by substituting these values into Eq. (11) and performing a computer program. We have developed a program for this purpose. The results are tabulated in Table 1, in which N_o and N_1 are obtained by the rigorous ray technique for a Gaussian profile with $n_b=2.20$, $\Delta n=0.05$ and various values of diffusion depth d. Note that our results agree well with those by ray technique, and are more accurate than the results by other approximations.

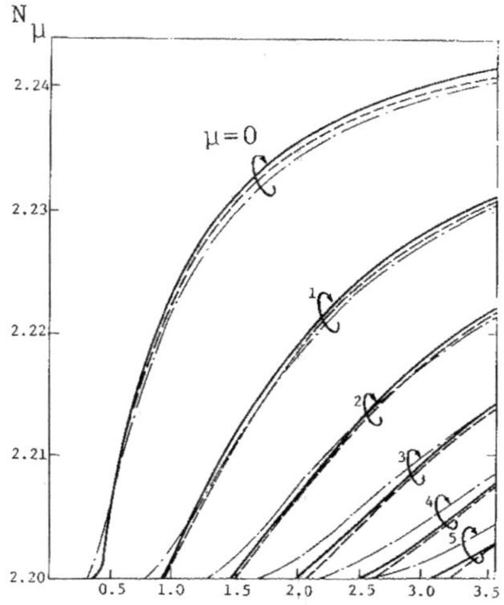

Fig.1 Solid curves, by the rigorous ray technique; dashed curves, by Eq.(11); dot-and-dash curves, by assuming hyperbolic profile.

Table 1. Waveguide Parameters Calculated from Mode Indices (d is in μm)

Mode Number	Mode Effective Indices	
	N_o	N_1
2	2.22385	2.20129
3	2.23157	2.21158
3	2.23580	2.21919
4	2.23843	2.22441

Mode Number	Ray Technique		Asymptotic Solution		Secant Hyperbolic		Parabolic	
	d	Δn	d	Δn	d	Δn	d	Δn
2	1.0	0.050	1.1	0.050	0.8	0.059	1.7	0.041
3	1.5	0.050	1.5	0.051	1.4	0.053	2.1	0.047
3	2.0	0.050	2.0	0.051	2.0	0.052	2.5	0.049
4	2.5	0.050	2.5	0.051	2.5	0.051	3.0	0.049

REFERENCES

1. R. D. Standly and V. Ramaswamy, Appl. Phys. Lett., Vol. 25, P. 711, 1974.
2. A. Kumar and E. Khular, Opt. Commun., Vol. 27, P. 349, 1978.
3. J. M. Arnold and L. B. Felsen, J. Acoust. Soc. Am., Vol. 67 No. 3, P. 757, 1979.
4. H. Ikuno, IEEE Trans. Microwave Theory Tech., MTT-26, No. 4, P. 261, 1978.

PART FIVE

PLANAR OPTICAL WAVEGUIDES

CARRIER-INDUCED INDEX CHANGE IN AlGaAs AND InGaAsP LASERS AND ITS INFLUENCE ON OPTICAL WAVEGUIDING

JOANNE S. MANNING AND ROBERT OLSHANSKY

ABSTRACT

A very large carrier-induced index change is reported for broad area AlGaAs and InGaAsP lasers. This index change has a strong influence on the waveguiding characteristics of stripe-geometry lasers and other laser operating characteristics.

INTRODUCTION

The carrier-induced index change affects such diverse aspects of diode laser behavior as waveguiding,[1] frequency modulation,[2] spectral linewidth,[3] and the effects of photon injection.[4] This paper presents the most extensive measurement yet reported of the carrier-induced index change. The observed index change with injected carrier density is $-(1.2\pm0.2)\times10^{-20}\,cm^3$ for AlGaAs and $-(2.3\pm0.6)\times10^{-20}\,cm^3$ for 1.3 μm InGaAsP. This corresponds to an index change at threshold of -0.03 to -0.06 for AlGaAs and -0.04 to -0.10 for InGaAsP lasers. The observed change at threshold is much larger than previously reported for AlGaAs lasers[5-7] and comparable to values recently reported by Turley[8] for InGaAsP.

MEASUREMENT OF CARRIER-INDUCED INDEX CHANGE

The index change is measured from the wavelength shift of a longitudinal mode as the pulsed bias current is varied below threshold: $\delta n=\delta\lambda(\bar{n}/\Gamma\lambda)$, where $\bar{n}$ is the effective index found from the longitudinal mode spacing and Γ is the optical confinement factor. The wavelength shift is observed to be linearly related to the carrier density for both AlGaAs and InGaAsP lasers, as shown in Figs. 1a and 1b. The relation between the index change at threshold and the threshold carrier density is shown in Figs. 2a and 2b for AlGaAs and InGaAsP lasers.

While previous studies included one or two lasers, this study reports on a large number of devices with a variety of active layer thicknesses and doping levels. It includes eight AlGaAs lasers from three different laboratories, grown by both LPE and MO-CVD processes, as well as nine 1.3 μm InGaAsP lasers. Device parameters and index changes are summarized for each diode in Table I.

Another significant advantage over previous studies is the direct experimental determination of the injected carrier density. The use of broad area lasers combined with measurements of the differential carrier lifetime permits an accurate determination of the injected carrier density as a function of current.[9]

The observed index change is an order of magnitude larger than that predicted for free-carrier absorption.[5] The index change is attributed primarily to the carrier-induced shift of the absorption edge, as suggested recently by several authors.[5-10] A simple model is presented which incorporates the absorption edge shift and free-carrier absorption to explain the magnitude of the index change as well as its observed wavelength dependence. The model also correctly predicts the current-produced change in dispersion $\lambda dn/d\lambda$ as deduced from the wavelength spacing between successive longitudinal modes.

WAVEGUIDING IN STRIPE GEOMETRY LASERS

These very large refractive index changes play a key role in determining the waveguiding characteristics of stripe-geometry lasers. A study of the near fields and far fields of 8-μm stripe oxide-isolated AlGaAs lasers shows that both gain-guiding and index anti-guiding must be included in an accurate model of waveguiding. The dielectric profile is determined experimentally[10] from measurements of the spontaneous emission profile at the laser facet, the carrier-induced index change, and the increase in gain with current. As shown in Fig. 4a, the observed spontaneous profile is well described by

$$S(x) \propto 1-\left(\frac{x}{w}\right)^{\alpha} \tag{1}$$

where α and w are fitted parameters whose values are shown in Table II. Since the spontaneous emission S(x) is proportional to the square of the carrier distribution N(x),[9]

$$N(x) \propto 1-\frac{1}{2}\left(\frac{x}{w}\right)^{\alpha}. \tag{2}$$

The dielectric profile in the junction plane contains real and imaginary parts:

$$\varepsilon(x)=n\ \Delta n(x)+i\frac{n}{k}\ \Delta g(x). \tag{3}$$

The index profile $\Delta n(x)$ is found from N(x) and the total observed index change at threshold, δn:

$$\Delta n(x)=\delta n(1-\frac{1}{2}(\frac{x}{w})^{\alpha}). \quad (4)$$

Similarly, the gain profile $\Delta g(x)$ is found from $N(x)$ and δg, the total current-produced change in gain measured at lasing wavelength from the longitudinal mode spectrum below threshold.[11] As shown in Fig. 3, the gain at lasing wavelength is observed to be linear with current, and thus linear with $(N(x))^2$. The gain profile can therefore be written as

$$\Delta g(x)=\delta g(1-(\frac{x}{w})^{\alpha}). \quad (5)$$

The dielectric profile determined in this way is then used to solve the waveguide problem for the stimulated emission near field and far field.[12] As shown in Table II and Figs. 4b and 4c, the predicted near field and far field profiles are in excellent agreement with the observed patterns.

SUMMARY

This work differs from earlier studies in several important ways: the current-induced index change of the active layer is measured to be very large; the dielectric profile shape is determined empirically rather than theoretically; the profile shape is fit to a power-law profile; and a closed-form expression is used to predict the properties of the guided wave.

ACKNOWLEDGEMENTS

We are grateful to D. Botez of RCA Laboratories and W. Streiffer and D. Scifres of Xerox PARC for generously providing us with their broad-area lasers.

REFERENCES

1. J. Manning and R. Olshansky, J. Appl. Phys. 53, 840 (1982).
2. S. Kobayashi, Y. Yamamoto, M. Ito, and T. Kimura, J. Quantum Electron. QE-18, 582 (1982).
3. C.H. Henry, IEEE J. Quantum Electron. QE-18, 259 (1982).
4. R. Lang, IEEE J. Quantum Electron. QE-18, 976 (1982).
5. C.H. Henry, R.A. Logan, and K.A. Bertness, J. Appl. Phys. 52, 4457 (1981).
6. S.E.H. Turley, G.H.B. Thompson, and D.F. Lovelace, Electron. Lett. 15, 356 (1979).
7. M. Ito and T. Kimura, IEEE J. Quantum Electron. QE-16, 910 (1980).
8. S.E.H. Turley, Electron. Lett. 18, 590 (1982).
9. C.B. Su and R. Olshansky, Appl. Phys. Lett. 41, 833 (1982).
10. J. Manning and R. Olshansky, Electron. Lett. 17, 506 (1981).
11. B. Hakki and T. Paoli, J. Appl. Phys. 44, 4113 (1973).
12. R. Olshansky, J. Appl. Phys. 53, 843 (1982).

TABLE I. Parameters and results for broad-area AlGaAs and InGaAsP lasers in study of carrier-induced index change.

	DIODE	d (μm)	p_ox10^{18} (cm^{-3})	λ (nm)	Γ	$\bar{n}$	J_{th} (kA/cm^2)	N_{th}x10^{18} (cm^{-3})	δλ (nm)	δn_{th}
	AlGaAs									
GTE	163-1	0.19	~0.1	830	0.60	4.67	2.3	3.2	-2.91	-0.028
	163-2	0.13	~0.1	829	0.38	4.47	1.6	3.2	-2.34	-0.034
	163-3	0.06	~0.1	849	0.17	4.10	1.2	4.1	-1.76	-0.050
	164-6	0.07	~0.1	838	0.20	4.08	1.2	3.7	-1.97	-0.048
RCA	R1	0.14	~0.1	839	0.44	4.41	1.2	2.7	-2.31	-0.028
	R2	0.08	~0.1	837	0.22	4.26	1.2	3.6	-1.65	-0.038
XEROX	X1	0.06	3	827	0.15	4.18	1.5	4.3	-1.79	-0.061
	X2	0.05	2	827	0.13	4.10	1.3	4.7	-1.61	-0.061
	1.3 μm InGaAsP									
GTE	192-4	0.11	1	1299	0.21	3.80	3.3	1.94	-3.94	-0.055
	192-5	0.12	1	1297	0.24	3.66	3.1	1.83	-4.37	-0.051
	192-6	0.10	1	1300	0.18	3.77	3.2	2.01	-4.41	-0.071
	192-9	0.12	1	1297	0.24	3.91	3.2	1.85	-5.89	-0.074
	194-1	0.09	~0.2	1290	0.15	3.78	2.1	3.33	-4.09	-0.080
	194-4	0.11	~0.2	1300	0.21	3.80	2.1	3.39	-7.02	-0.098
	194-8	0.09	~0.2	1289	0.15	3.72	2.2	3.33	-4.97	-0.096
	194-10	0.10	~0.2	1289	0.18	3.78	2.3	3.31	-5.63	-0.092
	236-6	0.33	~0.2	1309	0.20	3.78	2.9	2.48	-10.50	-0.043

TABLE II. Parameters for 8 μm stripe AlGaAs lasers: α and w are fitted to the spontaneous near field; δn and δg are the observed current-produced changes in index and gain at threshold; FWHM is the lasing near-field width; and FAHM is the lasing far-field width.

Laser no.	α	w(μm)	δn	δg(cm^{-1})	FWHM(μm) Exp	FWHM(μm) Theory	FAHM Exp	FAHM Theory
1	4.3	5.8	-0.062	1010	5.0	4.7	7.5°	7.5°
2	3.5	8.4	-0.044	960	6.4	6.4	6.1°	6.4°
3	4.5	7.5	-0.047	760	6.1	5.9	6.1°	5.8°

a

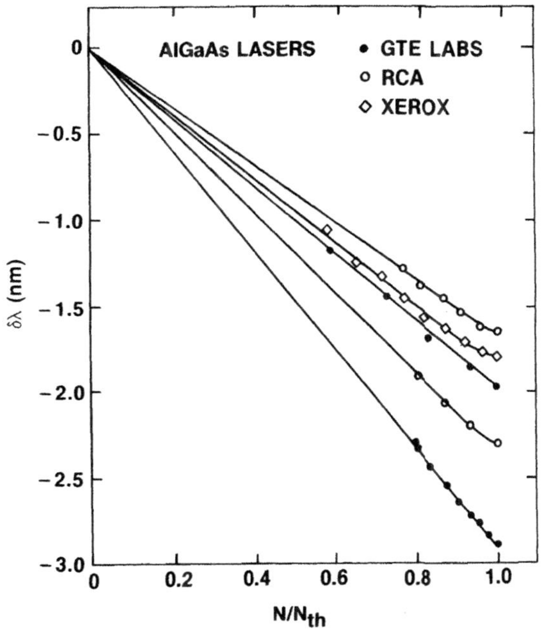

b

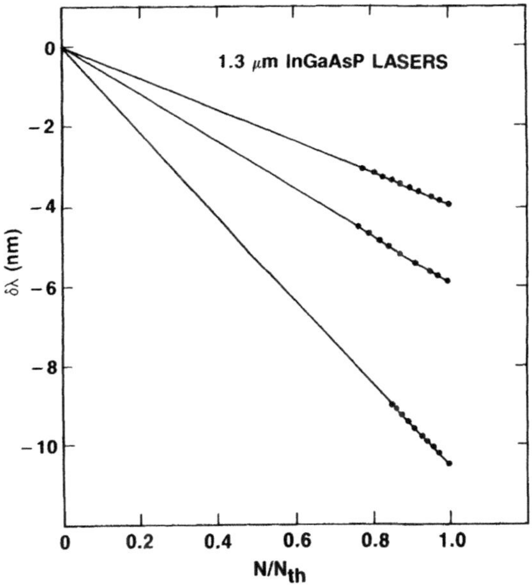

FIGURE 1. Longitudinal mode wavelength shift vs. normalized carrier density for broad-area a) AlGaAs lasers, and b) 1.3 μm InGaAsP lasers.

a

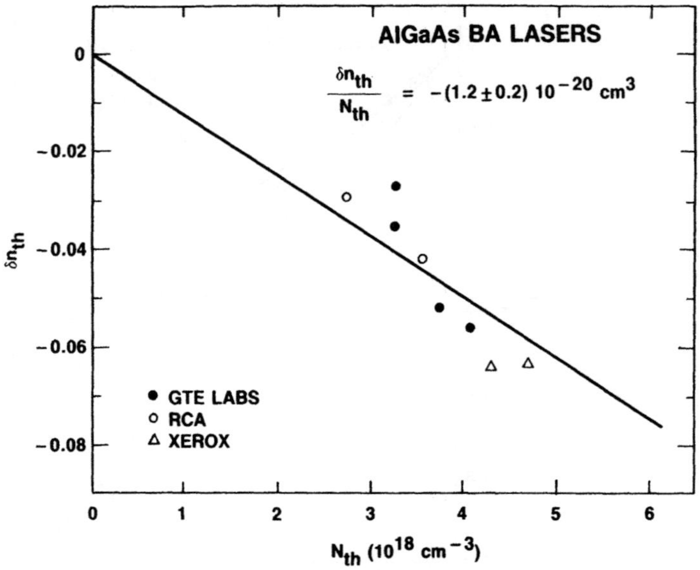

b

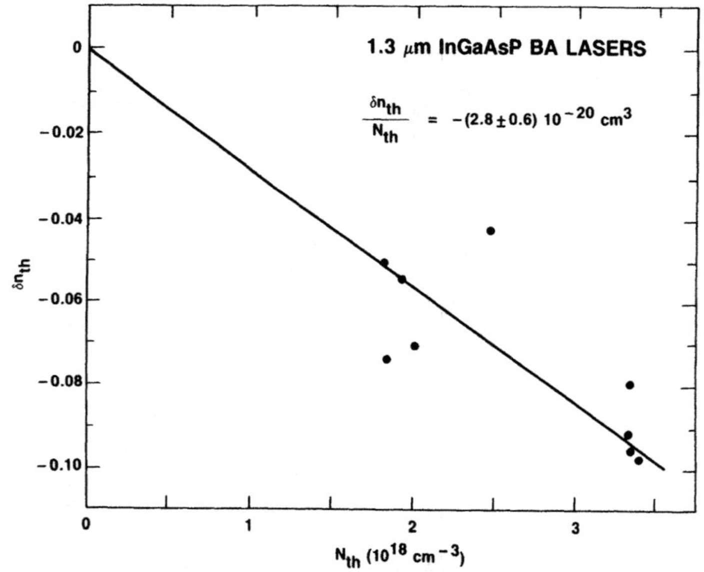

FIGURE 2. Index change at threshold vs. carrier density at threshold for a) AlGaAs lasers, and b) 1.3 μm InGaAsP lasers.

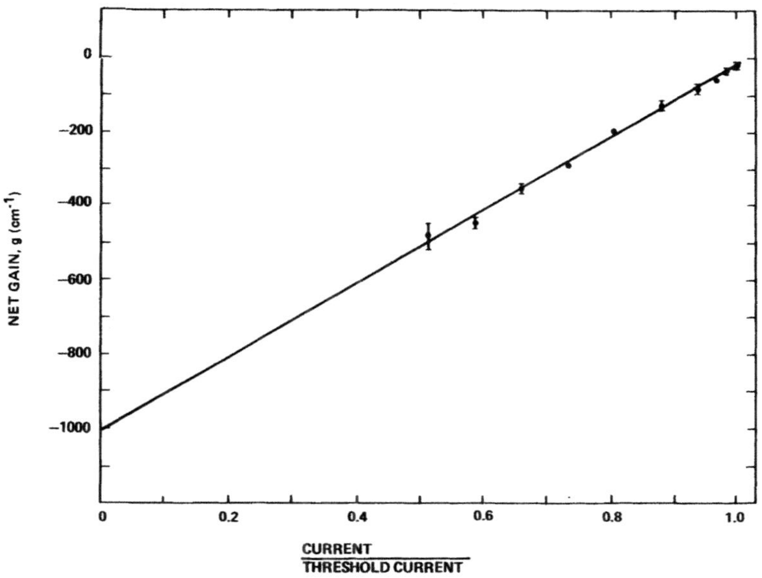

FIGURE 3. Modal gain of the active layer measured at lasing wavelength as a function of current for an 8-μm stripe AlGaAs laser in pulsed operation.

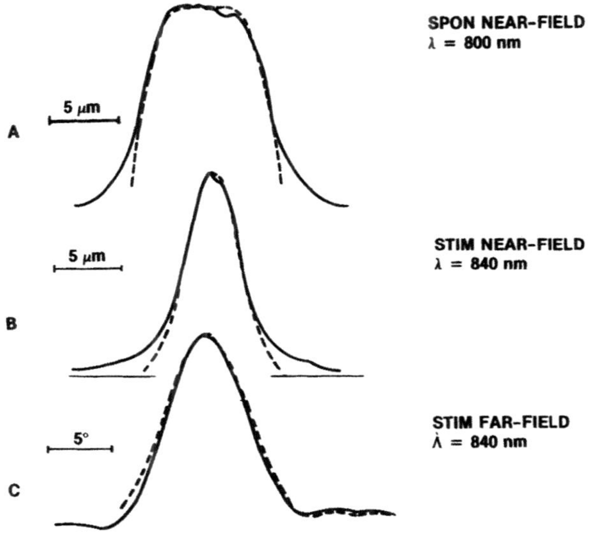

FIGURE 4. a) Observed (solid line) and fitted (dashed line) spontaneous emission profile for an 8-μm stripe AlGaAs laser; b) and c) Near and far fields of an 8-μm stripe AlGaAs laser. The solid line is observed profiles, and the dashed line is predicted from waveguiding model including index antiguiding factor.

WAVEGUIDE TYPE OPTICAL ISOLATOR AND CIRCULATOR USING MAGNETO-OPTIC GARNET THIN FILMS

Yasumitsu MIYAZAKI

Department of Information Engineering
Toyohashi University of Technology,
Tempaku-cho, Toyohashi 440 Japan

1. INTRODUCTION

Propagation and mode conversion properties of optical waves in film waveguides made of magneto-optic materials such as garnet crystals are studied, in order to construct optical thin film isolators and circulators using magneto-optic effects.

Waveguide type optical isolators are very important as uni-directional waveguides and are indispensable for stabilization of laser oscillation and separation of forward and backward optical waves in integrated optical circuits. However, waveguide type optical isolators with sufficient properties have not been realized,[1] because there are some difficulties to the fabrications and material properties. The author has shown the analysis on hybrid modes for the case of optical modulators using the Kerr effects and non-reciprocal circuits of optical isolators using the Faraday effects in gyrotropic garnet films with isotropic ZnO and anisotropic $LiIO_3$ top layers for accomplishing mode matching.[2,3] In order to construct optical isolators, in these constructions anisotropic crystals are required for reciprocal circuits. However, it is difficult to grow anisotropic single crystals on garnet films and substrates that are almost isotropic. It is found that, from a fabricational point of view, the Cotton-Mouton effects of gyrotropic films are very useful for reciprocal circuits.

The author has fabricated YIG and Bi-substituted YIG (Bi:YIG) single crystal films of good quality on GGG substrates by RF sputtering method. In this paper, propagation properties of optical waves in the garnet films are investigated. TE-TM mode conversions due to the Faraday and the Cotton-Mouton effects are measured in Bi:YIG single crystals sputtered on GGG substrates.

Birefringence effects in sputtered garnet films that are very useful for mode matching are found. From the experimental results, the design data for non-reciprocal circuits using the Faraday effect and reciprocal circuits using the Cotton-Mouton effect of waveguide type optical isolator are determined, such as propagation length and applied magnetic fields. By using ferrite magnets a waveguide type optical isolator is constructed practically with short propagation length and small magnetic external fields.

2. FABRICATION OF MAGNETO-OPTIC GARNET FILMS

Optical isolators consist of non-reciprocal part, reciprocal part and mode filter. For non-reciprocal part, magneto-optic garnet films are useful and fabricated by RF sputtering method. Substrates of GGG crystals are heated to 750°C and RF power is 85W. The composition of the sputtering target is $Y_3(Al,F)_5O_{12}$ and $Bi_1Y_2Fe_4Al_1O_{12}$.

Sputtered Bi:YIG thin-films on GGG substrates are single crystals. Film conpositions of thickness 1.68μm are measured by IMA. Saturation magnetizations are 90.5G for Bi:YIG and 1700G for YIG. Bi:YIG thin films fabricated by RF sputtering method have magnetic anisotropy with the easy axis parallel to the film plane and their coercive force is 3 Oe. Birefringenc

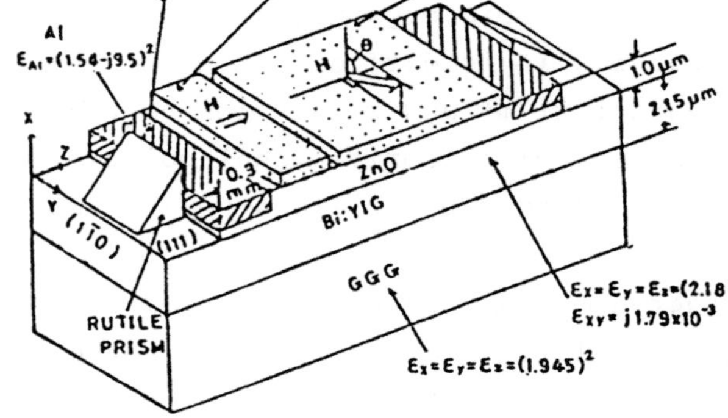

FIGURE 1. Optical film isolator

is $\Delta n=1.17x10^{-3}$. The refractive index of Bi:YIG film is 2.18 for the wavelengh $\lambda=1.15\mu m$. The spectific Faraday rotation and the absorption coefficient are 300-800deg/cm and $2cm^{-1}$. Faraday rotations of Bi:YIG films are by a few times larger than those of YIG.

3. OPTICAL FIELDS IN GARNET FILMS

The structure of optical isolator is shown in Fig.1. The propagation loss is 2dB/mm. Hybrid eigen modes in magneto-optic film waveguides with Faraday effects are analyzed and their dispersion curves are shown in Fig.2. Fig.3 shows propagation length dependencies of mode conversions when maximum mode conversions are 50%.

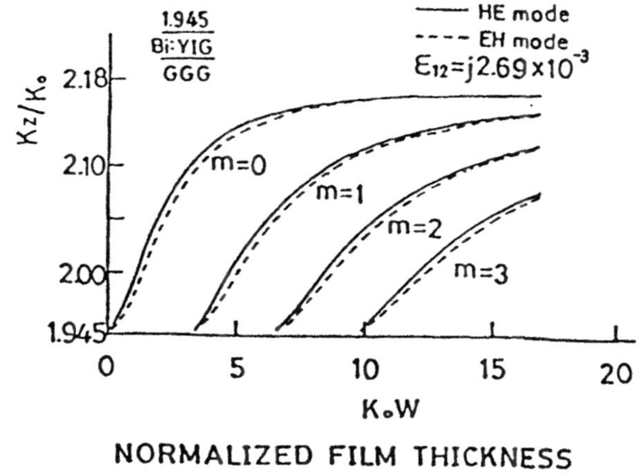

FIGURE 2. Normalized propagation constant constant and dispersion

Non-reciprocal parts using Faraday effects are constructed when top layer is ZnO ($n_3=1.945$), film thickness of Bi:YIG is 1.7m and film length is 0.32mm. Experimental TE-TM mode conversions are found as shown for several external magnetic fields parallel to propagation axis on the film surface. These efficient conversions are obtained by mode matching due to birefringences of Bi:YIG films. These birefringence effects lead to non-reciprocal parts of 50% conversions consisting of film thickness 2.13μm, length 1mm and air top, as in Fig.4.

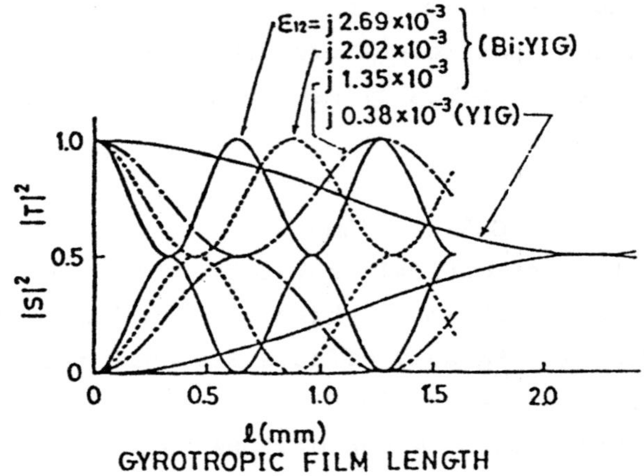

FIGURE 3. TE-TM mode conversion ratio

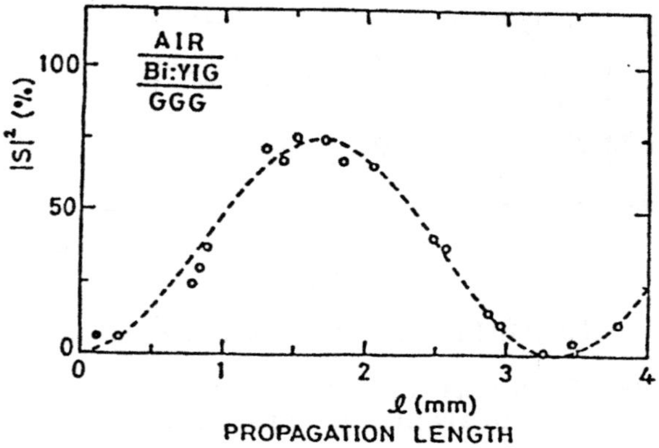

FIGURE 4. TE-TM power conversion efficiency

When magnetic field is vertical to propagation axis and oriented away from the line normal to the film plane, mode conversions due to Cotton-Mouton effects are obtained for reciprocal parts. 50% TE-TM conversions are obtained when $\theta=30^{o}$ for Bi:YIG films of thickness 2.13μm and length 2.3mm. The magnetization and conversions are saturated over 70 Oe.

From the experimental results, it is found that sputtered Bi:YIG films on GGG substrated are optical uniaxial anitropic crystals, where the optical axis is normal to the film surface. The birefringence is $\Delta n=1.17\times10^{-3}$. Theoretical evaluation of maximum TE-TM mode conversion due to Faraday and Cotton-Mouton effects are shown.

4. ISOLATOR AND CIRCULATOR

The compact structures of thin film type isolators of total lengths of 2.3mm, consisting of Bi-substituted YIG films of thickness2.3μm on GGG substrates are shown in Fig.5.

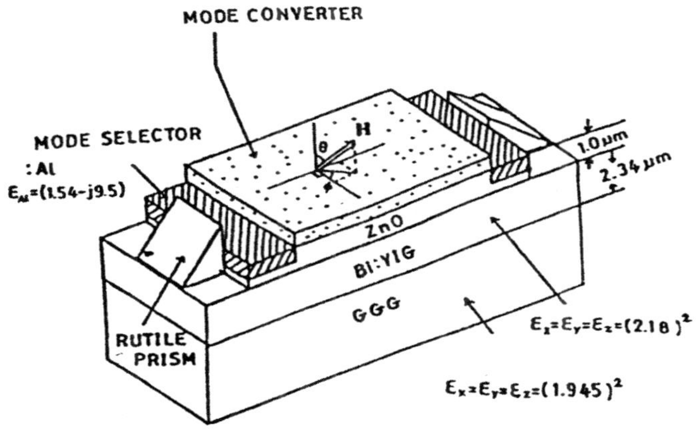

FIGURE 5. Compact integrated optical isolator

Isolation ratio of the isolator for forward and backward optical waves is 9-13dB with insertion loss of 3-4dB for the TEo mode. Optical circulators are constructed combining the 3 or 4 isolators of this type as in Fig.6.

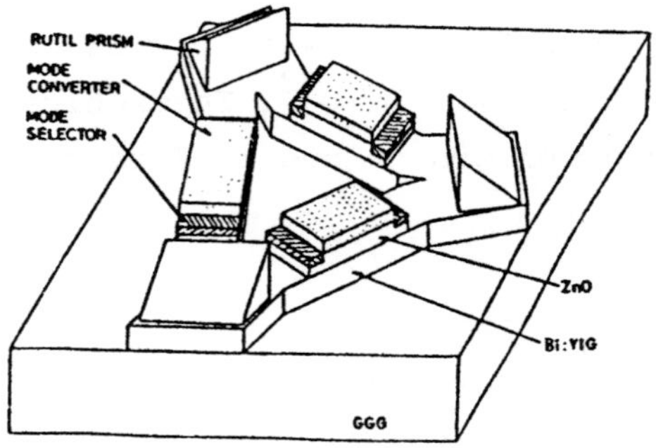

FIGURE 6. 3 ports circulator

REFERENCES

1. Warner,J. 1975. Nonreciprocal magnetooptic waveguides. IEEE Trans. Microwave Theory & Tech., **MTT-23**, pp70-78.
2. Agrawal,V.K.,Miyazaki,Y.and Akao,Y. 1975. Waveguide type optical modulator using Kerr magnetooptic effects in Ni-Fe thin films. J.Appl.Phys., **14**, pp1313-1322.
3. Taki,K.,Miyazaki,Y.and Akao,Y. 1980. Optical propagation and conversion properties of hybrid modes in YIG thin-film waveguides on GGG substrates using Faraday effects with isotropic top layer. Trans. IECE Japan,**E63**, pp754-761.

MODE COUPLING COEFFICIENTS AND BREWSTER'S LAW OF WAVEGUIDE GRATINGS

Pan Xiuwen and Tao Shangping
Beijing Institute of Posts and Telecommunications

ABSTRACT

We propose a method for calculating mode coupling coefficients of a guided wave obliquely incident on a grating with a rather large depth and an arbitrary index. It is shown that the null coupling incident angle depends on the parameters of the grating, according to the Modified Brewster's Law.

Introduction

The problem of a guided wave obliquely incident on a waveguide diffraction grating has been a topic of increasing importance in integrated optics because of its various applications. An effective method delling with this problem is the coupled mode theory.

Using Marcuse's formula (3.2-44) and replacing its K_{rs}^{+-} by a spatial average over a period D of the grating, D.G.Hall improved K.Wagatsuma's calculation of coupling coefficients and derived the so-called Waveguide Grating Brewster's Law, i.e. TE_o--TE_o mode coupling vanishes at an incident angle $\phi_{oc}=tg^{-1}[2/(n^2f/n_a^2+n_a^2/n^2f)]^{1/2}$, independently of any of the grating parameters marked on Fig.1.

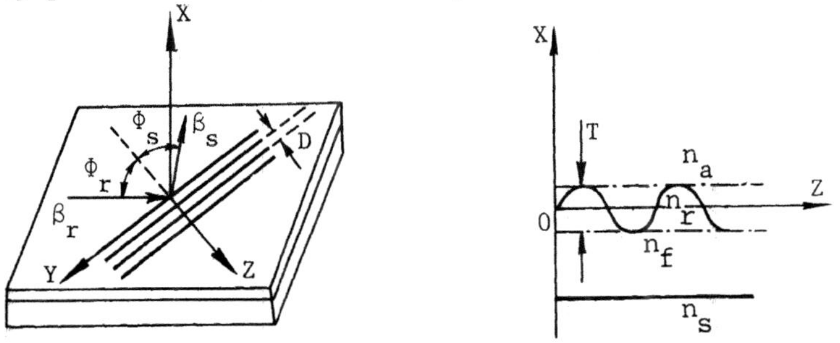

Fig.1 Schematic diagram of a waveguide grating with a profile x=f(z). T is depth of the grating; n_a, n_r, n_f, n_s are indices of the superstract, the grating, the film and the substract, respectively; f(z)=-f(z+D/2).

We have found that Hall's calculation is not valid for a grating with a large depth and with a refractive index quite different from that of the film, such as the waveguide grating Bragg deflector made from photoresist and used in our experiment (See Fig.2). The reason is that the quoted paper expanded

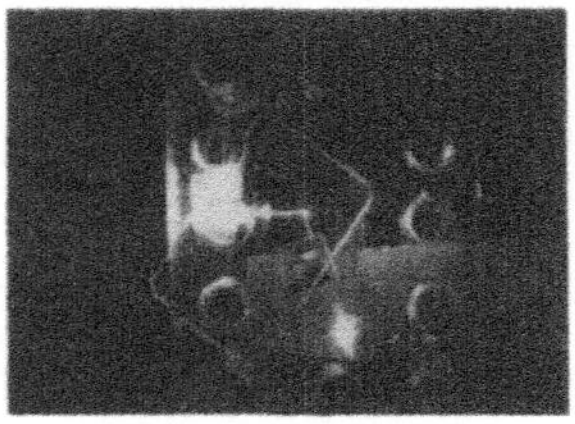

Fig.2 Waveguide grating Bragg deflector made from photoresist.

the fields in terms of a three-layer ideal waveguide and replaced the fields in the perturbed region by the fields at X=0. In order to make the calculation of K_{rs}^{+-} valid for the above kind of grating we make the following suggestions: 1. Take the perturbed region as an individual dielectric layer with an index $n_\sigma=(1/DT\int_0^D\int_0^r n^2(x,z)dxdz)^{\frac{1}{2}}$; 2. Expand the perturbed fields in terms of a four-layer ideal waveguide; 3. Use exact field expressions in calculation; and 4. Take the first order Fourier coefficient of K_{rs}^{+-} as the required coupling coefficient.

Mode Coupling Coefficients and Brewster's Law

In deriving K_{rs}^{+-} and null coupling incident angle ϕ_{rc}, we further assume: 1) each mode is normalized by P=1, which is the power per unit length along the Y direction. 2) the perturbation caused by the grating is sufficiently small so that the effective coupling takes place only between rth and sth modes, which satisfy the Bragg condition, while the other mode couplings can be neglected. The coupled wave equations in the grating coordinates are simplified as

$$\frac{dC_r^+(z)}{dz}=K_{rs}^{+-}C_s^-(z)e^{j\frac{2\pi}{D}z}$$

$$\frac{dC_s^-(z)}{dz}=K_{sr}^{-+}C_r^+(z)e^{-j\frac{2\pi}{D}z} \qquad (1)$$

And we have derived expressions involved in Eq.(1) in the form:

$$K^{+-}_{rs(TE_r-TE_s)}=\frac{\omega\varepsilon_0}{4jD}(n_r^2-n_a^2)\left(\cos\phi_r\cdot\cos\phi_s-\frac{n_g^4}{n_r^2n_a^2}\sin\phi_r\cdot\sin\phi_s\right)$$

$$\cdot\int_0^D\int_0^{f(z)+T/2}E_r^*(x)\cdot E_s(x)\cdot\exp(j2\pi Z/D)\,dxdz \tag{2}$$

$$\phi_{rc(TE_r-TE_s)}=tg^{-1}\frac{n_r^2n_a^2\beta_s}{n_g^4\beta_r}\sqrt{\frac{1}{2}\left(1-\frac{\beta_r^2}{\beta_s^2}+\sqrt{\left(1-\frac{\beta_r^2}{\beta_s^2}\right)^2+4\frac{n_g^8\beta_r^2}{n_r^4n_a^4\beta_s^2}}\right)} \tag{3}$$

$$\phi_{oc(TE_o-TE_o)}=tg^{-1}(n_r\cdot n_a/n_g^2)$$

$$K^{+-}_{rs(TM_r-TM_s)}=\frac{(n_r^2-n_a^2)}{4jD\omega\varepsilon_0n_g^4}\left(\left(\sin\phi_r\cdot\sin\phi_s-\frac{n_g^4\cos\phi_r\cdot\cos\phi_s}{n_r^2\cdot n_a^2}\right)\right.$$

$$\int_0^D\int_0^{f(z)+T/2}\frac{dH_r^*(x)}{dx}\cdot\frac{dH_s(x)}{dx}\cdot\exp(j2\pi z/D)\,dxdz$$

$$\left.+\beta_r\cdot\beta_s\int_0^D\int_0^{f(z)+T/2}H_r^*(x)\cdot H_s(x)\cdot\exp(j2\pi z/D)\,dxdz\right) \tag{4}$$

$\phi_{rc(TM_r-TM_s)}$ satisfies $K^{+-}_{rs(TM_r-TM_s)}=0$

$$K^{+-}_{rc(TE_r-TM_s)}=\frac{(n_r^2-n_a^2)}{4D}\left(\frac{\cos\phi_r\cdot\sin\phi_s}{n_g^2}+\frac{n_g^2\sin\phi_r\cdot\cos\phi_s}{n_r^2\cdot n_a^2}\right)$$

$$\cdot\int_0^D\int_0^{f(z)+T/2}E_r^*(x)\cdot\frac{dH_s(x)}{dx}e^{j\frac{2\pi}{D}z}\,dxdz \tag{5}$$

$$\phi_{rc(TE_r-TM_s)}\equiv 0 \tag{6}$$

Our results differ from Hall's (3) in several points. We have found that ϕ_{rc} for TE_r-TE_s null coupling depends not only on the parameters of the waveguide but also on those of the grating. Formula (4) shows that ϕ_{rc} for TM_r-TM_s null coupling does not always exist, which strongly depends on the parameters of both the waveguide and the grating. Formula (6) shows that as long as the incident angle $\phi_r=0$, the TE_r-TM_s coupling always exists.

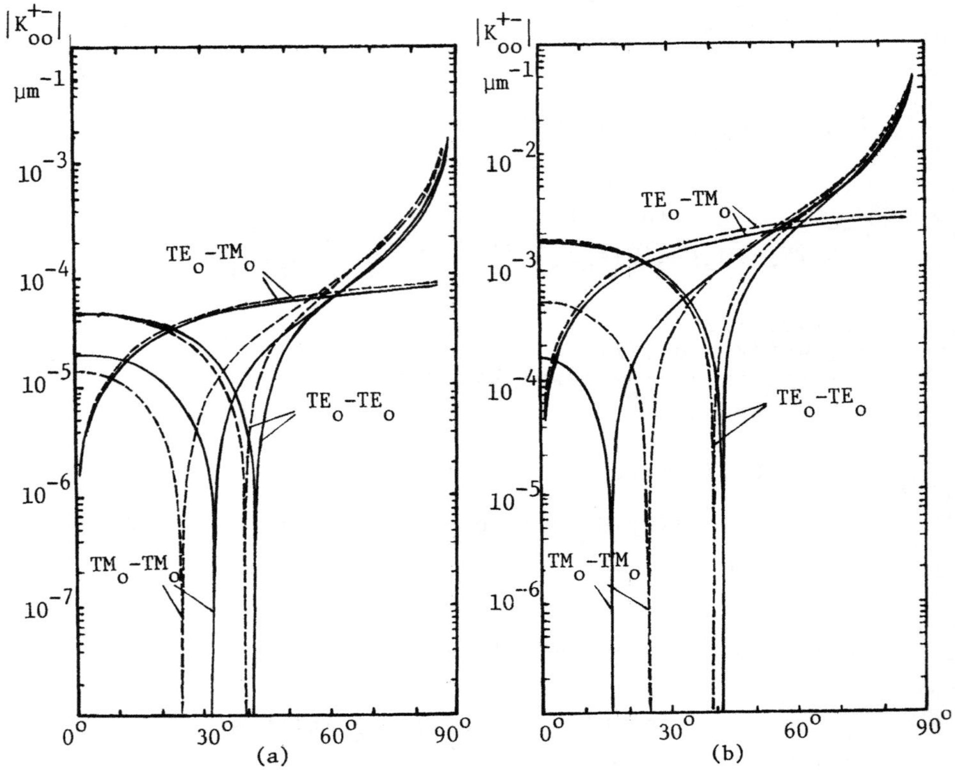

Fig.2 Curves of K_{rs}^{+-} - ϕ_r (a) T=10Å (b) T=382Å. Dashed curves indicate the results by Hall's method. Solid curves indicate the results by our method.

In Fig.2 we present plots of K_{oo}^{+-} as a function of ϕ_r for typical parameters. Note that the calculated values of mode coupling coefficients using two different methods are quite different even for a small depth of the grating. Both calculations are approximate. In Ref. 4 Y.Yamamoto proposes an improved coupled mode theory for calculation of corrugated waveguides, which is based on the selection of a proper boundary for the unperturbed waveguide to make the self-coupling coefficient smallest. We consider this theory also valid for improving our calculation.

Conclusion

1. The Waveguide Grating Brewster's Law derived here shows that a TE_r-TE_s null coupling incident angle exists, which strongly depends on the parameters of both the waveguide and the grating. This result is of value in practical design, as in many applications such as the chirped demultiplexer, it is desirable to eliminate some mode couplings so as to secure a fine wavelength resolution.

2. Mode coupling coefficients calculated by using different methods are quite different. Our method is perhaps more accurate for gratings with a large depth and an arbitrary index as we use less approximation. However, further sophisticated experiment is needed to test its accuracy.

REFERENCES

1. D.Marcuse, "Theory of Dielectric Optical Waveguide," Academic Press, New York, Chap. 3, (1974).
2. K.Wagatsuma, H.Sakaki and S.Saito, Mode Conversion and Optical Filtering of Obliquely Incident Waves in Corrugated Waveguide Filters, IEEE. QE-15. p.632 (1979).
3. D.G.Hall, Diffraction Efficiency of Waveguide Grating: Brewster's Law, Optics Lett, No.7, p.315 (1980).
4. Y.Yamamoto, T.Kamiya and H.Yanai, Improved Coupled Mode Analysis of Corrugated Waveguides and Lasers, IEEE, QE-14, No.4, p.245 (1978).

NUMERICAL CALCULATION OF THE DIFFRACTION PATTERN ON THE FOCAL LINE OF A CHIRPED GRATING LENS ON OPTICAL WAVEGUIDES

Shu-Tong Zhou (Fudan University, Shanghai)
Zong-Qi Lin (Jiao Tong University, Shanghai)

ABSTRACT

The principle and method are presented for computation of two dimensional wave fields in the focal region of a chirped and curved grating lens on optical waveguides. The phase and amplitude of the wave at the exit boundary of the lens are calculated by using the generalized coupled mode theory and the focal pattern has been calculated by means of Kirchhoff's integral. A numerical example and its computer algorithm are discussed.

INTRODUCTION

Recently there is a growing interest in the use of chirped grating structures for transforming one form of a guided-wave beam into another(1-5). The chirped grating lens is a typical application in which a planar guided wave has been diffracted into a focused guided-wave beam. Theoretical analysis and prediction of characteristics and limitation of the chirped grating lens are indispensable to its design and improvement.

The light diffraction properties of the chirped and curved grating have been analysed to various degrees of accuracy via several approaches(6-8). However, they can only be used to analyze structures that have a very limited chirping rate or curvature with no phase mismatch or a constant periodicity. In this paper, we describe the numerical calculation of the focal pattern of a chirped and curved grating lens using a generalized two dimensional coupled mode approach(9) which can take into account the effect of the phase mismatch as well as the effect of the curved guided-wave fronts and the curved grating grooves. We can obtain information such as the diffracted focus patterns, the diffracted efficiency, the angular range of the incident beam propagation within which an effective diffraction can take place, etc.

A two dimensional wave in a planer optical waveguide shown in Fig.1 can be expressed by a scalar function $\psi(x,y)$, such that the electromagnetic fields of a TE mode are

$$\vec{E}_E(x,y,z) = [\nabla \times (\psi(x,y)\ \vec{i}_z)]\ F_E \qquad (1a)$$

$$\vec{H}_E(x,y,z) = \frac{j}{\omega\mu_0\mu}\ \nabla \times \vec{E}_E \ , \qquad (1b)$$

where $\vec{i}_z$ is the unit vector along z axis, F_E is a TE eigen function of the differential equation

$$\frac{\partial^2}{\partial z^2} + k_0^2\varepsilon(z) - k_0^2 n_e)\ F_E(z) = 0, \qquad (2)$$

with the conventional boundary conditions of the guided mode. In this equation, $\varepsilon(z)$ is the relative dielectric constant of the waveguide and n_e is the eigenvalue representing the effective index of the guided-wave mode. $\psi(x,y)$, which represents the horizontal variation of the generalized TE mode, is a solution of the following differential equation:

$$(\frac{\partial^2}{\partial x^2} + \frac{\partial^2}{\partial y^2} + k_0^2 n_e^2)\ \psi(x,y) = 0 \qquad (3)$$

The TM modes have similar relations. Therefore the focal pattern can be obtained by calculating the scalar function ψ near the focus of the chirped grating lens via a Kirchhaff integral once the phase and amplitude of the field at the exit boundary of the lens are known.

GEOMETRIC RELATION OF CHIRPED GRATING LENS

As the dimension of the lens is much larger than the wavelength, it is reasonable to write the scalar functions of both the incident and diffracted waves within the grating region in the Sommerfeld-Runge form[10].

$$\psi_i(x,y) = A_i(x,y)\exp[-jk_0 n_e\ \Phi_i(x,y)] \qquad (4)$$

with $i = 1,2$ for the incident and diffracted waves respectively. According to the Generalized Coupled-mode Theory[9], the phase matching condition of the volume interaction is

$$\Phi_2 = \Phi_1 - (\Gamma/k_0 n_e) \qquad (5)$$

where $\Gamma(x,y)$ is the grating phase function describing the periodic variation of the grating. $\nabla\Gamma(x,y)$ is a vector perpendicular to the local grooves with a magnitud equal to $2\pi/\lambda$, where λ is the local periodicity of the grating.

Consider the case of a thin linear chirped grating with horizontal grooves as in Fig.2. The phase function $\Gamma(x,y)$ of the grating is designed to be

$$\Gamma = -(\frac{2\pi}{\lambda_e})\ 2\sin\Theta_0\ (y-\frac{RH}{2})\ [1 - C(y-\frac{RH}{2})] \tag{6}$$

where RH is the hight of the grating, $\frac{RH}{2}$ is the ordinate of its middle point. Θ_0 is the Brogg angle at the middle point, λ_e is the wave length of the guided mode $(2\pi/\lambda_e) = k_0 n_e$, and C is the chirping rate of the grating.

When the incident wave is a plane wave with an incident angle Θ_i equal to Θ_0

$$\Phi_1 = x\cos\Theta_i - y\sin\Theta_i \tag{7}$$

substituting (6),(7) in (5), we have

$$\nabla\Phi_2 = \vec{i}_x \cos\Theta_0 + \vec{i}_y \sin\Theta_0\ [1 - 4C\ (y-\frac{RH}{2})]\ , \tag{8}$$

where y is the coordinate of a point in the thin grating. As $\nabla\Phi_2$ denotes the direction of the eikonal line, the direction of the diffracted line will be given by

$$\tan\Theta_d = [1 - 4C(y-\frac{RH}{2})]\ \tan\Theta_0\ , \tag{9}$$

where Θ_d is a function of y. It is easy to prove by geometry that all the diffracted lines from different points of the grating will converge to a single focus so far as the grating is thin enough. Suppose the coordinates of this focus to be (XF, YF). As the diffracted line from the middle point of the grating must also pass through it, therefore

$$YF = XF\ \tan\Theta_0 + \frac{RH}{2} \tag{10}$$

Meanwhile the inclination of the line connecting the focus to any point on the y axis is

$$\tan\Theta_L = \frac{XF\ \tan\Theta_0 + (\frac{RH}{2} - y)}{XF} \tag{11}$$

Equating (9) and (11), we obtain

$$XF = \frac{1}{4C\tan\Theta_0} , \tag{12}$$

and from equation (10)

$$C = \frac{1}{4(YF - RH/2)} \tag{13}$$

$$YF = \frac{1}{4C} + \frac{RH}{2} \tag{14}$$

When the incident angle Θ_i deviates from Θ_0, then, from simple geometric relation, the new focus point will be located at (XFM, YFM)

$$XFM = XF - BO\ (\Theta_0 - \Theta_i)\ \sin\Theta_0 \tag{15a}$$

$$YFM = YF + BO\ (\Theta_0 - \Theta_i)\ \cos\Theta_0 \tag{15b}$$

where BO is the distance between the original focus point and the middle point of the grating.

In practice an autual grating lens has a finite thickness D as in Fig.3, so, strictly speaking, there will be no focal point in the sense of geometric optics and we can only define a geometric focal point corresponding to the middle vertical line of the grating:

$$XF = \frac{1}{4C\tan\Theta_0} + \frac{D}{2} \tag{16}$$

$$BO = \{(XF - \frac{D}{2})^2 + (YF - \frac{RH}{2})^2\}^{\frac{1}{2}} , \tag{17}$$

with YF unchanged as in equation (14).

MESH POINTS AND FIELD DIFFERENCE EQUATIONS

The incident angle Θ_i is assumed to be constant throughout the grating region, the eikonal lines of the incident wave can be easily defined if we devide the incident boundary of the grating into RN segments as in Fig.3. Each incident eikonal line is labeled with a number J, J = 0,1,2...RN. The diffracted eikonal lines have different directions at different points. They should be determined segment by segment. In the actual case, Θ_0 is rather small, we can neglect the contribution of the diffracted eikonal lines starting from the topedge of the grating lens. Along each incident eikonal line of number J, there will be a series of mesh points with label number I, I = 1,2,3··· . As in Fig.3, the position of mesh point J=1, I=2 can be determined by mesh points J=0, I=1 and J=1, I=1. Generally, the position of a mesh point J, I can be determined using the positions of mesn points [(J-1), (I-1)] and [J, (I-1)] simple geometric relations

$$\tan\Theta_S = \frac{(y - PY)}{[x - PX(I)]} \tag{18}$$

$$y = J(\frac{RH}{RN}) - x\tan\Theta_i \ , \tag{19}$$

where [PY, PX (I)] is the position of mesh point (J-1), (I-1) and

$$PY = (J-1)\ (\frac{RH}{RN}) - PX(I)\ \tan\Theta_i \tag{20}$$

$\tan\Theta_S$ in equation (18) is the direction of a diffracted line at a point (x,y) in the grating. From equation (5),(7), we have

$$\tan\Theta_S = \frac{(\nabla\Phi_2)\cdot\vec{i}_y}{(\nabla\Phi_2)\cdot\vec{i}_x} = \frac{2\sin\Theta_0 - \sin\Theta_i - 4C(y - \frac{RH}{2})\sin\Theta_0}{\cos\Theta_i} \tag{21}$$

From equations (18,(19), we have

$$X(I) = PX(I) + RH/RN\ (\tan\Theta_S + \tan\Theta_i) \tag{22}$$

$$Y(I) = J(RH/RN) - X(I)\tan\Theta_i \tag{23}$$

According to the generalized coupled mode theory, if the incident and diffracted waves are $P(x,y)A_1\exp(-jk_0n_e\Phi_1)\cdot \cdot F_E(\vec{i}_z \times \nabla\Phi_1)$ and $Q(x,y)A_2\exp(-jk_0n_e\Phi_2)FE(\vec{i}_z \times \nabla\Phi_2)$ respectively, then P(x,y), Q(x,y) are related by the difference equations

$$P(C) - P(A)\exp[-\alpha\,\overline{AC}\,] = -\frac{A_2}{A_1}K_E(\nabla\Phi_1\cdot\nabla\Phi_2)\ j\ Q(A)\overline{AC} \tag{24}$$

$$jQ(C) - jQ(B)\exp[-r] = = \frac{A_1}{A_2}K_E\frac{(\nabla\Phi_1\cdot\nabla\Phi_2)}{(\nabla\Phi_2\cdot\nabla\Phi_2)}P(B)\frac{[1-\exp(-r)]}{(\alpha + j\xi)} \ , \tag{25}$$

where

$$\xi = k_0n_e\ [1 - \nabla\Phi_2\cdot\ \nabla\Phi_2]/2$$

$$\gamma = (\alpha + j\xi)(\overline{BC})/(\nabla\Phi_2\cdot\nabla\Phi_2)^{\frac{1}{2}} \ ,$$

α is the waveguide attenuation coefficient, B,A,C denote the

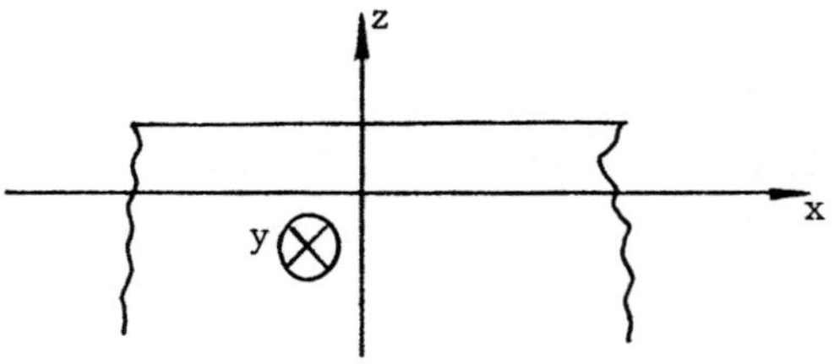

Fig.1 Coordinates of the planar waveguides

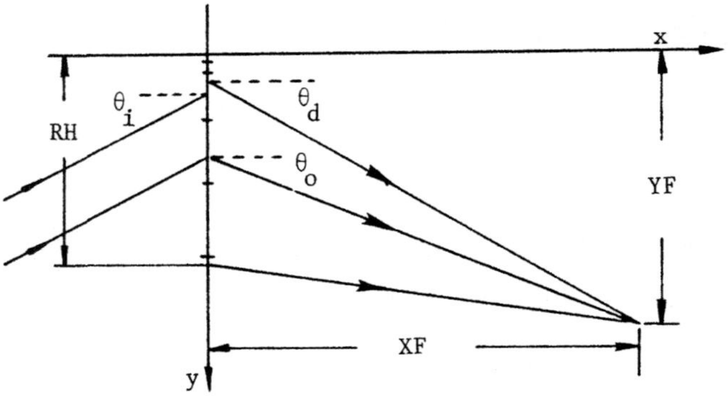

Fig.2 Geometric relation of a thin grating lens

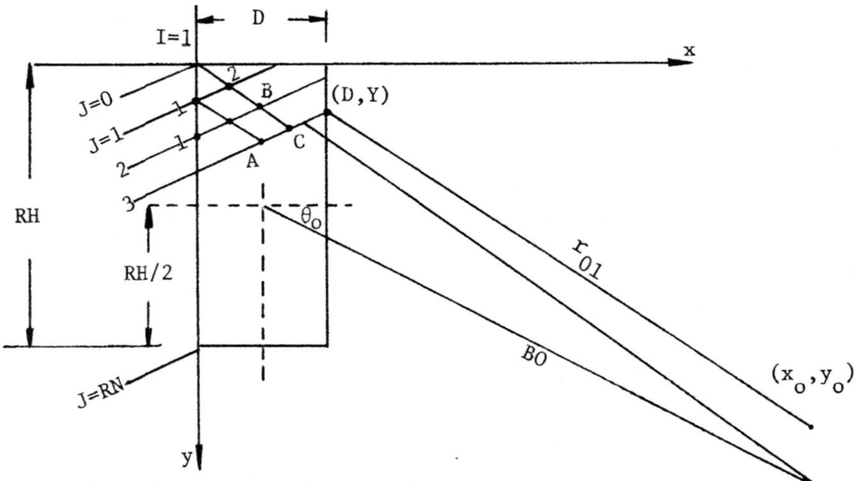

Fig.3 Geometric relation of a thick grating lens

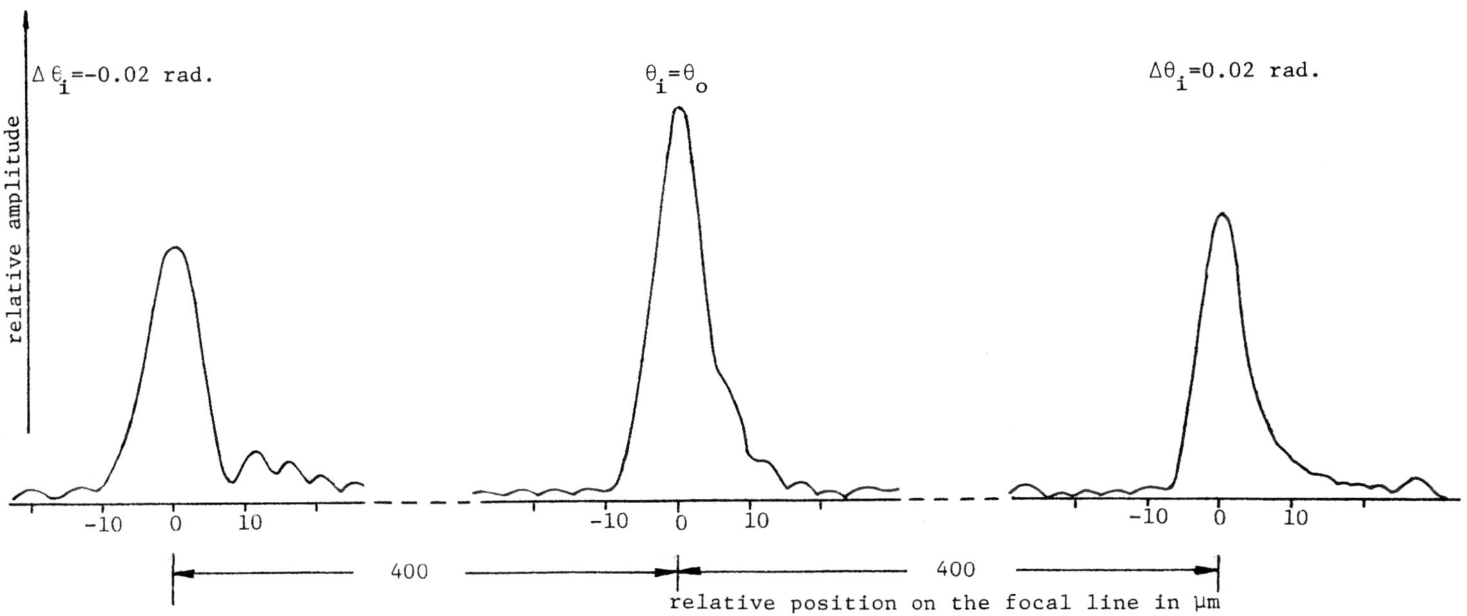

Fig.4 Field amplitude on the focal line with different angles of incidence

points [(J-1), (I-1)], [J,(I-1)], [J,I] respectively and $\overline{CA}$, $\overline{BC}$ are the distances between the corresponding points which can be calculated from the positions of these points, K_E is the coupling coefficient of the grating.

Calculation proceeds along each incident eikonal line from the left edge of the grating to the exit edge. When one eikonal line number J has been calculated, store the data of all the mesh points and begin the next eikonal line J+1. It is not necessary to retain the data of (J-2)th eikonal line while calculating the Jth eikonal line. The data which should be retained is the complex values P(I), Q(I) at each point (D,Y(I)) at the exit boundary. P(I), Q(I), Y(I) are then stored as R(I), S(I) and YA(I).

FOCUS PATTERN CALCULATION

We take the exit boundary of the grating lens as the diffraction aperture, the focal pattern can then be calculated by using an integral with a proper Green's function. As shown in Fig.3, the amplitude of the field at the point (x_0, y_0) is

$$U(x_0,y_0) = \frac{1}{4\pi} \int_{\text{Aperture}} h(x_0,y_0;\, x_1,y_1) \Big|_{x_1=D} f(y_1)\cdot dy_1 \qquad (26)$$

where

$$h(x_0,y_0;\, x_1,y_1) = 2\left(\frac{2\pi k_0 n_e}{r_{01}}\right)^{\frac{1}{2}} e^{j\frac{\pi}{4}} e^{-jk_0 n_e r_{01}} \cos\Theta$$

$$\simeq 4\pi\left(\frac{1}{\lambda_e r_{01}}\right)^{\frac{1}{2}} e^{j\frac{\pi}{4}} e^{-jk_0 n_e r_{01}} \qquad (27)$$

in which r_{01} is the distance between the point of integration (D, YA(I)) and the point (x_0, y_0).

$$r_{01} = \{(XF - D)^2 + (y_0 - YA(I))^2\}^{\frac{1}{2}} \qquad (28)$$

$f(y_1)$ in equation (26) is the field at point (D,YA(I)) on the exit boundary of the grating lens

$$f(y_1) = S(I)\, A_2 e^{-jk_0 n_e \Phi_2} \simeq \frac{S(I)}{(r_{01})^{\frac{1}{2}}} e^{-jk_0 n_e \Phi_2} \qquad (29)$$

as it is easy to prove that $A_2 \simeq \frac{1}{(r_{01})^{\frac{1}{2}}}$. Φ_2 is the phase function of the diffracted wave at the point (D,YA(I)) and can be obtained from equation[5]

$$\Phi_2 = \Phi_1 - \left(\frac{\Gamma}{k_0 n_e}\right) = D\cos\Theta_i - YA(I)\sin\Theta_i + $$
$$+ 2\sin\Theta_0 \left(YA(I) - \frac{RH}{2}\right)\left[1 - C\left(YA(I) - \frac{RH}{2}\right)\right] \tag{30}$$

Our calculated results are shown in Fig.4, the parameters of the grating lens being $D = 65\mu m$ (grating thickness), $YF = 2500\mu m$, $XF = 20032.5\mu m$, $RH = 2000\mu m$, coupling coefficient $K_E = 0.5\pi/D$, $\lambda_e = 0.4125\mu m$. The incident beam is assumed to be a $1/e^2$ truncated Gaussian distributed quasi-plane wave. To the authors' knowledge, calculation of the focal pattern of a grating lens has not appeared in the literature using a method other than the authors' approach.

This method can also be used to calculate the diffracted wave from a curved line grating lens. We have calculated grating lenses with parabolically curved grooves, which differ from linear chirped grating lenses only in the grating phase function $\Gamma(x,y)$. Since the methods are the same, we will not discuss it in this paper.

CONCLUSION

We have presented in this paper the principle and method for computation of the focal patterns of chirped and curved grating lenses with a certain amount of phase mismatch. This method is useful where the grating is fabricated on a single mode planar waveguide and the directions of propagation of the incident and diffracted beams are approximately parallel or antiparallel to each other so that very little mode conversion between TE and TM waves will occur. It can also be used in other cases such as grating reflectors, beam expanders, thin film spectrography etc.

REFERENCES

(1) S. K. Yao and D. E. Thompson, "Chirped grating lens for guided-wave optics", Appl. Phys. Lett., Vol.33, p635, 1978.

(2) P. K. Tien, "Method of forming novel curved line grating and their use as reflectors and resonators in integrated optics", Opt. Lett., Vol.1, p.64, 1977.

(3) P. K. Tien and R. J. Capik, "A thin film spectrography for guided-waves", in Tech. Dig., IEEE/OSA Topical Meet. Integrated and Guided-wave Optics (Incline Village, NV), 1980.

(4) P. R. Ashley and W. S. C. Chang, "Fresnel lens in a thin-film waveguide", Appl. Phys. Lett. 33(6), 15 Sept. p.490, 1978.

(5) G. Hatakosi and S. Tanaka, "Grating lenses for integrated optics" Opt. Lett., Vol.2, p.142, 1978.

(6) W. Streifer et al., "Perturbation analysis of nonuniform almost periodic Bragg reflectors", J. Opt. Soc. Am., Vol.66, No.12, p.1359, 1976.

(7) A. Hardy and W. Streifer, "Analysis of waveguided Gaussian beams coupled by misaligned or curved gratings", J. Opt. Soc. Am., Vol.69, p.1235, 1979.

(8) J. Van Roey and P. E. Lagasse, "Coupled wave analysis of obliquely incident waves in thin film gratings", Appl. Opt., Vol.20, p.423, 1981.

(9) Z. Q. Lin, S. T. Zhou, et al., "A generalized two-dimensional coupled-mode analysis of curved and chirped periodic structures in open dielectric waveguides", IEEE, Trans. Microwave Theory Tech. MTT-29, No.9, Sept. p881, 1981.

(10) M. Kline and I. W. Kay, Electromagnetic Theory and Geometrical Optics. New York: Interscience, 1965.

WAVEGUIDE OPTICAL PLANAR LENS —— A NEW DESIGN

Zang De-yu

Shanghai Institute of Semiconductor Devices

ABSTRACT

This paper reports, probably for the first time, a simple and efficient method for fabricating a "real" waveguide optical palanar lens (WOPL), which is useful in many places in integrated optics. The method is based upon the TI (Ti-Indiffusion) technique[1], PE (Proton Exchange) technique[2] and TIPE (Ti-Indiffusion Proton Exchange) technique[3] on a $LiNbO_3$ substrate. The effective index of a TI waveguide is quite different from that of a TIPE waveguide. Our design starts from the Fermat principle. Experimental results of a lens with F=15mm are described. The focal characteristics of the lens is found excellent.

There has been a variety of optical waveguide lenses constructed by different methods, such as the thin-film Luneburg lens, SAW diffraction lens, Fresnel diffraction lens and geodesic waveguide lens, of which the last one is mostcommonly used.

While the making of TI diffusion $LiNbO_3$ waveguide is a well known technique, the art of making PE and TIPE $LiNbO_3$ waveguides are only recently under development. By an integration of the TI and TIPE (or TI and PE) techniques, a real waveguide optical planar lens can be made. The design is based on the Fermat principle.

For a Y-cut $LiNbO_3$ substrate, the manufacturing process is as follows: a sample with a deposition of Ti about 400 Å is diffused at 1010°C for 6.5 hours in a LiO_2-rich atmosphere. Three TE modes and two TM modes can exist at λ=6328 Å. The effective index of TE_0 is 2.2172. After Ti-indiffusion, a material is deposited onto the sample, and the mask is made for the selective PE technique by using the standard photolithograph technique. The sample is then placed in a cell containing benzoic acid and heated to 200°C in a period of two hours, which is the desired proton exchange time. In the TIPE region, five TE modes appear while no TM modes can be supported, the effective index of TE_0 being 2.3266.

Consider a lens with an aspherical contour used for paraxial light. As shown in Fig. 1

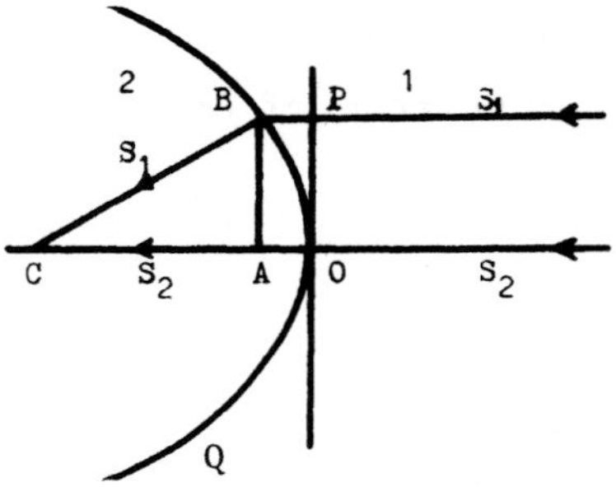

Fig.1 Scheme of lens design

TI waveguide region 1 and TIPE region 2 with effective indices Ne and Né for the TE_0 mode are separated by a curve BOQ. We start from the Fermat principle $\delta\int nds=0$. For two beams of paraxial light S_1 and S_2, as shown in Fig. 1, Fermat principle is expressed as

$$\overline{BP}\cdot Ne+\overline{BC}\cdot N\acute{e}-\overline{OC}\cdot N\acute{e}=0 \tag{1}$$

With the notations $\overline{BP}=x$, $\overline{OC}=F$, $\overline{BA}=y$, we have

$$\frac{[x-F/(1+Ne/N\acute{e})]^2}{[F/(1+Ne/N\acute{e})]^2}\quad\frac{y^2}{[F\cdot\sqrt{1-Ne^2/N\acute{e}^2}/(1+Ne/N\acute{e})]^2}=1 \tag{2}$$

Substitution of Ne=2.2172, Né=2.3266 in Eq. (3) yields

$$\frac{(x-0.512F)^2}{(0.512F)^2}+\frac{y^2}{(0.156F)^2}=1 \tag{3}$$

Eq. (4) represents an ellipse having a semimajor axis a=0.512F and a semiminor axis b=0.156F, where F is the focal length. The possible effective aperture D of the loss can be obtained as:

$$D=2\cdot b=0.312F \tag{4}$$

For an integrated optical spectrum analyzer (IOSA), which requires D-4mm, it is appropriate to take F=15mm. For an ideal lens, whose focal spot size is $\sim\lambda F/D(4)$, where λ is the wavelength in the medium, D is the aperture of the lens, according to Eq. (4), the largest aperture, D, of our lens is 0.312F; the smallest focal spot size is obviously λ_0 (the (wavelength in vacuum), which is a constant.

For experimental work, we designed a mask for a F=15mm dual lens as shown in Fig. (2).

Fig.2 Mask of a dual lens with F=15 mm. White, TIRE region; Black, TI region.

The first lens is used to collimate light and the second to focus light. In making the sample, we noted that Benzoic acid does not attack most metals and therefore we used some appropriate material as the mask for the selective PE technique. The results of the experiment are shown in Figs. (3) and (4).

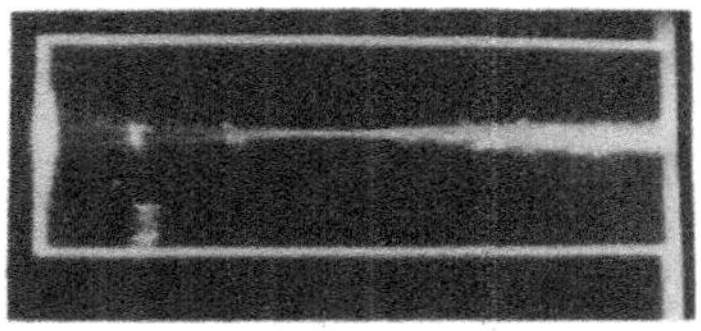

Fig.3 Experimental results. Light coupled into TE_o mode by a rutile prism.

As expected, the guided light is very well focused by the second lens. The focal spot is of good quality. As can be seen from Fig. 4, the test arrangement is similar to that used in Ref. 5. In our experiment, the reimaging lens used was not of high quality , which introduced aberrations. By means of prism coupling technique, in the neighbourhood of the focus, most of the light is in the TE_0 mode, only a small part of it is in the TE_3 and TE_4 modes. The effective indices

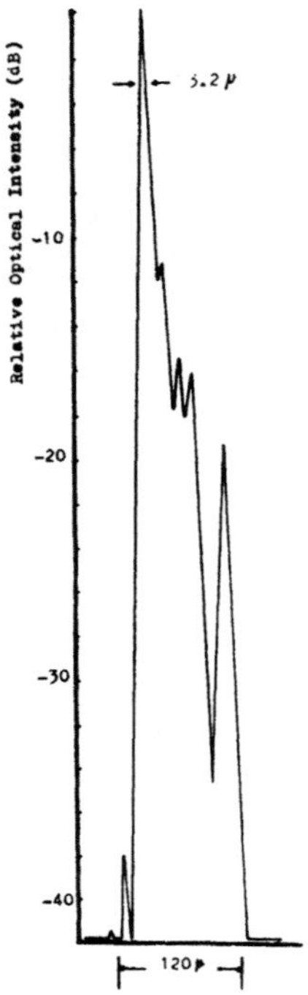

Fig.4 Focal intensity distribution of a F=15 mm dual lens through a F=95.8mm reimaging lens.

of TE_3 and TE_4 in the TIPE region are very close to those of the TE_0 and TE_1 modes in the TI region. It is very likely that there exists a TI sublayer in the TIPE region.

The WOPL described here has many advantages, such as, easy fabrication, flexibility to design for various uses, aberration-free in the case of paraxial rays. In addition, its focal spot size is theoretically independent of its focal length F as mentioned above. These characteristics are very interesting, especially for an integrated optical spectrum analyzer (IOSA).

A CKNOWLEDGEMENTS

The author wishes to thank Prof. D. B . Ostrowsky, l'Université de Nice, France for his encouragement and helpful suggestions. He would like to thank Profs. Y. Chen and Z. Xu, Shanghai Jiao Tong University for useful discussions. Thanks also go to his colleagues I. Chou, X. Chen,

K. Wan, P. Xu, J. Ma and Z. Zhou for their help in doing this research.

REFERENCES

1. R. V. Shmidt, I. P. Kaminow, Appl. Phys. Lett. 25 458 (1974)
2. J. L. Jackel, C. E. Rice & J. J. Veselka, Post Deadline Paper, PDP 1, Topical Meeting on Integrated and Guide Wave Optics, Cal. (Jan. 1982)
3. M. De Micheli et al, unpublished communication
4. R. Shubert & J. H. Harris, J. Opt. Soc. Am. 61 154 (1971)
5. S. K. Yao et al, Appl. Opt. 18 446 (1979)
6. V. R. Checcacci, "Fiber and Integrated Optics" P. 345, edited by D. B. Ostrowsky, Plenum Press (1979)
7. D. C. Foster et al, Appl. Opt. 9 2154 (1970)
8. W. S. C. Chang et al, Third International Conference on IOOC, Technical Digest, P. 114, San Francisco, Cal. (1981)

ON A ZIG-ZAG RAY PICTURE IN A PLANAR WAVEGUIDE

K. YASUMOTO

1. INTRODUCTION

Zig-zag ray picture is a simple and a useful concept to study the characteristics of various guided-wave structures. In this ray optical treatment, the Goos-Hänchen shift and the associated time delay of a wave packet that travels along a zig-zag path, being totally reflected successively at the boundaries of a waveguide, play a significant role to determine the modal properties of the waveguide. Kogelnik and Weber[1] have developed a description for the zig-zag ray model in a planar dielectric waveguide, based on the stationary-phase argument[2] for the evaluation of the Goos-Hänchen shift and the time delay. In this paper, a more useful ray picture, which enables us to understand intuitively the interconnections between the zig-zag ray, group velocity, power flow, and stored energy in a planar waveguide, is presented based on the energy-flux conservation argument.

2. GOOS-HÄNCHEN SHIFT AND TIME DELAY

Consider a plane interface $x = 0$ separating medium 1 ($x > 0$) from medium 2 ($x < 0$), as is shown in Fig. 1. When a space-like and time-like wave packet, which is uniform in the y direction, is incident on the interface at an angle greater than the critical angle for total reflection, the reflected wave packet undergoes the Goos-Hänchen shift z_s and experiences the associated time delay τ_s. It is well-known[1,2] that z_s and τ_s are related to the phase shift 2ϕ on total reflection by

$$z_s = 2\frac{\partial\phi}{\partial\beta_z}, \qquad \tau_s = -2\frac{\partial\phi}{\partial\omega}, \tag{1}$$

where ω and β_z are, respectively, the central angular frequency

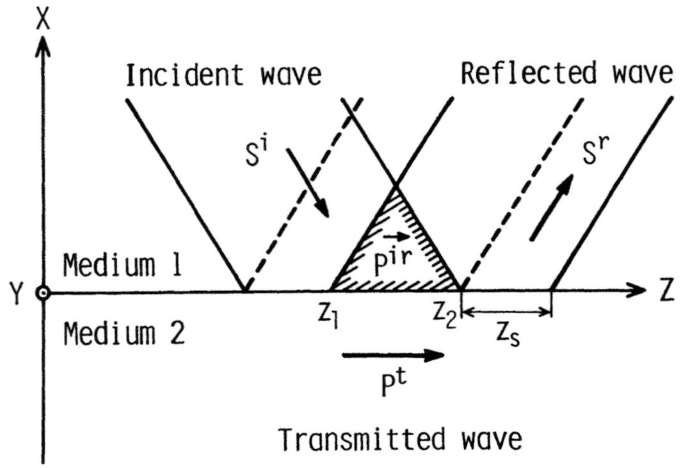

FIGURE 1. Illustration of the Goos-Hänchen shift at a plane reflecting interface.

and the z component of the central wavenumber of the spectrum for the incident wave packet.

We shall reexamine this phenomenon from the point of view of energy-flux conservation. For this purpose, we assume in Fig. 1 that the incident wave packet consists of a finite plane wave and the reflected wave packet is displaced from its expected position by the distance z_s. It is noted that the reflected wave has the same energy-flux as the incident wave, but there arise two additional energy-fluxes through this reflection process. One is a net energy-flux P^t inside the medium 2 carried by the transmitted evanescent wave in the z direction. The other is a energy-flux P^{ir} inside the medium 1 produced by the interaction between the incident and reflected waves, which exists in the region of overlap of the two wave packets and is directed in the z direction. The energy-flux P^t and associated stored-energy W^t are given by

$$P^t = \int_{-\infty}^{0} S_z^t \, dx, \tag{2}$$

$$W^t = \int_{-\infty}^{0} w^t \, dx, \tag{3}$$

where S_z^t and w^t are, respectively, the z component of the time-averaged Poynting vector and the time-averaged stored-energy density for the transmitted plane wave. The energy-flux P^{ir} can be evaluated by introducing an appropriate averaging process into the integral of the energy-flux density over the region of overlap of the two wave packets where $z_1 \leq z \leq z_2$ and $0 \leq x \leq x(z)$ as shown in Fig. 1. Let S_z^{ir} be the z component of the time-averaged Poynting vector

$$S_z^{ir} = \frac{1}{2} Re[\bar{E}^i \times \bar{H}^{r*} + \bar{E}^r \times \bar{H}^{i*}]_z \tag{4}$$

produced in this region by the interaction between the two wave packets, where $Re[\ ]$ and the asterisk denote the real part in the bracket and the complex conjugate, and the superscripts i and r refer to the incident and reflected waves. Then we first integrate S_z^{ir} over x at a fixed point of z to obtain

$$p^{ir}(z) = \int_0^{x(z)} S_z^{ir}\, dx, \tag{5}$$

where $p^{ir}(z)$ indicates the energy-flux of the interaction carried across a particular plane perpendicular to the interface and is of course a function of z. We next calculate an average of $p^{ir}(z)$ over z from z_1 to z_2 and finally take the limit where $(z_2 - z_1) \to \infty$. Thus we have the following expression:

$$P^{ir} = \lim_{(z_2 - z_1) \to \infty} \int_{z_1}^{z_2} dz \int_0^{x(z)} dx\, S_z^{ir}/(z_2 - z_1). \tag{6}$$

We assume that P^{ir} is the averaged energy-flux, parallel to the interface, carried by the interaction between the incident and reflected wave packets. Similarly, the averaged stored-energy W^{ir} associated with the energy-flux P^{ir} may be expressed in the form

$$W^{ir} = \lim_{(z_2 - z_1) \to \infty} \int_{z_1}^{z_2} dz \int_0^{x(z)} dx\, w^{ir}/(z_2 - z_1), \tag{7}$$

where w^{ir} is the time-averaged stored-energy density of the interaction of the two wave packets. If the medium 1 is a dielectric with permittivity ε_1 and permeability μ_1, w^{ir} is given as follows:

$$w^{ir} = \frac{1}{2} Re\,[\varepsilon_1 \bar{E}^i \cdot \bar{E}^{r*} + \mu_1 \bar{H}^i \cdot \bar{H}^{r*}]. \tag{8}$$

It should be noted that the averaging and limiting process introduced above implies that P^{ir} and W^{ir} may be evaluated approximately in terms of infinite plane waves. This view is obviously true, since a finite plane wave behaves in its central part, in a first approximation, like an infinite plane wave.

Now, based on the consideration of energy-flux conservation, it is argued that first, the total energy-flux $P^t + P^{ir}$ should be equal to the energy-flux of the incident plane wave in a strip of oblique width z_s and of course also to the energy-flux of the reflected plane wave in a strip of the same width, and second, the delay τ_s in the reflection process should be equal to the time needed for the total energy $W^t + W^{ir}$ to travel along the interface for the distance z_s. This argument provides the expressions for the Goos-Hänchen shift z_s and the associated time delay τ_s as follows:

$$z_s = \frac{P^t + P^{ir}}{S_x^r}, \tag{9}$$

$$\tau_s = \frac{W^t + W^{ir}}{S_x^r}, \tag{10}$$

where S_x^r is the x component of the time-averaged Poynting vector of the reflected plane wave. Equations (9) and (10) are new expressions. It must be stressed that the energy portions P^{ir} and W^{ir} arising from the interaction of the incident and reflected waves have not been included in the corresponding equations derived by using the conventional energy-flux conservation argument[3]. In the recent study[4], we have applied Eqs.(9) and (10) to evaluate the Goos-Hänchen shift and associated time delay at several types of reflecting interface and have shown that the results are in complete agreement with those obtained from Eq.(1).

3. ZIG-ZAG RAY PICTURE

For the sake of simplicity, we consider a symmetric-slab waveguide of medium 1 ($-a/2 \leq x \leq a/2$) with the substrate and cover of medium 2. Figure 2 shows the zig-zag ray path of the successively and totally reflected homogeneous plane wave inside the slab.

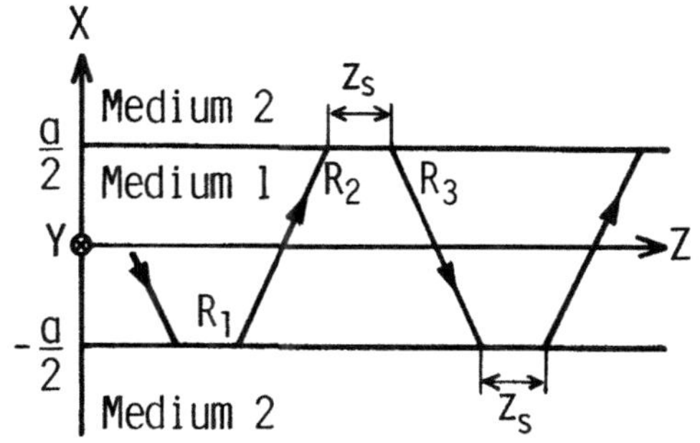

FIGURE 2. Zig-zag ray picture in a symmetric-slab waveguide.

After the passage through the slab from R_1 to R_2, the ray reappears at R_3, shifted by z_s and delayed by τ_s. From the symmetry, $R_1R_2R_3$ may be considered to form a unit cell in the periodic pattern of the zig-zag ray path. The total time taken for the ray to travel from R_1 to R_3 is given by $a/v_{gx} + \tau_s$ and, in this time, the ray associated with the guided wave advances in the z direction from R_1 to R_3 a distance of $(a/v_{gx})v_{gz} + z_s$. Thus the zig-zag ray description gives the group velocity V_g of the guided wave as follows:

$$V_g = \frac{(a/v_{gx})v_{gz} + z_s}{a/v_{gx} + \tau_s}, \tag{11}$$

where v_{gx} and v_{gz} are the x and z components of the group velocity of the plane wave in the medium 1.

The group velocity of the guided wave can be determined exactly by two different wave-theoretical approaches, which are refered to, for the sake of convenience, as a kinematic method and a dynamic method. In the kinematic method, the dispersion relation of the waveguide

$$2\phi(\omega,\beta_z) + a\beta_x(\omega,\beta_z) = m\pi \tag{12}$$

is used and the group velocity is determined from the relation

$$V_g = \frac{d\omega}{d\beta_z}, \tag{13}$$

where ϕ is half the phase shift on total reflection at the boundary surfaces $x = a/2$ and $x = -a/2$, β_x is the x component of the wavenumber of the plane wave inside the slab, and m is an integer which identifies the mode number of the guided wave. The dynamic method, on the other hand, is based on the consideration of the power flow and stored energy in the waveguide and gives the group velocity as

$$V_g = \frac{P_f + 2P_s}{W_f + 2W_s}, \tag{14}$$

where, as is shown in Fig. 3, P_f and W_f are the total power flow and the total stored-energy inside the slab, $2P_s$ and $2W_s$ are those in both the substrate and cover regions, and W_f and W_s stand for per unit length in the z direction. Using Eqs.(1) and (12), Kogelnik and Weber[1] have shown that the group velocity (11) predicted by the zig-zag ray model agrees with the group velocity (13) obtained by the kinematic method. We shall prove here, with the help of our new expressions (9) and (10) for the Goos-Hänchen shift and time delay, the group velocity (11) agrees with the group velocity (14) obtained by the dynamic method. This will help us to have an intuitive understanding for the interconnections between the zig-zag ray, group velocity, power flow, and stored-energy in the waveguide.

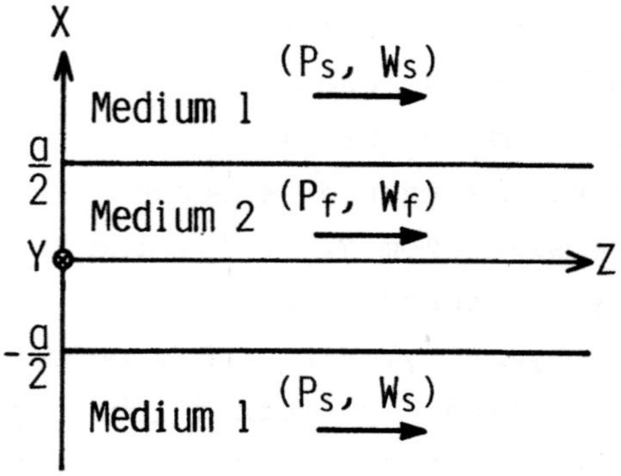

FIGURE 3. Illustration of the power flow and stored-energy in a symmetric-slab waveguide.

From a brief comparison of Fig. 3 with Fig. 1, we first find the following relations:

$$P_f = a(S_z^i + S_z^r) + 2P^{ir}, \quad W_f = a(w^i + w^r) + 2W^{ir}, \tag{15}$$

$$P_s = P^t, \quad W_s = W^t, \tag{16}$$

where S_z^i and w^i are the z component of the time-averaged Poynting vector and the time-averaged stored-energy density for the incident plane wave and S_z^r and w^r are those for the reflected plane wave. The terms of $2P^{ir}$ and $2W^{ir}$ in Eq.(15) imply that the energy portions of interaction between the incident and reflected waves inside the slab are produced through two reflection procesis at the interfaces $x = a/2$ and $x = -a/2$. Substituting Eqs.(15) and (16) into Eq.(14), we have

$$V_g = \frac{aS_z^r + P^t + P^{ir}}{aw^r + W^t + W^{ir}}, \tag{17}$$

where we have used the relations $S_z^i = S_z^r$ and $w^i = w^r$ which hold under the total reflection. Taking into account that $S_z^r = w^r v_{gz}$ and $S_x^r = w^r v_{gx}$, Eq.(17) can be rewritten in the form

$$V_g = \frac{(a/v_{gx})\, v_{gz} S_x^r + P^t + P^{ir}}{(a/v_{gx})\, S_x^r + W^t + W^{ir}}, \tag{18}$$

Using Eqs.(9) and (10), it is readily shown that Eq.(18) agrees with Eq.(11).

4. CONCLUSIONS

Using a new energy-flux conservation argument for the evaluation of the Goos-Hänchen shift and associated time delay, we have developed a useful zig-zag ray picture which connects directly the ray optical description for guided waves in a planar waveguide with the corresponding dynamic description based on the wave-theoretical approach. The present ray picture can be easily extended to the case of asymmetric structures.

REFERENCES

1. H. Kogelnik and H.P. Weber, J. Opt. Soc. Am., vol.64, pp.174-185, 1974.
2. K. Artmann, Ann. Phys. (Leipz.), vol.2, pp.87-102, 1948.
3. R.H. Renard, J. Opt. Soc. Am., vol.54, pp.1190-1197, 1964.
4. K. Yasumoto and Y. Oishi, to be published in J. Appl. Phys., vol.54, no.5, 1983.

[illegible]

4. CONCLUSION

[illegible]

REFERENCES

[illegible]

SCATTERING OF SURFACE WAVES ON TRANSVERSE DISCONTINUITIES IN SYMMETRICAL MULTI-LAYER DIELECTRIC WAVEGUIDES

K. UCHIDA and K. AOKI

1. INTRODUCTION

Open dielectric waveguides have become increasingly important in the past few years, particularly in connection with the areas of integrated optics and millimeter-wave integrated circuits. Much attention has also been given to the surface wave scattering at the discontinuities in dielectric waveguides, such as steps, changes in material properties, and others. In this paper, we present a rigorous analysis, based on the Wiener-Hopf technique, of step discontinuities in relative permittivities for an open-type planar symmetrical multi-layer dielectric waveguide.

2. FREE WAVES

The geometry of the problem is shown in Fig.1, where the structure is uniform in the z-direction. The magnetic parameters are introduced only for the duality between TE and TM waves. If we assume the time dependence $e^{i\omega t}$, the wave equation for TE case becomes

$$(1)\qquad \left[\frac{\partial^2}{\partial x^2} + \frac{\partial^2}{\partial y^2} + \kappa_j^2(y) \right] E_z(x,y) = 0, \quad j=1,2$$

with

$$(2)\qquad \begin{aligned} \kappa_j^2(y) &= \kappa_0^2 = \omega^2 \varepsilon_0 \mu_0 \qquad (y>a_N) \\ &= \kappa_0^2 \varepsilon_j(y) \mu_j(y) = \kappa_0^2 \varepsilon_{mj} \mu_{mj} \quad (a_{m-1}<y<a_m,\ m=1\sim N). \end{aligned}$$

Other field components are given by

$$(3)\qquad H_x = \frac{-1}{i\omega\mu_0\mu_j(y)} \cdot \frac{\partial E_z}{\partial y}, \quad H_y = \frac{1}{i\omega\mu_0\mu_j(y)} \cdot \frac{\partial E_z}{\partial x}.$$

The boundary conditions are summarized as follows:

(B1) radiation condition,

(B2) continuity of E_z at $y=a_m$ $(m=1\sim N)$,

(B3) continuity of H_x at $y=a_m$ $(m=1\sim N-1)$,

(B4) continuity of H_x at $y=a_N$,

(B5) continuity of E_z and H_y at $x=0$ and $0<y<a_N$,

(B6) edge condition.

First, we consider free waves along a uniform dielectric waveguide, that is, $\varepsilon_1(y)=\varepsilon_2(y)$ and $\mu_1(y)=\mu_2(y)$ in Fig.1. Let x-dependence be $e^{-i\zeta x}$, then the solution to eq.(1) can be expressed as

$$(4)\qquad \begin{aligned} E_z(\zeta,y) &= Ue^{-ik_0(y-a_N)} && (y>a_N) \\ &= UF_j(\zeta,y) && (0<y<a_N) \end{aligned}$$

where U is a constant, and

$$(5)\qquad k_0=\sqrt{\kappa_0^2-\zeta^2}$$

and

$$(6)\qquad \begin{aligned} F_j(\zeta,y) &= [\prod_{q=m}^{N} G_{qj}(\zeta)]\frac{\sin k_{mj}(y-a_{m-1})+G_{m-1}(\zeta)\sin k_{mj}(a_m-y)}{\sin k_{mj}d_m} \\ &\qquad (a_{m-1}<y<a_m,\ m=2\sim N) \\ &= [\prod_{q=1}^{N} G_{qj}(\zeta)]\frac{\cos k_{1j}y}{\cos k_{1j}a_1} \qquad (0<y<a_1). \end{aligned}$$

The branch of k_0 is chosen as Im $k_0>0$ to satisfy (B1), and

$$G_{1j}(\zeta)=\frac{k_{2j}}{\mu_{2j}}\operatorname{cosec}k_{2j}d_2\{\frac{k_{2j}}{\mu_{2j}}\cot k_{2j}d_2-\frac{k_{1j}}{\mu_{1j}}\tan k_{1j}a_1\}^{-1},$$

$$(7)\qquad \begin{aligned} G_{mj}(\zeta) &= \frac{k_{m+1j}}{\mu_{m+1j}}\operatorname{cosec}k_{m+1j}d_{m+1}\{\frac{k_{m+1j}}{\mu_{m+1j}}\cot k_{m+1j}d_{m+1} \\ &\quad +\frac{k_{mj}}{\mu_{mj}}[\cot k_{mj}d_m-\operatorname{cosec}k_{mj}d_m G_{m-1j}(\zeta)]\}^{-1} \qquad (m=2\sim N-1), \end{aligned}$$

$$G_{Nj}(\zeta)=1$$

where

$$(8)\qquad \begin{aligned} k_{mj}^2 &= \kappa_0^2\varepsilon_{mj}\mu_{mj}-\zeta^2 && (m=1\sim N), \\ d_m &= a_m-a_{m-1} && (m=2\sim N). \end{aligned}$$

These recurrence relations ensure the boundary conditions (B2) and (B3). The remaining boundary condition (B4) can be written in the form

$$UG_j(\zeta) = 0 \tag{9}$$

where the kernel function is defined by

$$G_j(\zeta) = ik_0 + \tilde{G}_j(\zeta), \tag{10}$$

$$\tilde{G}_j(\zeta) = \frac{k_{Nj}}{\mu_{Nj}}[\cot k_{Nj}d_N - \operatorname{cosec} k_{Nj}d_N G_{N-1j}(\zeta)]. \tag{11}$$

Since the boundary conditions (B5) and (B6) are unnecessary for a uniform waveguide, the zero $s_{j\nu}$ of the kernel function corresponds to the propagation constant of the ν-th surface wave supported on this system:

$$G_j(\pm s_{j\nu}) = 0 \quad (\nu = 1 \sim M_j). \tag{12}$$

On the other hand, its pole $\zeta_{j\nu}$ is the propagation constant of the parallel plate waveguide mode:

$$1/\tilde{G}_j(\pm\zeta_{j\nu}) = 0 \quad (\nu = 1, 2, \cdots). \tag{13}$$

Now we define the following functions by residue calculus:

$$\tilde{F}_{j\nu}(y) = \lim_{\zeta \to \zeta_{j\nu}} (\zeta - \zeta_{j\nu}) F_j(\zeta, y). \tag{14}$$

These functions constitute an orthogonal set, that is,

$$\int_0^{a_N} \tilde{F}_{j\nu}(y) \tilde{F}_{j\xi}(y) \frac{dy}{\mu_j(y)} = D_{j\nu} \quad (\nu = \xi) \qquad = 0 \quad (\nu \neq \xi). \tag{15}$$

3. WIENER-HOPF EQUATION

We define the scattered fields (superscript s) by

$$(E^t, H^t) = (E^i, H^i) + (E^s, H^s) \tag{16}$$

where superscripts t and i denote total and incident fields respectively. The incident surface wave, as was discussed in the previous section, is given by

$$E_z^i(y) = N_1(s_{11}) e^{-\sqrt{s_{11}^2 - \kappa_0^2}(y - a_N)} \quad (a_N < y) \qquad = N_1(s_{11}) F_1(s_{11}, y) \quad (0 < y < a_N) \tag{17}$$

where $N_1(s_{11})$ is the normalization factor; unit incidence from the left is assumed. For analytical convenience, we divide the the scattered fields into two parts as

$$(E^s, H^s) = (E_1^s, H_1^s) + (E_2^s, H_2^s) \tag{18}$$

where each field is defined for $y<a_N$ by

$$\begin{aligned} (E_1^s, H_1^s) &= (E^s, H^s)L(-x) - (E^i, H^i)L(x), \\ (E_2^s, H_2^s) &= (E^t, H^t)L(x). \end{aligned} \tag{19}$$

$L(x)$ is a step function such as

$$\begin{aligned} L(x) &= 1 \qquad (x>0) \\ &= 0 \qquad (x<0). \end{aligned} \tag{20}$$

In the following analysis, we use the Fourier and its inverse transformation defined by

$$\begin{aligned} f(\zeta) &= F[f(x)] = \int_{-\infty}^{\infty} f(x) e^{i\zeta x} dx, \\ f(x) &= F^{-1}[f(\zeta)] = \frac{1}{2\pi} \int_c f(\zeta) e^{-i\zeta x} d\zeta \end{aligned} \tag{21}$$

where the contour c must lie in the overlapped strip ($\mathrm{Im}\ \kappa_0 < \mathrm{Im}\ \zeta < \mathrm{Im}\ -\kappa_0$); when the analysis is complete, we put $\mathrm{Im}\ \kappa_0 \to 0$.

For $y>a_N$ (in the free space), direct Fourier transformation of the wave equation (1) yields

$$\left(\frac{d^2}{dy^2} + k_0^2 \right) E_z^s(\zeta, y) = 0. \tag{22}$$

Taking the boundary conditions (B1) and (B2) into consideration, we can express this solution as

$$E_z^s(\zeta, y) = \frac{U_1(\zeta) + U_2(\zeta)}{\zeta - s_{11}} e^{-ik_0(y-a_N)} \tag{23}$$

where

$$\begin{aligned} U_1(\zeta)/(\zeta - s_{11}) &= F[E_{z1}^s(x, a_N)], \quad U_1(s_{11}) = -iN_1(s_{11}), \\ U_2(\zeta)/(\zeta - s_{11}) &= F[E_{z2}^s(x, a_N)], \quad U_2(s_{11}) = 0. \end{aligned} \tag{24}$$

For $0<y<a_N$ (in the dielectric waveguides), direct Fourier transformation of the wave equation(1) leads to the following nonhomogenious differential equations due to the discontinuities of the media at x=0:

$$(25)\qquad \left(\frac{d^2}{dy^2}+k_{mj}^2\right)E_{zj}^s(\zeta,y)=(-1)^j\left(\frac{\partial}{\partial x}-i\zeta\right)E_z^t(0,y).$$

$$(a_{m-1}<y<a_m,\ m=1\sim N,\ j=1,2)$$

Now we expand the right-hand sides of eq.(25) as follows:

$$(26)\qquad \partial E_z^t(0,y)/\partial x=-\sum_\nu \alpha_{j\nu}\tilde{F}_{j\nu}(y),\quad E_z^t(0,y)=-\sum_\nu \beta_{j\nu}\tilde{F}_{j\nu}(y)$$

together with the boundary condition (B5)

$$(27)\qquad \begin{aligned}\sum_\nu \alpha_{1\nu}\tilde{F}_{1\nu}(y)/\mu_1(y) &= \sum_\xi \alpha_{2\xi}\tilde{F}_{2\xi}(y)/\mu_2(y),\\ \sum_\nu \beta_{1\nu}\tilde{F}_{1\nu}(y) &= \sum_\xi \beta_{2\xi}\tilde{F}_{2\xi}(y).\end{aligned}\qquad (\nu,\xi=1,2,\cdots)$$

Then the solution to eq.(25) can be expressed as

$$(28)\qquad E_{zj}^s(\zeta,y)=\frac{U_j(\zeta)}{\zeta-s_{11}}F_j(\zeta,y)+(-1)^j\sum_\nu\frac{\alpha_{j\nu}-i\zeta\beta_{j\nu}}{\zeta^2-\zeta_{j\nu}^2}\tilde{F}_{j\nu}(y).\qquad (j=1,2)$$

It should be noted that eq.(28) satisfies the boundary conditions (B2) and (B3) as well as (B5). Moreover, coefficients $\alpha_{j\nu}$ and $\beta_{j\nu}$ should be related to $U_j(\zeta)$ in order to cancel poles in appropriate half-planes:

$$(29)\qquad \begin{aligned}\alpha_{1\nu}-i\zeta_{1\nu}\beta_{1\nu} &= 2\zeta_{1\nu}U_1(\zeta_{1\nu})/(\zeta_{1\nu}-s_{11}),\\ \alpha_{2\xi}+i\zeta_{2\xi}\beta_{2\xi} &= 2\zeta_{2\xi}U_2(-\zeta_{2\xi})/(\zeta_{2\xi}+s_{11}).\end{aligned}$$

Imposing the boundary condition (B4) on eqs.(23) and (28), we obtain after some manipulation

$$(30)\qquad \frac{U_1(\zeta)}{\zeta-s_{11}}G_1(\zeta)+\frac{U_2(\zeta)}{\zeta-s_{11}}G_2(\zeta)+\sum_\nu\frac{\alpha_{1\nu}-i\zeta\beta_{1\nu}}{\zeta^2-\zeta_{1\nu}^2}\mathrm{Res}[G_1(-\zeta_{1\nu})]$$
$$+\sum_\xi\frac{\alpha_{2\xi}-i\zeta\beta_{2\xi}}{\zeta^2-\zeta_{2\xi}^2}\mathrm{Res}[G_2(\zeta_{2\xi})]=0.$$

This is the Wiener-Hopf equation; $U_1(\zeta)$ is regular in the lower half-plane and so is $U_2(\zeta)$ in the upper, and this relation holds in the overlapped strip ($\mathrm{Im}\ \kappa_0<\mathrm{Im}\ \zeta<\mathrm{Im}\ -\kappa_0$).

4. FORMAL SOLUTION

First we factorize the kernel functions in eq.(30) as

$$(31)\qquad G_j(\zeta)=G_j^+(\zeta)/G_j^-(\zeta)=G_j^+(\zeta)G_j^+(-\zeta)$$

where $G_j^+(\zeta)$ is a regular and nonzero function in the upper half-plane. Following the conventional procedure together with the edge condition (B6), we can determine the formal solution as

$$(32)\qquad \begin{aligned} U_1(\zeta) &= G_1^-(\zeta)[U_1(s_{11})/G_1^-(s_{11}) + I_1(\zeta) + (\zeta - s_{11})S_1(\zeta)], \\ U_2(\zeta) &= [I_2(\zeta) + (\zeta - s_{11})S_2(\zeta)]/G_2^+(\zeta) \end{aligned}$$

with

$$(33)\qquad I_{\substack{1\\2}}(\zeta) = \frac{\zeta - s_{11}}{2\pi i}\int_\Gamma \frac{ik_0 G_{\substack{1\\2}}^-(t)U_{\substack{2\\1}}(\mp t)}{(t\pm\zeta)(t\pm s_{11})}dt,$$

$$(34)\qquad S_{\substack{1\\2}}(\zeta) = \sum_\nu \frac{U_{\substack{1\\2}}(\mp\zeta_{\substack{1\\2}\nu})\,\mathrm{Res}[G_{\substack{1\\2}}(\zeta_{\substack{1\\2}\nu})]}{(\zeta\pm\zeta_{\substack{1\\2}\nu})(\zeta_{\substack{1\\2}\nu}\pm s_{11})G_{\substack{1\\2}}^+(-\zeta_{\substack{1\\2}\nu})}$$

where Γ is the contour in clockwise direction along the branch cut which starts from κ_0 to infinity in the lower half-plane, and $\frac{1}{2}$ and $\pm$ signs go together. Moreover, $U_1(-\zeta_{1\nu})$ and $U_2(\zeta_{2\xi})$ in eq.(34) should be related to $U_1(\zeta_{1\nu})$ and $U_2(-\zeta_{2\xi})$ for numercal convenience. Combination of eqs.(15),(27) and (29) leads to

$$(35)\qquad \begin{aligned} \sum_\nu P_{\nu\xi}^{(2)}\frac{U_1(-\zeta_{1\nu})}{\zeta_{1\nu}+s_{11}} &= -2\zeta_{2\xi}D_{2\xi}\frac{U_2(-\zeta_{2\xi})}{\zeta_{2\xi}+s_{11}} + \sum_\nu P_{\nu\xi}^{(1)}\frac{U_1(\zeta_{1\nu})}{\zeta_{1\nu}-s_{11}}, \\ \sum_\xi P_{\nu\xi}^{(2)}\frac{U_2(\zeta_{2\xi})}{\zeta_{2\xi}-s_{11}} &= -2\zeta_{1\nu}D_{1\nu}\frac{U_1(\zeta_{1\nu})}{\zeta_{1\nu}-s_{11}} - \sum_\xi P_{\nu\xi}^{(1)}\frac{U_2(-\zeta_{2\xi})}{\zeta_{2\xi}+s_{11}} \end{aligned}$$

where

$$(36)\qquad P_{\nu\xi}^{(j)} = \int_0^{a_N}\left[\frac{\zeta_{1\nu}}{\mu_1(y)} + (-1)^j\frac{\zeta_{2\xi}}{\mu_2(y)}\right]\tilde{F}_{1\nu}(y)\tilde{F}_{2\xi}(y)dt. \quad (j=1,2$$

5. CONCLUSION

We have analyzed the problem of two symmetrical multi-layer dielectric waveguide junctions by the Wiener-Hopf technique, and obtained formal solutions only for TE even modes. However, the exchange of functions $(\cos k_{1j}a_1, \sin k_{1j}a_1)\rightarrow(\sin k_{1j}a_1, -\cos k_{1j}a$ is enough to obtain final results for odd modes. Moreover, extension to TM modes is possible with the exchange $(E,H,\varepsilon,\mu)\rightarrow(H,-E,\mu,\varepsilon)$. To obtain physical quantities such as reflected, transmitted, and radiated fields, we have to determine eqs.(33) and (34) numerically. An iterative method will be effectively applicable for this type of formal solutions.

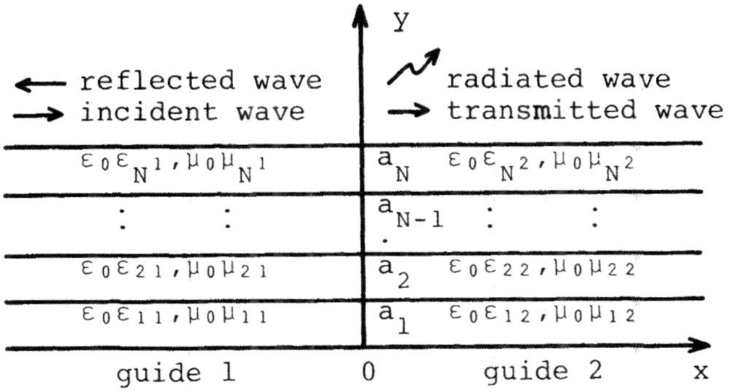

Fig.1 Geometry of the problem.

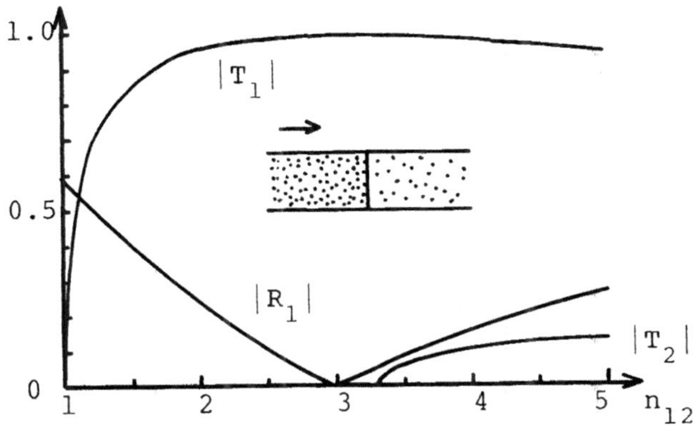

Fig.2 Reflection and transmission coefficients for single layer. (TE incidence; $n_{11}=3, \kappa_0 a_1=1$)

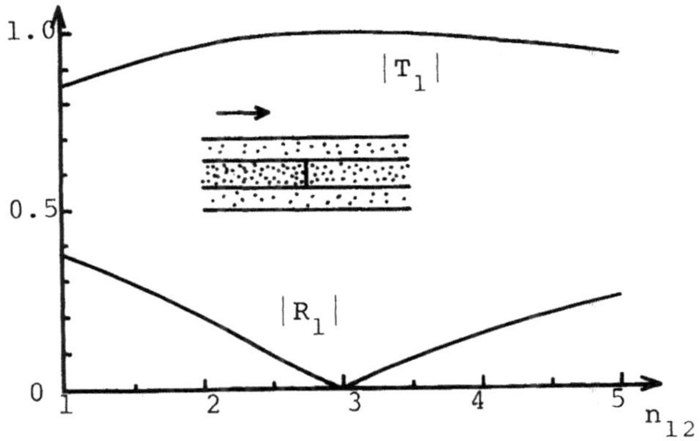

Fig.3 Reflection and transmission coefficients for 3-layer. (TE incidence; $n_{11}=3, n_{21}=n_{22}=2, a_1/a_2=0.5, \kappa_0 a_2=1$)

CUTOFF FREQUENCIES OF OPTICAL PLANAR/CYLINDRICAL STRUCTURES USING THE RESONANCE TECHNIQUE

C. D. Papageorgiou

J. D. Kanellopoulos

Department of Electrical Engineering, National Technical University of Athens, Athens 147, Greece

A. C. Boucouvalas

Department of Electrical Engineering Imperial College, of Science and Technology, London, SW7 2BT, U.K.

ABSTRACT

We present a simple, accurate, and fast numerical method for calculating the cutoff frequencies of optical planar dielectric guiding structures. From Maxwell's equations we derive a transmission-line equivalent circuit for any mode of the guiding structure. The problem of finding the cutoff frequencies is transformed into finding the resonance frequency of the equivalent circuit. The algorithm uses no Bessel or other complicated functions and could be implemented on small microcomputers. The technique can also be used with data from measurements.

1. INTRODUCTION

The fields of microwaves, integrated optics and Laser Technology are areas sharing a common interest in optical planar layered structures. Knowledge of the propagation constants and cutoff frequencies of the various modes in such waveguides is important for design work[1,2].

In this paper we propose a new method based on transmission-line theory for obtaining the mode cutoffs of planar layered structures and cylindrical dielectric waveguides. It uses no Bessel functions and eliminates "small argument" approximations. To devise such a method, time and space Fourier transforms when properly applied to Maxwell equations for the planar geometry yield two separately excited circuits, one electric and the other magnetic, related to the transverse electric and transverse magnetic fields supported by a planar dielectric layer.

For this geometry the two circuits are not coupled. On the

other hand, the space Fourier transformations of the Maxewell's equations for the cylindrical layer lead generally to two interlinked transmission lines which correspond to the HE or EH hybrid modes, except for the symmetrical case (axial wavenumber $\ell=0$) where the transverse electric or transverse magnetic modes exist. For each case, equivalent T-circuits for planar/cylindrical dielectric layers are derived. The entire planar optical structure or a fiber can be approximated by a large number of such thin layers of uniform refractive index and correspondingly by a large number of equivalent T-circuits in tandem. When the whole circuit resonates, the propagation constants and the cutoff frequencies can be found at the resonance condition.

2. THE TRANSMISSION LINE MODEL

2.1 Planar structure

Maxwell's equations for source free, time dependent fields are

$$\nabla x\tilde{E} = -\frac{\partial \bar{B}}{\partial t}, \qquad \nabla x\tilde{H} = \frac{\partial \bar{D}}{\partial t} \tag{1}$$

where $\tilde{E}(t)$, $\tilde{H}(t)$, $\bar{B}(t)$ and $\bar{D}(t)$ are, as usual, time dependent field vectors.

We assume fields with periodic time dependence of the form $\exp(j\omega t)$ and direction of propagation along the y axis of an anisotropic medium with dielectric constants εy along the x and y directions and εz along the z direction. Applying space Fourier transforms to the field components of equations (1), along the x and y axes with respective wavenumbers ℓ and β, the following equations can be derived.

$$\begin{aligned}
&\frac{j\beta\bar{B}_z}{\mu_o} - \frac{\partial\bar{H}_y}{\partial z} = j\omega\varepsilon_o\varepsilon_y\bar{E}_x \\
&\frac{\partial\bar{H}_x}{\partial z} - \frac{j1\bar{B}_z}{\mu_o} = j\omega\varepsilon_o\varepsilon_y\bar{E}_y \\
&\ell\bar{H}_y - \beta\bar{H}_x = \omega\bar{D}_z \\
&\frac{j\beta\bar{D}_z}{\varepsilon_o\varepsilon_z} - \frac{\partial\bar{E}_y}{\partial z} = -j\omega\mu_o\bar{H}_x \\
&\frac{\partial\bar{E}_x}{\partial z} - \frac{j\ell\bar{D}_z}{\varepsilon_o\varepsilon_z} = -j\omega\mu_o\bar{H}_y
\end{aligned} \tag{2}$$

$$\ell\bar{E}_y - \beta\bar{E}_x = -\omega\bar{B}_z$$

where $\bar{\Psi}_i$ is the spatial Fourier transform of the field component Ψ_i i.e.

$$\bar{\Psi}_i(\ell,\beta,z) = \int\int_{-\infty}^{\infty}\Psi_i(x,y,z)\exp\{-j(\ell_x+\beta_y)\}\, d_x d_y$$

$(i=x,y,z)$

Defing the following variables

$V_{TE} = \ell\bar{H}_x + \beta\bar{H}_y$ as TE voltage

$V_{TM} = \ell\bar{E}_x + \beta\bar{E}_y$ as TM voltage

$I_{TE} = \omega\bar{B}_z = \omega\mu_o\bar{H}_z = \beta\bar{E}_x - \ell\bar{E}_y$ as TE current

$I_{TM} = \omega\bar{D}_z = \omega\varepsilon_z\bar{E}_z = \ell\bar{H}_y - \beta H_x$ as TM current

we deduce from equations (2) the following relations:

$$\frac{\partial V_{TE}}{\partial z} = -\frac{\gamma^2{}_y}{j\omega\mu_o}I_{TE}, \qquad \frac{\partial I_{TE}}{\partial z} = -j\omega\mu_o V_{TE} \tag{3}$$

where $\gamma^2{}_y = \ell^2 + \beta^2 - n_y^2k_o^2$ and

$$\frac{\partial V_{TM}}{\partial z} = -\frac{\gamma^2{}_z}{j\omega\varepsilon_o n_z^2} I_{TM}, \qquad \frac{\partial I_{TM}}{\partial z} = -j\omega\varepsilon_o n_y^2 V_{TM} \tag{4}$$

where $\gamma_z^2 = \ell^2+\beta^2-n_z^2k_o^2$, $k_o = 2\P/\lambda_o$, λ_o being the free space wavelength, $\varepsilon_y = n_y^2$, $\varepsilon_z = n_z^2$, n_i being the corresponding refractive index.

The two equivalent transmission lines have different propagation factors (γ_y, γ_z) and different characteristic impedances:

$$Z_{TE} = \frac{\gamma_y}{j\omega\mu_o}, \qquad Z_{TM} = \frac{\gamma_z}{j\omega\varepsilon_o n_y n_z} \tag{5}$$

Let us consider a dielectric planar layer of thickness d and refractive indices n_y, n_z along the y and z directions, as shown in Fig.1. With the z axis selected to be normal to the layer boundaries, we assume wave propagation along the y axis. For the proper choice of cartesian coordinates $\frac{\partial}{\partial x} = 0$ and thus $\ell = 0$. Hence $\gamma_y^2 = \beta^2 - n_y^2k_o^2$; $\gamma_z^2 = \beta^2 - n_z^2k_o^2$.

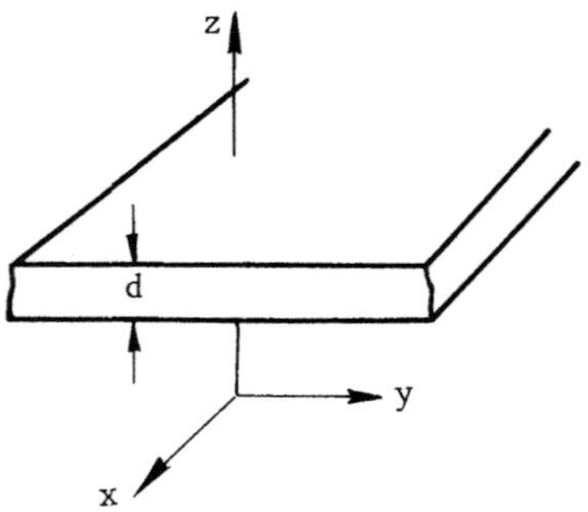

Fig.1 Dielectric planar layer and reference coordinates

The planar layer can be represented by two transmission lines and consequently by two equivalent T-circuits (fig.2) with elements given by

$$Z_{s,TE} = Z_{TE} \cdot \tanh(\gamma_y d/2)$$
$$Z_{p,TE} = Z_{TE}/\sinh \ (\gamma_y d) \qquad (6)$$

$$Z_{s,TM} = Z_{TM} \cdot \tanh(\gamma_z d/2)$$
$$Z_{p,TM} = Z_{TM}/\sinh \ (\gamma_z d) \qquad (7)$$

An infinitely thick layer will be represented by its characteristic impedances Z_{TE} and Z_{TM} as they are shown in Eqs (5). A set of successive planar layers can now be represented by a set of equivalent circuits connected in tandem with their elements given by equations (6), and a similar set by using equations (7). For certain β values, it is possible that some of the elements are of the "capacitive" type, while the remainder, of the "inductive" type. In such cases the condition of resonance gives the propagation constants of the TE modes using the magnetic equivalent circuits, and of the TM modes using the electric equivalent circuits. This technique will be illustrated by the following example of an inhomogeneous anisotropic layer between two uniform anisotropic layers (Fig.3). The top layer has indices n_{YT}, n_{ZT} and the bottom layer n_{YB}, n_{ZB}, along the Y and Z directions, respectively. Propagation is in the Y direction. The equivalent circuit of the composite structure is shown in Fig. 3b.

For TM modes the characteristic impedances of the top layer

Z_T and the bottom layer Z_B are given by

$$Z_T = \frac{-j(\beta^2 - n_{ZT}^2 k_o^2)^{\frac{1}{2}}}{\omega\varepsilon_o n_{YT} n_{ZT}} \qquad Z_B = \frac{-j(\beta^2 - n_{ZB}^2 k_o^2)^{\frac{1}{2}}}{\omega\varepsilon_o n_{YB} n_{ZB}}$$

and for TE modes

$$Z_T = \frac{-j(\beta^2 - n_{YT}^2 k_o^2)^{\frac{1}{2}}}{\omega\mu_o} \qquad Z_B = \frac{-j(\beta^2 - n_{YB}^2 k_o^2)^{\frac{1}{2}}}{\omega\mu_o}$$

respectively.

The middle layer with an arbitrary profile is represented by a large number, N, of uniform thin layers in a stair-like fashion. The equivalent circuits of the successive layers are connected in tandem as in Fig. 3b, and are terminated by Z_T and Z_B. Using elementary circuit theory, the impedance of the nth layer can be expressed in terms of the (n-1)th layer:

$$Z_n = \frac{(Z_{n-1} + Z_{s,n}) Z_{p,n}}{Z_{n-1} + Z_{s,n} + Z_{p,n}} + Z_{s,n} \tag{8}$$

n=1,2,3........., N; where $Z_{s,n}$, $Z_{p,n}$ are the series and parallel elements of the T-circuit representation of the nth layer.

When the layers are thin, then $Z_{s,n}$ and $Z_{p,n}$ simplify to the following formulae:

For TM modes

$$Z_{s,TM} = \frac{\gamma'^2_z (\delta z')^2}{2} Z_{P,TM}$$

$$Z_{P,TM} = \frac{Z_o}{n_y n_z (\delta z')}$$

For TE modes

$$Z_{s,TE} = \frac{\gamma'^2_Y (\delta Z')^2}{2} Z_{P,TE}$$

$$Z_{P,TE} = \frac{1}{Z_o (\delta z')}$$

where $\gamma'^2_z = (\beta/k_o)^2 - n_z^2$, $\gamma'^2_Y = (\beta/k_o)^2 - n_y^2$, δz is the thickness of the layer, $Z_o = 120\pi$, and $\delta z' = \delta z k_o$.

To achieve resonance, the sum of Z_N (impedance of the inhomogeneous and bottom layers) and Z_T (characteristic impedance of the top layer) should be zero. Z_N is computed by using Eq.(8) with Z_B as Z_{n-1} for n = 1,2...N. The resonance condition is satisfied when

$$Z_N + Z_T = 0 \tag{9}$$

Thus, β can be found by using a root searching technique.

2.2 Cylindrical model

An Optical fiber, with an arbitrary refractive index profile, could be considered as consisting of a large number of thin, cylindrical homogeneous dielectric layers of thickness δr, dielectric constant $\varepsilon = n^2\varepsilon_o$, permeability μ and conductivity $\sigma = 0$. With reference to a previous paper by the present author[5], the electromagnetic behaviour of a cylindrical dielectric layer can be similarly described by T-circuits with series impedance Z_B and parallel Z_P, where

$$Z_B = Z\tanh(\gamma\delta r/2), \quad Z_P = Z/\sinh(\gamma\delta r) \tag{10}$$

and Z, γ, δr are, respectively, the characteristic impedance, the propagation factor and the length of the transmission line. Since $\frac{\delta r}{r} \ll 1$, Z_B and Z_P of the transmission lines can be approximated as

$$Z_B = \frac{1}{2}(\delta r')^2\gamma'^2 Z_P' \tag{11}$$

$$Z_P' = \frac{Z_o}{jr'\delta r' n(n_2^2 + (1/r')^2)} \tag{12}$$

where $r' = rk_o$, $\delta r' = \delta r k_o$, $Z_P' = Z_P k_o$, $n_2 = \beta/k_o$ and

$$\gamma'^2 = n_2^2 + \left(\frac{1}{r'}\right)^2 - n^2 \pm \frac{2nn_2\ell}{\ell^2 + (n_2 r')^2} \tag{13}$$

where - and + before the last term are for HE mode and EH mode, respectively.

3. CUTOFF FREQUENCIES

3.1 Planar structure

Cutoff frequencies can be easily obtained with the resonance technique. It is best illustrated in the case of a guiding layer with $n_1k_o > \beta > n_2k_o$.

The cutoff frequencies of the various modes are obtained when $\beta = n_2k_c$. The propagation factors γ_Y^2 and γ_Z^2 of equations (3) and (4) will now become

$$\gamma_{Yc}^2 = (n_2^2 - n_Y^2)k_c^2, \quad \gamma_{Zc}^2 = (n_2^2 - n_Z^2)k_c^2$$

and $\omega = k_c c$ where $c = 1/(\mu_o\varepsilon_o)^{\frac{1}{2}}$, $k_c = 2\pi/\lambda_c$, λ_c being the cutoff wavelength.

The transmission lines will now have characteristic impedances

$$\bar{Z}_{TE} = \frac{\gamma_{Yc}}{jw_c\mu_o}$$
$$\bar{Z}_{TM} = \frac{\gamma_{zc}}{jw_c\varepsilon_o n_Y n_Z} \tag{14}$$

Finally for TE and TM modes the equivalent circuit for the cutoff frequencies of a uniformly anisotropic layer of thickness d will have series and parallel elements defined by

$$Z_{s,TE} = Z_{TE}\cdot\tanh(\gamma_{Yc}d/2); \quad Z_{s,TM} = Z_{TM}\cdot\tanh(\gamma_{Zc}d/2)$$
$$Z_{P,TE} = Z_{TE}/\sinh(\gamma_{Yc}d) \quad ; \quad Z_{P,TM} = Z_{TM}/\sinh(\gamma_{Zc}d) \tag{15}$$

respectively, where $\bar{Z}_{TE}$ and $\bar{Z}_{TM}$ are given by Eqs.(14).

An infinitely thick uniformly anisotropic layer can be represented in a similar manner as in section 2.1 by Z_{TM} and Z_{TE}, Eqs.(14). In the case of an inhomogeneous anisotropic layer between two uniform anisotropic layers, represented in section 2.1. for illustration of the method, Eq.(8) holds true for evaluation of the cutoff frequencies, and λ_c is obtained when Eq.9 is satisfied, with variable λ_c and

$$Z_{s,TM} = \frac{\gamma'^2_{Zc}\,(\delta Z')^2}{2}\, Z'_{P,TM}$$
$$Z'_{P,TM} = \frac{Z_o}{n_Y n_Z(\delta Z')} \tag{16}$$

for TM modes and

$$Z'_{s,TE} = \frac{\gamma'^2_{Yc}(\delta Z')^2}{2}\, Z'_{P,TE}$$
$$Z'_{P,TE} = \frac{1}{Z_o(\delta Z')} \tag{17}$$

for TE modes, where $\gamma'^2_{Yc} = (n_2^2 - n_Y^2)$, $\gamma'^2_{Zc} = (n_2^2 - n_Z^2)$, and $\delta Z' = \delta Z K_c$.

As an indication of the convergence and accuracy of this method, the parabolic profile was chosen for computation

$$n^2(Z) = n_1^2(1 - \Delta(\frac{Z}{\alpha})^2) \qquad Z \leq \alpha$$
$$= n_2^2 \qquad Z > \alpha \tag{18}$$

where α is the radius of the parabolic layer and $\Delta = (n_1^2 - n_2^2)/n_1^2$.

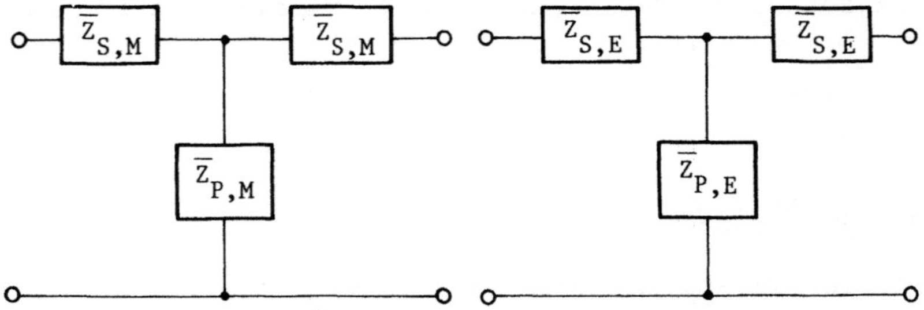

Fig.2 Equivalent T-circuit of the transmission lines representing the planar layer of Fig.1

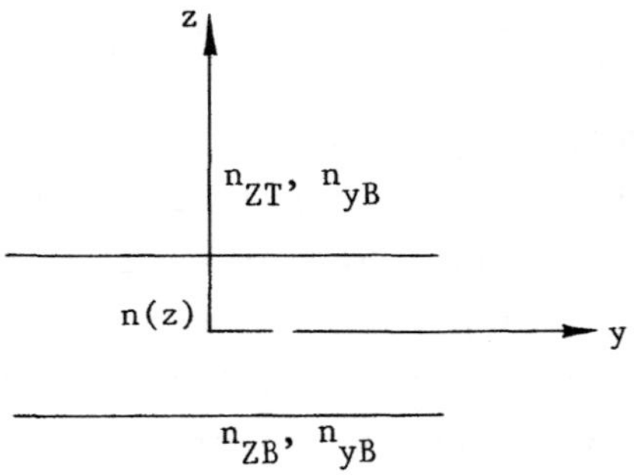

(a)

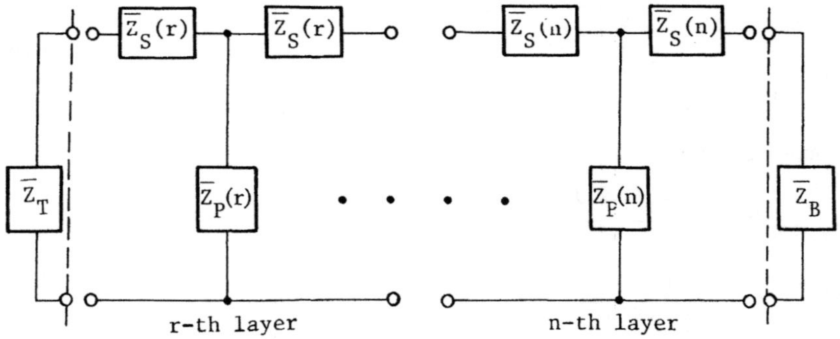

Fig.3 (a) Inhomogeneous anisotropic planar waveguide
(b) Equivalent transmission line consisting of a large number of equivalent T-circuits of thin layers in tandom

(b)

The derived results for 500 layers were in excellent agreement with preexisting results.

3.2 Cylindrical structure

In order to obtain the cutoff equivalent circuits of the transmission lines, it is appropriate, at this stage, to introduce

$$\beta = n_2 k_C$$

where $k_C = 2\pi/\lambda c$, λc being the cutoff wavelength, n_2 the cladding refractive index.

The series and parallel elements Z_B, Z_P of the equivalent T-circuits of such a transmission line are given by Eqs. (10)-(13) with r', $\delta r'$, Z_P' substituted by

$$r' = rk_C, \quad \delta r' = \delta r k_C, \quad Z_P' = Z_P k_C$$

Using again the elementary circuit theory as described in section 2.2, one obtains the cutoff frequency of any mode as the corresponding resonant frequency of the equivalent circuit.

As an indication of the convergence and accuracy of the method, the graded index fiber profile was chosen for computation. We define

$$\begin{aligned} n^2(r) &= n_1^2 \left(1 - \Delta\left(\frac{r}{\alpha}\right)q\right) & r \leq \alpha \\ &= n_2^2 & r > \alpha \end{aligned} \qquad (19)$$

where $\Delta = (n_1^2 - n_2^2)/n_1^2$, α is the core radius, and q is the profiling factor.

For all computations we have chosen the thickness of the layers, inside and outside the core, to be $\delta r/r = 0.02$. Depending on the required accuracy of the result, 400 to 500 layers are in general sufficient for this $\delta r/r$ ratio, where an error of $\pm 1 \times 10^{-3}$ in V_C is acceptable. Increasing the number of layers and decreasing the $\delta r/r$ ratio will improve the accuracy at the expense of computation time. Finally, a very important factor for improving the speed of the method is to use an efficient root searching technique.

REFERENCES

1. Hartog, A.H. and Adams, M.J., On the accuracy of the W.K.B. approximations in optical dielectric waveguides, Optical and Quantum Electronics, Vol.9, pp.223-232, 1977.
2. Adams, MJ., An Introduction to Optical waveguides, Chichester, J. Wiley, 1981, pp.156-166.
3. Gambling, W. A., Payne, D. H., and Matsumura, H., Cut-off frequency in radially inhomogeneous single mode fibre, Electron. Lett., 1977, 13 pp.139-140.
4. Sharma, E. K., Goyal, I. C., and Ghatak, A.K. : Calculation of cutoff frequencies in optical fibers for arbitrary profiles using the Matrix Method, I•E•E•E• J. Quant. Electr., 1981, GE-17, pp.2317-2321.
5. Papageogiou, C. D. and Boucouvalas, A.C., Propagation constants of cylindrical dielectric waveguides with arbitrary refractive-index profile using resonance technique, Electron. Lett., 1982, Vol. 18, No.18, pp.786-788.

PART SIX

SYMPOSIUM BRIEFS

THROUGHPUT AND COUPLING IN OPTICAL FIBERS

Henri Hodara
Tetra Tech, Inc., Pasadena, California

ABSTRACT

We show how to apply the concept of optical throughput, product of area times solid angle to emitters and fibers, in order to simplify calculations of coupling efficiency. We show that throughput conservation yields readily the maximum theoretical coupling efficiency, independently of intervening optics.

SUMMARY

Coupling of light to an optical fiber depends on the relative area and numerical aperture of both emitter and fiber. It follows that calculations of the light fraction coupled from emitter to fiber must take into account four possible geometries. Additional cases arise if matching optics are introduced to increase the power coupled into the fiber.

In this paper, we show that power coupling calculations for each of the four cases can be carried out very simply by using the concept of optical throughput conservation. Throughput of any optical component, defined as the product of area times its subtended solid angle, is proportional to the number of modes supported by the structure. We also show that throughput conservation yields readily the maximum theoretical power coupled into a fiber, without attention to the details of the matching optics.

BEAM PROPAGATION IN THE PARAXIAL APPROXIMATION

Reynaldo Castillo, A. K. Ghatak*, and H. Hora
Department of Theoretical Physics, University of New South Wales,
Kensington - Sydney, Australia.

Abstract and Summary

We present a new method for studying the propagation of beams in the paraxial approximation.

We start with the exact scalar wave equation corresponding to propagation in a homogeneous medium of refractive index n:

$$[\,\nabla^2 - \frac{n^2}{c^2}\frac{\partial^2}{\partial t^2}\,]\,\Psi(x,y,z,t) = 0 \qquad (1)$$

We assume propagation in the z-direction and write the solution in the form

$$\Psi = \psi(x,z)\, e^{i(kz-\omega t)} \qquad (2)$$

where $k = nk_o$, $k_o = \omega/c$ being the free space wave number. We substitute (2) in (1) to obtain

$$[\,\frac{\partial^2}{\partial x^2} + 2ik\frac{\partial}{\partial z}\,]\,\psi(x,z) = 0 \qquad (3)$$

where we have made use of the paraxial approximation i.e., we have assumed

$$\left|\frac{\partial^2\psi}{\partial z^2}\right| << \left|k\frac{\partial\psi}{\partial z}\right| \qquad (4)$$

Physically, Eq. (4) implies that ψ does not vary appreciably in a distance $\sim\lambda$ along the z direction. Eq. (3) is identical in form to the one dimensional Schrodinger equation for a free particle in one dimension:

$$[\,\frac{\partial^2}{\partial x^2} + \frac{2mi}{\hbar}\frac{\partial}{\partial t}\,]\,\psi(x,t) = 0 \qquad (5)$$

It is interesting to note that a first integral for Eq. (5) is given by

$$\Delta = x - \frac{i\hbar t}{m}\frac{\partial}{\partial x} \qquad (6)$$

Which is nothing but the operator $x + \frac{p_x}{m}t$. We may mention that Δ is a

* On leave of absence from Department of Physics, Indian Institute of Technology, New Delhi, India.

first integral in the sense that if ψ is a solution of Eq. (5) then $\Delta\psi$ is also a solution. Now, corresponding to Eq. (3) the first integral is given by

$$\Delta = x - i\,\kappa z \frac{\partial}{\partial z} \tag{7}$$

where $\kappa \equiv 1/k$. One can now develop an operator algebra for Δ as in quantum mechanics.

We return to Eq. (3), multiply it by $e^{i\lambda x}\,dx$ and integrate from $-\infty$ to $+\infty$ to obtain

$$\frac{\partial u(\lambda,z)}{\partial z} = -\frac{i\lambda^2}{2k}\,u(\lambda,z) \tag{8}$$

where

$$u(\lambda,z) = \int_{-\infty}^{+\infty} \psi(x,z)e^{i\lambda x}\,dx \tag{9}$$

is the Fourier transform of $\psi(x,z)$, and in writing Eq. (8) we have done two partial integrations and assumed ψ to vanish at $x = \pm\infty$. The solution of Eq. (8) is obvious:

$$u(\lambda,z) = u_o(\lambda)\; e^{-(i\lambda^2/2k)z} \tag{10}$$

where

$$u_o(\lambda) = \int_{-\infty}^{+\infty} \psi(x,o)\; e^{i\lambda x}\,dx \tag{11}$$

We not take the inverse Fourier transform to obtain

$$\psi(x,z) = \frac{1}{2\pi}\int_{-\infty}^{+\infty} u_o(\lambda)\; e^{-i(\lambda^2/2k)z}\; e^{-i\lambda x}d\lambda \tag{12}$$

which is nothing but the "diffraction pattern". From the known form of $\psi(x,o)$ one can calculate $u_o(\lambda)$ and hence $\psi(x,z)$. For example, for $u_o(\lambda) = \lambda^n$ we have

$$\psi(x,z) = \frac{1}{2}\sqrt{\frac{2k}{\pi z}}\,\frac{d^n}{dx^n}\left[\, e^{ikx^2/2z}\,\right] \tag{13}$$

Although the final result is known in optics, what we have given here is a new method to study the propagation of beams of finite extent in the paraxial approximation. We are now attempting to extend the method to inhomogeneous media.

LEAKY-WAVE MODES AND THEIR ROLE IN THE NUMERICAL EVALUATION OF THE FIELD EXCITED BY A LINE SOURCE IN A NON-SYMMETRIC, INHOMOGENEOUSLY LAYERED, SLAB WAVEGUIDE

H. Blok

Laboratory of Electromagnetic Research, Department of Electrical Engineering, Delft University of Technology, The Netherlands

ABSTRACT & SUMMARY

In scattering problems in open-waveguide configuations source-type representations for the fieldcomponents in the cross sections of the waveguides are needed. The relevant representations can be found using a special approach with respect to time and with respect to the longitudinal coordinate. In the spectral representation we consider those spectral constituents that can be excited by a source of finite extent. Such an approach naturally leads to the concept of modes and it also gives a better understanding of the role of leaky-wave modes, the latter being part of the radiation field. In piecewise homogeneously layered (multi-step-index) waveguides, the spectrum can be constructed from analytical representations of the field, whereas in the more general, inhomogeneously layered (graded-index) waveguides the field representations as well as the spectrum has to be found numerically. In this context, we have investigated numerically the canonical problem of the line-source excitation of planar, open-waveguide configurations. In the subsequent computations, the spatial Fourier contour is deformed into a path of steepest descent that passes through a saddle point of the argument of the propagation factor of the field in the complex Fourier-transform variable plane. The latter path is situated on various sheets of a multi-sheeted Riemann-surface. The contribution from all poles (surface-wave as well as leaky-wave poles) that are circumvented in this process have to be determined. These are the surface-wave poles that are located on the physically acceptable Riemann sheet and two types of leaky-wave poles (air leaky-wave poles and substrate leaky-wave poles) that are located on other Riemann sheets. The properties of the field distributions of the corresponding surface-wave and leaky-wave modes, and where possible their interrelationship, are discussed for various symmetric and non-symmetric planar-waveguide configurations.

The path of steepest descent is approximated by a contour of finite length. Its contribution to the excited electromagnetic field is computed using standard integration techniques.

Numerical results are presented for various configurations. In particular, the excitation of a non-symmetric planar waveguide and the role of the leaky-wave modes in the excited field are discussed in detail.

OPTICAL IMAGE TRANSMISSION THROUGH A SINGLE FIBER BY CHROMATIC CODING

T. C. Yang*

C. Froehly

Laboratoire d'Optique, University of LIMOGES, France

ABSTRACT

We have devised a new system for real-time, direct transmission of self-luminous (white light) images through a single fiber with a 600 transmitted points and a loss of 40 dB. Color and 3D information transmission are under study in our laboratory.

SUMMARY

Real-time transmission of two-dimensional self-luminous black/white images through a single fiber is demonstrated. It is based on a new two-dimensional chromatic coding of the illuminance distribution to be transmitted. The coding and decoding devices consist of combinations of two spectroscopes working with crossed dispersions, one of them being in the first order of interference (prism spectroscope), the other in the kth order of interference (echelette grating spectroscope). The image available at the fiber output is a two-dimensional "rainbow" spectrum which consists of the juxtaposition of about $m \simeq k/2$ sub-spectra (spectral raster with m lines) modulated by the image intensity pattern (see fig).†

The number of the points simultaneously imaged through the fiber is about $N^2=600 (\simeq 25 \times 25)$ in the experiment presented. We observed no degradation of the image sharpness by mode couplings, microbends of the guide and other usual causes of signal distortion.

The energy loss, measured

as the ratio τ' of the image emittance L' to the source emittance L, may be shown equal to τ/N^2, where τ accounts for the overall attenuation of the radiation through the fiber and the spectroscopes. The measured value of $10 \log(\tau/N^2)$ was of the order of 40 dB.

The optical power available per pixel is given by $d\Phi = LK\lambda_o^2\tau'/N^2$, where:

L: source emittance (here, $L \simeq 1W.mm^{-2}.sr^{-1}$: object illuminated by halogen glow lamp),

K: number of the excited modes of the fiber (here $N \simeq 10^4$),

λ_o: average wavelength of the spectral carrier (here $\lambda_o = 0{,}5$ μm),

τ', N^2 having been defined above ($\tau' \simeq 10^{-4}$, $N^2 \simeq 6.10^2$ in this experiment), thus $d\Phi \simeq 4.10^{-10}$ W/pixel $\simeq 10^{10}$ photons/pixel in the case of our 600 pixel transmission.

This allows the picture information flow to be transmitted at rates as high as 10^7 images per second, that is 6.10^9 bits per second, if the image detector is a two dimensional photodiode matrix, the sensitivity of each of the photocells being 10^3 photons per bit.

Generalization to two dimensional color transmission is then demonstrated experimentally (with lower number of transmitted points).

Application to real-time stereoscopic imaging of three-dimensional coloured pictures through a single fiber, now under study in our laboratory, will be discussed.

* On leave from the 210 Electronics Institute, Shanxi, PRC.

† The original "rainbow" spectrum photo is in color. Black & white one is presented here to simplify printing processes.

TRANSIENT STIMULATED RAMAN SCATTERING IN GLASS FIBER

Gan Fuxi, Jin Deyun, Huang Guosong and Chen Shizheng
Shanghai Institute of Optics and Fine Mechanics, Academia Sinica

ABSTRACT

Using a train of 0.53μm pulses from a mode-locked laser with a peak power of 10^7 W or more to stimulate a 20 m long GeO_2-P_2O_5-SiO_2 glass fiber, stable transient stimulated Raman scatterings were observed, of which 15 orders of Stokes waves and 8 orders of anti-stokes waves are involved. The Raman frequency shift is 430 cm^{-1} in wave number, the frequency spectrum covers from 0.45μm to 1.00μm, and the conversion efficiency is 80%.

A passive mode-locked Nd: YAG laser is used in the experimental arrangement. It produces a stable light pulse train of 1.06μm wavelength at a repetition rate of 10-12 Hz. The pulse width measured by TPF method is 23 ps, the peak power is larger than $5x10^7$ W, and the divergent angle of the light beam is less than 1 mrad. By the use of a KDP frequency-doubling crystal in type II phase-matching, a light pulse train of 0.53μm wavelength with a peak power of 10^7 W or more is obtained. It is focused on the end of a 20 m long GeO_2-P_2O_5-SiO_2 glass fiber. The output stimulated Raman scattering light is detected on the other end of the fiber.

Stable transient high-order Stokes and anti-Stokes stimulated Raman scattering light has been observed. The scattering frequency spectrum ranges from 9.9x3 to $22x10^3$ cm^{-1}. We have not observed any scattering light at the range of 0.838-0.925μm, because there are stronger absorption losses in glass fiber at this band. The spectrum of the stimulated Raman scattering in the glass fiber is shown in Fig.1. The central frequency, line-width and time-width of lower Stokes scatterings are given in Table 1.

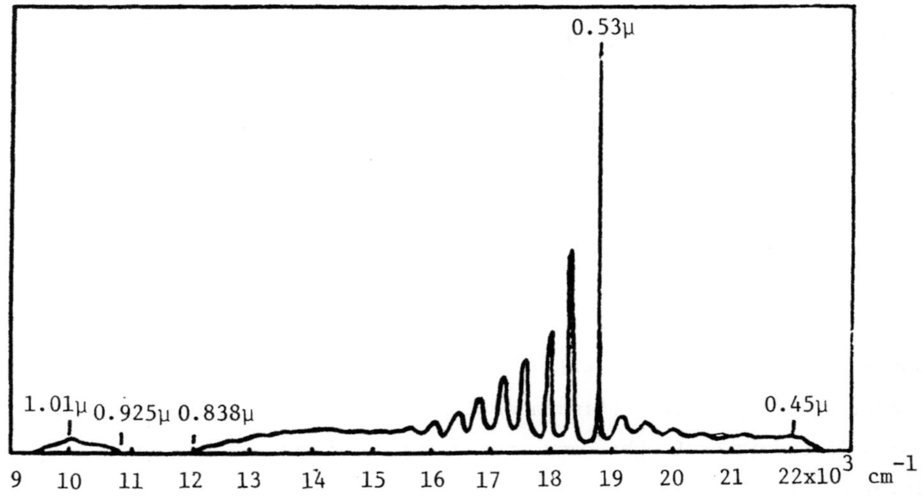

FIGURE 1. Spectrum of SRS in Glass Fiber

TABLE 1. The central frequency, line-width and time-width of lower order Stokes waves

	Frequency (cm^{-1})	Line-width (cm^{-1})	Time-width (ps)	Frequency shift (cm^{-1})
SRS-0	18793	44	23	0
SRS-1	18363	96	10.6	430
SRS-2	17933	128	7.9	430
SRS-3	17503	176	5.8	430

RESEARCH ON INTENSITY DISTRIBUTION OF MULTI-ORDER SRS IN AN OPTICAL FIBER

Yang Tian-long
Shanghai University of Science and Technology

Gao Pei-juan & Zhao Dong-huan
Shanghai Institute of Optics & Fine Mechanics, Academia Sinica

ABSTRACT

This paper presents a theoretical analysis as well as an experimental investigation on the relative intensities of multi-order stimulated Raman scattering (SRS) in an optical fiber. Our analysis shows that the Raman mixing is a principal factor which affects the relative intensities of multi-order SRS. We also show by experiment that there are simultaneously two kinds of scattering elements corresponding to the deformation vibration of SiO_2 and GeO_2 tetrahedra with frequencies around 1.3×10^{13} Hz.

EXPERIMENTAL INVESTIGATION OF PRISM-GRATING COUPLERS

Xu Dewei & Wang Xianxiu
Changchun Institute of Physics, Academia Sinica

Zhang Qinying
Changchun Institute of Optics & Fine Mechanics,
Academia Sinica

ABSTRACT AND SUMMARY

We have proposed and produced a novel structure, so-called "prism-grating" coupler, to excite guided modes in film waveguide by an incident beam of laser. The coupler is a composite structure consisting of a prism and a grating, with the peculiarity that the grating is not cut on the planar film, but on the bottom surface of the prism.

In the coupler, manufactured in our laboratory, we cut a grating of 1200 lines/mm on the surface of a polished optical glass (ZF_7) prism by a diffraction grating ruling machine. A film of ZnS was deposited on a glass substrate by vacuum evaporation. The prism grating was placed on top the ZnS film waveguide by means of a clamping apparatus.

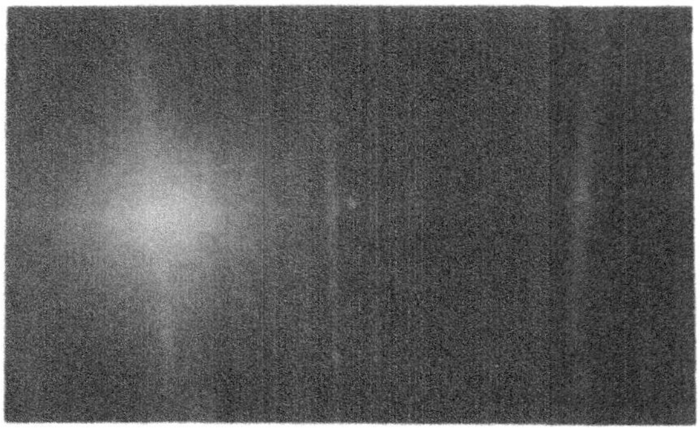

Fig. 1. Excitation of waveguide modes using a "prism-grating" coupler. The three modes shown corresponding to m=0,1,2.

A 6328 Å He-Ne- laser was used as the source in our experiment. Three waveguide modes excited by our prism-grating coupler were observed as shown in Fig. 1.

Further work on this coupler, including especially a quantitative determination of the coupling efficiency, continues at our laboratory in order to address its feasibility as a practical optical waveguiding device.

TRANSMISSION OF HIGH LASER POWER BY LARGE-CORE OPTICAL FIBER

Ye Biqing

Ma Zhonglin

Shanghai Institute of Optics and Fine Mechanics, Academia Sinica

ABSTRACT AND SUMMARY

We report the experiments made in our laboratory on the launching of a high power (> 100 w, CW) laser beam at 1.06 μ m into a large-core (∿ 400 μm diameter) quartz cable. With a fiber length of 1 ∿ 2 km, the coupling-transmission efficiency achieved is larger than 85%.

Optical fibers carrying high power laser light appear potentially useful in certain applications, especially in medicine, in industry, as well as in basic research. A high power CW Nd:YAG laser, with a series of jointed arms as the "conductor" of the laser beam, was successfully used in liver cancer surgery (1980, Reference 1), the first attempt of such kind reported in the literature.

Recently, we replace the jointed-arms by a large-core fiber cable to transmit the high power laser beam, so that the laser scalpel becomes as flexible as an endoscopic instrument.

Our light source is a high power segmented CW Nd:YAG - laser, with continuously variable output 0 - 250 w, beam divergence less than 10 mrad. We use China-made quartz fibers cladded by silicon rubber (diameters of the inner core, the outer core and the cladding being 250 - 400 μm, 650 -700 μm and 1 mm, respectively), with an numerical aperture around 0.17. We use different optical focusers to achieve optimal coupling. One typical example of our experimental results is shown in the figure.

Fig.4 Coupling-transmission efficiency vs injected laser power for an optical fiber with 400 μm inner core diameter and 1 m length. Laser power at 1.06 μm is injected to the fiber by three different focusing devices.

+ - single lens (f_c=30 mm)

o - telescopic combined focuser (M=2.8,f=65mm)

× - telescopic combined focuser (M=2.8,f=50mm)

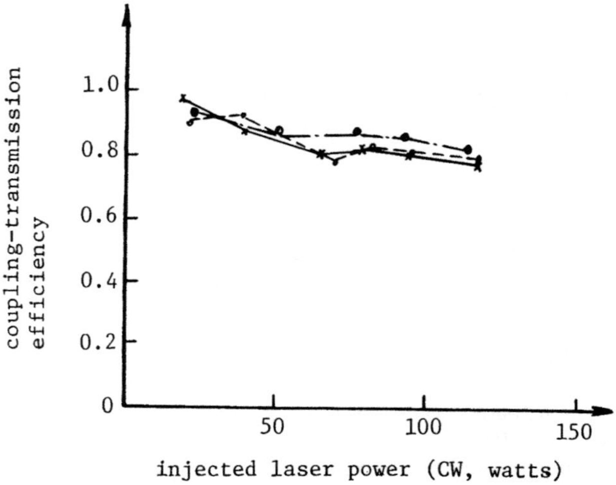

(Fig.1)

The authors wish to thank Messrs. Song Boqing and Shao Qing who provided the quartz fibers for our experiment.

REFERENCE

1. Yu Yeqin, Ye Biqing et al, Kexue Tongbao, 26 (1981) 826

COUPLINGS BETWEEN SINGLE-MODE LASERS AND SINGLE-MODE FIBERS

Zhu Yan-nian* and G. Wenke

Institute of Optical Communications, Heinrich Hertz Institute, Berlin, West Germany

ABSTRACT

This Symposium brief reports some preliminary results obtained in our laboratory in single-mode laser and single-mode fiber couplings, with a view to ascertaining a satisfactory coupling technique which possesses simultaneously a higher coupling efficiency and a weaker optical feedback. Measurements have been made on couplings between lasers and different structures of fiber ends, including plane end-face, micro-lens and a hemispherical lens at a tapered end (or TH-end fiber).

SUMMARY

We used the Japan-made CSP single-mode laser, with wavelength 0.83 μm, threshold current 62 mA, beam divergence angle 24° (horizontal) and 10° (vertical). Short wavelength laser produces larger dispersion, so as to make it easier to observe the frequency jitter caused by optical feedback. The laser is modulated by a 50 MHz square wave. A large area PIN photodiode measures the optical power, and a high bit rate APD measures the optical pulse group delay.

We used the Germany-made (SEL) single-mode fiber with outer diameter 125 μm, core diameter 7 μm, relative index difference 0.23%, attenuation 2.1 dB/km.

In our experiment, we found that the coupling efficiency reaches a maximum when the micro-lens radius is around 7-10 μm. For a TH-end fiber the tapered length is about 2∿3 times the fiber diameter to obtain high coupling efficiency. Our experiment also shows that when the diameter of the hemispherical lens is about 25-35 μm, higher coupling efficiency and smaller optical feedback are achievable at the same time.

Fig. 1 shows a periodic change of optical power with the separation z between the laser and the fiber. The period is $\lambda/2$, in conformity with the resonance-cavity theory.

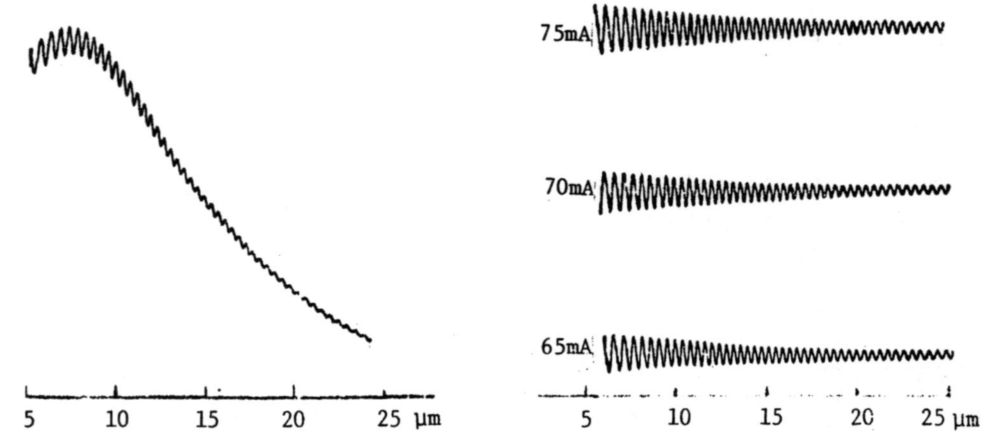

Fig. 1. Optical power change of a laser coupled to an optical fiber with a micro-lens end. Left: at the front end-face of laser. Right: at the back end-face of laser.

(a) (b)

Fig. 2. Phase jitter of a pulse signal passing through a 3.5 km single-mode fiber. Left: Plane end-face. Right: TH-end fiber.

In conclusion, our experiment shows that plane end-face fiber has a coupling loss as high as 6.4 dB, as well as a rather strong optical feedback. The coupling efficiency of a carefully made micro-lens and the TH-end fiber can be lower than 1.9 dB. The latter has an additional advantage of smaller optical feedback.

* on leave from the Kweilin Institute of Optical Communications

FARADAY ROTAION IN MONOMODE FIBER WITH AXIALLY VARYING MAGNETIC FIELD

Qian Jing-ren

University of Science and Technology of China, Hefei

ABSTRACT

A simple integral formula is derived for calculating the total Faraday rotation in a monomode optical fiber produced by an axially varying magnetic field. While Faraday rotation in coiled monomode fibers has been studied by the use of Jones calculus, the author has found that Miller's theory of periodic coupling particularly useful in calculating the Faraday rotation produced in a fiber of the helix form.

SUMMARY

Under the action of an applied magnetic field, linearly polarized HE_{11} mode in a monomode optical fiber undergoes Faraday rotation, which can be utilized in making a sensor or a Faraday-effect current transducer. In order to ensure high accuracy, either a nearly perfect conventional fiber or a nonconventional fiber of the spun version should be employed. The Faraday rotation angle for invariant magnetic field has been found in the literature. In practice, however, the applied magnetic field may vary along the axial direction z. Use of the coupled-mode theory leads to an integral formula $\int_0^L K(z)dz$, for the total angle of Faraday rotation (K denoting coupling coefficient caused by axial magnetic field, L being the fiber length). To the author's knowledge, this formula has not yet appeared in the literature.

A useful criterion for determining the power transfer between the coupled modes is the coupling capacity Q[1]. For a monomode optical fiber with a linear birefrigence b, $Q=b/(2K)$, and hence, if K is not sufficiently large as compared with b, the Faraday rotation as expressed by the above integral formula cannot be preserved owing to mode conversion-reconversion effect.

It is found that the Faraday rotation can be preserved if the mode couplings become periodic. For example, the periodic-coupling capacity[2] for a spun fiber is $\Omega=b/(\tau-2K)$, where τ is the spinning speed during fiber drawing. The Faraday rotation is preserved as long as $(\tau-2K)>>b$, which is the actual case in practice.

Another example is a fiber helix. The magnetic field transversely applied to the helix axis does not vary in space. However, owing to the helix form of the fiber, the direction of the magnetic field relating to the successive half turns of the fiber becomes alternating, nearly according to a sinusoidal function. Couplings between the two linearly polarized modes become periodic, and the Faraday rotation is preserving. Maximum Faraday rotation occurs when the period $2\Pi R$ satisfies the condition $b=n/R$, $n=1,2,\ldots$, where b is the linear birefringence produced by bending. With the aid of Miller's theory(3), the author finds that a rotation of 45°is achievable under the condition $(\sin\frac{KL}{2})^2=\frac{1}{2}$ where K is the maximum value of the coupling coefficient. This result agrees with that obtained by using Jones calculus[4].

AKNOWLEDGEMENT

The author is indebted to Prof. Huang Hung-chia, Member of Academia Sinica, who suggested the use of periodic coupling for treating the problem concerned.

REFERENCES

1. Huang Hung-chia, Scientia Sinica, Vol. 9, No. 1, 1960, p.142
2. Huang Hung-chia, Radio Science, Vol. 16, No. 4, 1981, p.495
3. S. E. Miller, B.S.T.J., Vol. 33, No. 5, 1954, p.661; ibid, Vol. 47, No. 8, 1968, p.1800
4. A. J. Barlow, J. J. Ramskov-Hanson & D. N. Payne, Appl. Optics, Vol. 20, N0. 17, 1981, p.2962

$LiNbO_3$ Waveguides by Electrically Enhanced Ion Migration and a Comparison of Techniques

Y.X. Chen
Shanghai Jiao Tong University

Abstract and Summary

Recently we have investigated the electrically enhanced migration technique for fabricating $LiNbO_3$ waveguides. With the aid of an analysis by an X-ray photo-electron spectroscopy (XPS) and a second ion mass spectroscopy (SIMS), we have found that the formation of a waveguide is mainly due to the migration of the Li atoms from the surface layer of the $LiNbO_3$ substrate.

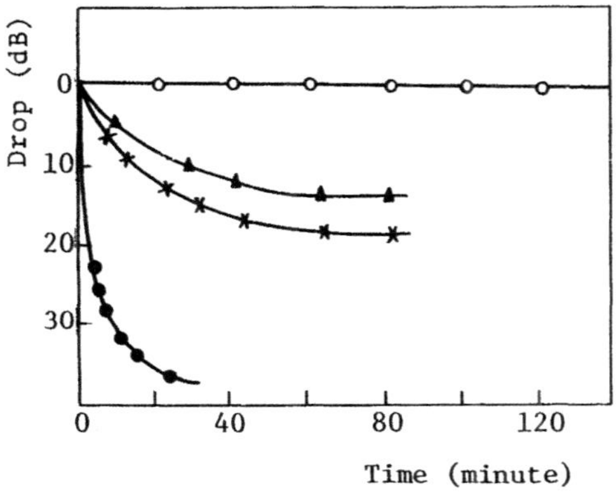

Fig.1
Time-dependence of the drop of main peak of m-line.
o ion migration
▲ Ti indiffusion in close crucible
x Ti indiffusion in atmosphere
● Ti indiffusion in wet oxygen

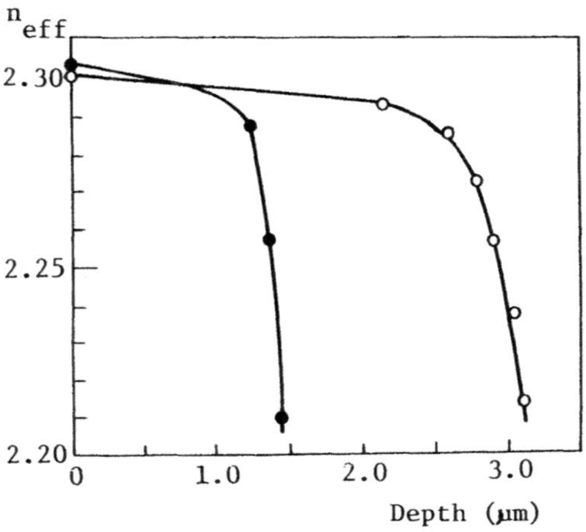

Fig.2
Effective indices profile of ion-migration waveguide. Sample No. 15, 260° C, 5v, 125 min.
o positive side
● negative side

TABLE 1. COMPARISON OF VARIOUS SINGLE-MODE $LiNbO_3$ WAVEGUIDES

FABRICATION TECHNIQUES	ION-MIGRATION	PORTON-EXCHANGE	Ti-IN-DIFFUSION
T (^oC)	260	200	1000
t (min)	50	10	360
Δn_{eff}	0.04	0.06	0.01
d_m (μm)	0.8-1.2	0.7	1.5-2.0
Attenuation (dB/cm)	0.8	0.5	0.8
Scattering (mr, at 40dB)	1.7	1.3	2.5
Optical damage (632.8 nm)	40mw, 5h, no damage	40mw, 5h, no damage	13mw, 2h, damage
n_{eff} profile	Nearly step	Nearly step	Complementary error function

The characteristics of three kinds of $LiNbO_3$ waveguide by different techniques have been measured for comparison, as given in table 1. It shows that $LiNbO_3$ waveguides by ion migration are more resistive to optical damage (Fig.1), with a higher index increase, a small mode depth and a nearly step-profile of effective index. (Fig.2).

It is also found that waveguides made by the ion-migration technique are compatible with Ti-indifussion waveguides. It is therefore possible to design waveguide devices with a multilayer structure and a certain planar pattern of various indices on a $LiNbO_3$ substrate.

GEODESIC LENSES FOR INTEGRATED OPTICS

W. Jiang

G. R. Dubois

V. M. Ristic

Department of Electrical Engineering, University of Toronto, Canada

ABSTRACT

The authors present a procedure for the design of an aspherical, rounding-edge geodesic lens for integrated optics. The design, which is based on geometrical optics, makes use of the annulus technique, and of the principle of equivalent lenses.

In practice, the geodesic lens necessarily has an appropriate rounding-edge in order to reduce the scattering loss due to the steep rim between the waveguide and the lens. A numerical procedure has been devised to find the joint point between the rounding-edge and the inner part of the lens. To facilitate the design, a relationship between the focal length, the effective aperture and the smoothness of the joint point has been found and represented in a graphical form. The tolerances achievable with the state-of-the-art technology are discussed.

Experimental results obtained in the authors' laboratory for evaluation of the produced lenses are in good agreement with the numerical calculation.

ZIRCONIA-CONTAINING OPTICAL FIBERS PULLED BY DOUBLE CRUCIBLE METHOD

Longin Kociszewski, Ryszard Stepien and Jan Buzniak

Institute of Electronic Material Technology
Warsaw, Poland

ABSTRACT AND SUMMARY

The double crucible method is attractive in its capacity of producing optical fibers with high numerical aperture and low cost. The loss of optical fibers produced by this method can be reduced if chemical purity and physical homogenity of the glass are carefully considered.

In addition to chemical purity and physical homogenity, the authors of this report also considered the chemical composition of the glass. Following are the subjects under investigation:

1. Differences in transmission of two kinds of quartz glass rods which produce fibers with losses of 10 and 100 dB/m.
2. Transmission of the F-2 type glass, which is generally used for producing optical fibers for intermediate-distance communications.
3. Transmission of the optical fibers produced by the zirconia-containing glass as compared with the fibers drawn from the F-2 type glass.

The above listed optical fibers have been investigated regarding transmission, chemical composition and refractive index. The relation between structural property of the glass and the optical property of the fiber is studied with special emphasis on zirconia-containing glass fibers. It is found that the zirconia-containing glass fibers have considerable lower losses, even though they have a higher level of impurities.

LAUNCHING LIGHT FROM SEMICONDUCTOR LASERS INTO SINGLE-MODE FIBER OPTICAL WAVEGUIDES

Harish R. D. Sunak
State University of Campinas (UNICAMP), Institute of Physics, Campinas, 13.100, São Paulo - Brazil

ABSTRACT

In this report, problems relating to couplings of semiconductor lasers with monomode optical fibers are considered. Particular emphasis will be placed on the techniques for increasing the launching efficiency η in presence of various misalignments of the sources and fibers.

SUMMARY

Our investigation includes the following topics:

1. Characteristics of the light output of semiconductor laser diodes and what limitations these characteristics impose on the coupling coefficient η. The LD's characteristics considered include the near-field intensity profiles and the far-field divergence angles of the emerging light.
2. The basic characteristics such as core diameters and numerical apertures of single-mode optical fibers, and how they influence η.
3. Calculation of η, with various theoretical models, based on the characteristics considered in the above two items.
4. Experimental results as compared with the above calculated results (or theoretical predictions) using, as a first example, plane-ended optical fibers. The sources that cause deviations of the experimental results from the calculation are discussed.
5. The above results provide a basis to consider various end-geometries of optical fibers for the purpose of increasing η, including hemispherical ended fibers, tapered fibers with plane ends, taper with hemisphere-ended fibers, etc. Mechanisms by which greater η is achievable are discussed.

   The effects of axial, lateral and anguler misalignments between lasers and fibers for various fiber-end configulations are compared using a 1 m and a 1 km fiber models.
6. Various techniques to make the fiber end geometries as discussed above, including CO_2 laser microtorch, photoresist, arc-discharge, etc. The relative advantages and disadvantages of each technique are described. Reproducibility is a main factor for consideration.
7. The design of transmitter modules based on the above investigation and the type of systems using the transmitted modules.

LIST AND INDEX OF AUTHORS

The last line under each name indicates the number (or numbers) of pages in this Symposium Proceedings on which the author's article (or articles) appears.

GPSR Compliance

The European Union's (EU) General Product Safety Regulation (GPSR) is a set of rules that requires consumer products to be safe and our obligations to ensure this.

If you have any concerns about our products, you can contact us on ProductSafety@springernature.com

In case Publisher is established outside the EU, the EU authorized representative is:

Springer Nature Customer Service Center GmbH
Europaplatz 3
69115 Heidelberg, Germany

Batch number: 10755284

Printed by Printforce, the Netherlands